LINEAR ALGEBRA
PROBLEM BOOK

Paul R. Halmos

THE
DOLCIANI MATHEMATICAL EXPOSITIONS

Published by
THE MATHEMATICAL ASSOCIATION OF AMERICA

———

The Dolciani Mathematical Expositions

NUMBER SIXTEEN

LINEAR ALGEBRA PROBLEM BOOK

PAUL R. HALMOS

Published and Distributed by
THE MATHEMATICAL ASSOCIATION OF AMERICA

©1995 by
The Mathematical Association of America (Incorporated)
Library of Congress Catalog Card Number 94-79588

Complete Set ISBN 0-88385-300-0
Vol. 16 ISBN 0-88385-322-1

Printed in the United States of America

Current printing (last digit):
10 9 8 7 6 5 4 3 2

The DOLCIANI MATHEMATICAL EXPOSITIONS series of the Mathematical Association of America was established through a generous gift to the Association from Mary P. Dolciani, Professor of Mathematics at Hunter College of the City University of New York. In making the gift, Professor Dolciani, herself an exceptionally talented and successful expositor of mathematics, had the purpose of furthering the ideal of excellence in mathematical exposition.

The Association, for its part, was delighted to accept the gracious gesture initiating the revolving fund for this series from one who has served the Association with distinction, both as a member of the Committee on Publications and as a member of the Board of Governors. It was with genuine pleasure that the Board chose to name the series in her honor.

The books in the series are selected for their lucid expository style and stimulating mathematical content. Typically, they contain an ample supply of exercises, many with accompanying solutions. They are intended to be sufficiently elementary for the undergraduate and even the mathematically inclined high-school student to understand and enjoy, but also to be interesting and sometimes challenging to the more advanced mathematician.

1. *Mathematical Gems,* Ross Honsberger
2. *Mathematical Gems II,* Ross Honsberger
3. *Mathematical Morsels,* Ross Honsberger
4. *Mathematical Plums,* Ross Honsberger (ed.)
5. *Great Moments in Mathematics (Before 1650),* Howard Eves
6. *Maxima and Minima without Calculus,* Ivan Niven
7. *Great Moments in Mathematics (After 1650),* Howard Eves
8. *Map Coloring, Polyhedra, and the Four-Color Problem,* David Barnette
9. *Mathematical Gems III,* Ross Honsberger
10. *More Mathematical Morsels,* Ross Honsberger
11. *Old and New Unsolved Problems in Plane Geometry and Number Theory,* Victor Klee and Stan Wagon
12. *Problems for Mathematicians, Young and Old,* Paul R. Halmos
13. *Excursions in Calculus: An Interplay of the Continuous and the Discrete,* Robert M. Young
14. *The Wohascum County Problem Book,* George T. Gilbert, Mark I. Krusemeyer, Loren C. Larson
15. *Lion Hunting and Other Mathematical Pursuits: A Collection of Mathematics, Verse, and Stories by Ralph P. Boas, Jr.,* Gerald L. Alexanderson and Dale H. Mugler (eds.)
16. *Linear Algebra Problem Book,* Paul R. Halmos
17. *From Erdös to Kiev: Problems of Olympiad Caliber,* Ross Honsberger
18. *Which Way Did the Bicycle Go? . . . and Other Intriguing Mathematical Mysteries,* Joseph D. E. Konhauser, Dan Velleman, and Stan Wagon

Mathematical Association of America
P. O. Box 91112
Washington, DC 20090-1112
1-800-331-1MAA   FAX: 1-301-206-9789

PREFACE

Is it fun to solve problems, and is solving problems about something a good way to learn something? The answers seem to be yes, provided the problems are neither too hard nor too easy.

The book is addressed to students (and teachers) of undergraduate linear algebra—it might supplement but not (I hope) replace my old *Finite-Dimensional Vector Spaces.* It largely follows that old book in organization and level and order—but only "largely"—the principle is often violated. This is not a step-by-step textbook—the problems vary back and forth between subjects, they vary back and forth from easy to hard and back again. The location of a problem is not always a hint to what methods might be appropriate to solve it or how hard it is.

Words like "hard" and "easy" are subjective of course. I tried to make some of the problems accessible to any interested grade school student, and at the same time to insert some that might stump even a professional expert (at least for a minute or two). Correspondingly, the statements of the problems, and the introductions that precede and the solutions that follow them sometimes laboriously explain elementary concepts, and, at other times assume that you are at home with the language and attitude of mathematics at the research level. Example: sometimes I assume that you know nothing, and carefully explain the associative law, but at other times I assume that the word "topology", while it may not refer to something that you are an expert in, refers to something that you have heard about.

The solutions are intrinsic parts of the exposition. You are urged to look at the solution of each problem even if you can solve the problem without doing so—the solution sometimes contains comments that couldn't be made in the statement of the problem, or even in the hint, without giving too much of the show away.

I hope you will enjoy trying to solve the problems, I hope you will learn something by doing so, and I hope you will have fun.

CONTENTS

Preface vi

Chapter 1. Scalars 1

    1. Double addition
    2. Half double addition
    3. Exponentiation
    4. Complex numbers
    5. Affine transformations
    6. Matrix multiplication
    7. Modular multiplication
    8. Small operations
    9. Identity elements
    10. Complex inverses
    11. Affine inverses
    12. Matrix inverses
    13. Abelian groups
    14. Groups
    15. Independent group axioms
    16. Fields
    17. Addition and multiplication in fields
    18. Distributive failure
    19. Finite fields

Chapter 2. Vectors 17

    20. Vector spaces
    21. Examples
    22. Linear combinations
    23. Subspaces
    24. Unions of subspaces

25. Spans
26. Equalities of spans
27. Some special spans
28. Sums of subspaces
29. Distributive subspaces
30. Total sets
31. Dependence
32. Independence

Chapter 3. Bases 39
33. Exchanging bases
34. Simultaneous complements
35. Examples of independence
36. Independence over $\mathbb{R}$ and $\mathbb{Q}$
37. Independence in $\mathbb{C}^2$
38. Vectors common to different bases
39. Bases in $\mathbb{C}^3$
40. Maximal independent sets
41. Complex as real
42. Subspaces of full dimension
43. Extended bases
44. Finite-dimensional subspaces
45. Minimal total sets
46. Existence of minimal total sets
47. Infinitely total sets
48. Relatively independent sets
49. Number of bases in a finite vector space
50. Direct sums
51. Quotient spaces
52. Dimension of a quotient space
53. Additivity of dimension

Chapter 4. Transformations 51
54. Linear transformations
55. Domain and range
56. Kernel
57. Composition
58. Range inclusion and factorization
59. Transformations as vectors
60. Invertibility
61. Invertibility examples
62. Determinants: 2×2

63. Determinants: $n \times n$
64. Zero-one matrices
65. Invertible matrix bases
66. Finite-dimensional invertibility
67. Matrices
68. Diagonal matrices
69. Universal commutativity
70. Invariance
71. Invariant complements
72. Projections
73. Sums of projections
74. not quite idempotence

Chapter 5. Duality 85

75. Linear functionals
76. Dual spaces
77. Solution of equations
78. Reflexivity
79. Annihilators
80. Double annihilators
81. Adjoints
82. Adjoints of projections
83. Matrices of adjoints

Chapter 6. Similarity 97

84. Change of basis: vectors
85. Change of basis: coordinates
86. Similarity: transformations
87. Similarity: matrices
88. Inherited similarity
89. Similarity: real and complex
90. Rank and nullity
91. Similarity and rank
92. Similarity of transposes
93. Ranks of sums
94. Ranks of products
95. Nullities of sums and products
96. Some similarities
97. Equivalence
98. Rank and equivalence

Chapter 7. Canonical Forms 107

   99. Eigenvalues
  100. Sums and products of eigenvalues
  101. Eigenvalues of products
  102. Polynomials in eigenvalues
  103. Diagonalizing permutations
  104. Polynomials in eigenvalues, converse
  105. Multiplicities
  106. Distinct eigenvalues
  107. Comparison of multiplicities
  108. Triangularization
  109. Complexification
  110. Unipotent transformation
  111. Nipotence
  112. Nilpotent products
  113. Nilpotent direct sums
  114. Jordan form
  115. Minimal polynomials
  116. Non-commutative Lagrange interpolation

Chapter 8. Inner Product Spaces 129

  117. Inner products
  118. Polarization
  119. The Pythagorean theorem
  120. The parallelogram law
  121. Complete orthonormal sets
  122. Schwarz inequality
  123. Orthogonal complements
  124. More linear functionals
  125. Adjoints on inner product spaces
  126. Quadratic forms
  127. Vanishing quadratic forms
  128. Hermitian transformations
  129. Skew transformations
  130. Real Hermitian forms
  131. Positive transformations
  132. positive inverses
  133. Perpendicular projections
  134. Projections on $\mathbb{C} \times \mathbb{C}$
  135. Projection order
  136. Orthogonal projections

137. Hermitian eigenvalues
138. Distinct eigenvalues

Chapter 9. Normality 149

139. Unitary transformations
140. Unitary matrices
141. Unitary involutions
142. Unitary triangles
143. Hermitian diagonalization
144. Square roots
145. Polar decomposition
146. Normal transformations
147. Normal diagonalizability
148. Normal commutativity
149. Adjoint commutativity
150. Adjoint intertwining
151. Normal products
152. Functions of transformations
153. Gramians
154. Monotone functions
155. Reducing ranges and kernels
156. Truncated shifts
157. Non-positive square roots
158. Similar normal transformations
159. Unitary equivalence of transposes
160. Unitary and orthogonal equivalence
161. Null convergent powers
162. Power boundedness
163. Reduction and index 2
164. Nilpotence and reduction

Hints 169

Solutions:

Chapter 1 185
Chapter 2 204
Chapter 3 216
Chapter 4 228
Chapter 5 252
Chapter 6 259
Chapter 7 277
Chapter 8 296
Chapter 9 310

SCALARS

1. Double addition

Is it obvious that

$$63 + 48 = 27 + 84?$$

It is a true and thoroughly uninteresting mathematical statement that can be verified in a few seconds—but is it obvious? If calling it obvious means that the reason for its truth is clearly understood, without even a single second's verification, then most people would probably say no.

What about

$$(27 + 36) + 48 = 27 + (36 + 48)$$

—is that obvious? Yes it is, for most people; the instinctive (and correct) reaction is that the way the terms of a sum are bunched together cannot affect the answer. The approved technical term is not "bunch together" but "associate"; the instinctive reaction is a readiness to accept what is called the **associative law** of addition for real numbers. (Surely every reader has noticed by now that the non-obvious statement and the obvious one are in some sense the same:

$$63 = 27 + 36 \quad \text{and} \quad 84 = 36 + 48.)$$

Linear algebra is concerned with several different kinds of operations (such as addition) on several different kinds of objects (not necessarily real numbers). To prepare the ground for the study of strange operations and to keep the associative law from being unjustly dismissed as a triviality, a little effort to consider some good examples and some bad ones is worthwhile.

Some of the examples will be useful in the sequel, and some won't—some are here to show that associativity can fail, and others are here to show that even when it holds it may be far from obvious. In the world of linear algebra non-associative operations are rare, but associative operations whose good behavior is not obvious are more frequently met.

Problem 1. *If a new addition for real numbers, denoted by the temporary symbol $\boxed{+}$, is defined by*

$$\alpha \boxed{+} \beta = 2\alpha + 2\beta,$$

is $\boxed{+}$ associative?

Comment. The plus sign on the right-hand side of the equation denotes ordinary addition.

Note: since ordinary addition is commutative, so that

$$2\alpha + 2\beta = 2\beta + 2\alpha,$$

it follows that

$$\alpha \boxed{+} \beta = \beta \boxed{+} \alpha.$$

Conclusion: the new addition is also commutative.

2   2. Half double addition

Problem 2. *If a new addition for real numbers, denoted by the temporary symbol $\boxed{+}$, is defined by*

$$\alpha \boxed{+} \beta = 2\alpha + \beta,$$

is $\boxed{+}$ associative?

Comment. Since $2\alpha + \beta$ is usually different from $2\beta + \alpha$, this $\boxed{+}$ is not commutative.

3   3. Exponentiation

Problem 3. *If an operation for positive integers, denoted by the temporary symbol $*$, is defined by*

$$\alpha * \beta = \alpha^{\beta},$$

is it commutative? Is it associative?

4. Complex numbers 4

Suppose that an operation $\boxed{+}$ is defined for ordered pairs of real numbers, that is for objects that look like $\langle \alpha, \beta \rangle$ with both α and β real, as follows:

$$(\alpha, \beta) \boxed{+} (\gamma, \delta) = (\alpha + \gamma, \beta + \delta).$$

Is it commutative? Sure, obviously—how could it miss? All it does is perform the known commutative operation of addition of real numbers twice, once for each of the two coordinates. Is it associative? Sure, obviously, for the same reason.

The double addition operations in Problems 1 and 2 are artificial; they were cooked up to make a point. The operation of exponentiation in Problem 3 is natural enough, and that is its point: "natural" operations can fail to be associative. The coordinatewise addition here defined for ordered pairs is a natural one also, but it is far from the only one that is useful.

> **Problem 4.** *If an operation for ordered pairs of real numbers, denoted by the temporary symbol $\boxed{\cdot}$, is defined by*
>
> $$(\alpha, \beta) \boxed{\cdot} (\gamma, \delta) = (\alpha\gamma - \beta\delta, \alpha\delta + \beta\gamma),$$
>
> *is it commutative? Is it associative?*

Comment. The reason for the use of the symbol $\boxed{\cdot}$ (instead of $\boxed{+}$) is twofold: it is reminiscent of multiplication (instead of addition), and it avoids confusion when the two operations are discussed simultaneously (as in many contexts they must be).

5. Affine transformations 5

Looking strange is not necessarily a sign of being artificial or useless.

> **Problem 5.** *If an operation for ordered pairs of real numbers, denoted by $\boxed{\cdot}$ again, is defined by*
>
> $$(\alpha, \beta) \boxed{\cdot} (\gamma, \delta) = (\alpha\gamma, \alpha\delta + \beta),$$
>
> *is it commutative? Is it associative?*

6. Matrix multiplication 6

The strange multiplication of Problem 5 is a special case of one that is more complicated but less strange.

Problem 6. *If an operation for ordered quadruples of real numbers, denoted by* ⊡ *, is defined by*

$$\langle \alpha, \beta, \gamma, \delta \rangle \boxdot \langle \alpha', \beta', \gamma', \delta' \rangle$$
$$= \langle \alpha\alpha' + \beta\gamma', \alpha\beta' + \beta\delta', \gamma\alpha' + \delta\gamma', \gamma\beta' + \delta\delta' \rangle,$$

is it commutative? Is it associative?

Comment. How is the multiplication of Problem 5 for ordered pairs a "special case" of this one? Easy: restrict attention to only those quadruples $\langle \alpha, \beta, \gamma, \delta \rangle$ for which $\gamma = 0$ and $\delta = 1$. The ⊡ product of two such special quadruples is again such a special one; indeed if $\gamma = \gamma' = 0$ and $\delta = \delta' = 1$, then $\gamma\alpha' + \delta\gamma' = 0$ and $\gamma\beta' + \delta\delta' = 1$. The first two coordinates of the product are $\alpha\alpha'$ and $\alpha\beta' + \beta$, and that's in harmony with Problem 5.

Another comment may come as an additional pleasant surprise: the multiplication of complex numbers discussed in Problem 4 is also a special case of the quadruple multiplication discussed here. Indeed: restrict attention to only those quadruples that are of the form

$$\langle \alpha, \beta, -\beta, \alpha \rangle,$$

and note that

$$\langle \alpha, \beta, -\beta, \alpha \rangle \boxdot \langle \gamma, \delta, -\gamma, \delta \rangle = \langle \alpha\gamma - \beta\delta, \alpha\delta + \beta\gamma, -\beta\gamma - \alpha\delta, -\beta\delta + \alpha\gamma \rangle$$

—in harmony with Problem 4.

7. Modular multiplication

Define an operation, denoted by ⊡, for the numbers 0, 1, 2, 3, 4, 5 as follows: multiply as usual and then throw away multiples of 6. (The technical expression is "multiply modulo 6".) Example: $4 \boxdot 5 = 2$ and $2 \boxdot 3 = 0$.

Problem 7. *Is multiplication modulo 6 commutative? Is it associative? What if 6 is replaced by 7: do the conclusions for 6 remain true or do they change?*

8. Small operations

Problem 7 shows that interesting operations can exist on small sets. Small sets have the added advantage that sometimes they can forewarn us about some dangers that become more complicated, and therefore harder to see, when the sets get larger. Another reason small sets are good is that operations on them can be defined in a tabular manner that is reassuringly explicit.

Consider, for instance, the table

×	0	1	2
0	0	0	0
1	0	1	2
2	0	2	1

which defines multiplication modulo 3 for the numbers 0, 1, 2. The information such tables are intended to communicate is that the product of the element at the left of a **row** by the element at the top of a **column**, in that order, is the element placed where that row and that column meet. Example: $2 \times 2 = 1$ modulo 3.

It might be worth remarking that there is also a useful concept of addition modulo 3; it is defined by the table

+	0	1	2
0	0	1	2
1	1	2	0
2	2	0	1

It's a remarkable fact that addition and multiplication modulo 3 possess all the usually taught properties of the arithmetic operations bearing the same names. They are, for instance, both commutative and associative, they conspire to satisfy the distributive law

$$\alpha \times (\beta + \gamma) = (\alpha \times \beta) + (\alpha \times \gamma),$$

they permit unrestricted subtraction (so that, for example, $1 - 2 = 2$), and they permit division restricted only by the exclusion of the denominator 0 (so that, for example, $\frac{1}{2} = 2$). In a word (officially to be introduced and studied later) the integers modulo 3 form a **field**.

Problem 1 is about an operation that is commutative but not associative. Can that phenomenon occur in small sets?

Problem 8. *Is there an operation in a set of three elements that is commutative but not associative?*

9   9. Identity elements

The commonly accepted attitudes toward the commutative law and the associative law are different. Many real life operations fail to commute; the mathematical community has learned to live with that fact and even to enjoy it. Violations of the associative law, on the other hand, are usually considered by specialists only. Having made the point that the associative law deserves respect, this book will concentrate in the sequel on associative operations only. The next job is to see what other laudable properties such operations can and should possess.

The sum of 0 and any real number α is α again; the product of 1 and any real number α is α again. The phenomenon is described by saying that 0 and 1 are **identity elements** (or zero elements, or unit elements, or neutral elements) for addition and multiplication respectively. An operation that has an identity element is better to work with than one that doesn't. Which ones do?

> **Problem 9.**   *Which of the operations*
> (1) *double addition,*
> (2) *half double addition,*
> (3) *exponentiation,*
> (4) *complex multiplication,*
> (5) *multiplication of affine transformations,*
> (6) *matrix multiplication,*
> *and*
> (7) *modular addition and multiplication*
> *have an identity element?*

In the discussion of operations, in Problems 1–8, the notation and the language were both additive (+, sum) and multiplicative (×, product). Technically there is no difference between the two, but traditionally multiplication is the more general concept. In the definition of groups, for instance (to be given soon), the notation and the language are usually multiplicative; the additive theory is included as a special case. A curious but firmly established part of the tradition is that multiplication may or may not be commutative, but addition always is. The tradition will be followed in this book, with no exceptions.

An important mini-theorem asserts that an operation can have at most one identity element. That is: if × is an operation and both ε and ε' are identity elements for it, so that

$$\varepsilon \times \alpha = \alpha \times \varepsilon = \alpha \qquad \text{and} \qquad \varepsilon' \times \alpha = \alpha \times \varepsilon' = \alpha$$

for all α, then

$$\varepsilon = \varepsilon'.$$

Proof. Use ε' itself for α in the equation involving ε, and use ε for α in the equation involving ε'. The conclusion is that $\varepsilon \times \varepsilon'$ is equal to both ε and ε', and hence that ε and ε' are equal to each other.

Comment. The proof just given is intended to emphasize that an identity is a two-sided concept: it works from both right and left.

10. Complex inverses

10

Is there a positive integer that can be added to 3 to yield 8? Yes.

Is there a positive integer that can be added to 8 to yield 3? No.

In the well-known language of elementary arithmetic: subtraction within the domain of positive integers is sometimes possible and sometimes not.

Is there a real number that can be added to 5 to yield 0? Yes, namely -5. Every real number has a negative, and that fact guarantees that within the domain of real numbers subtraction is always possible. (To find a number that can be added to 8 to yield 3, first find a number that can be added to 3 to yield 8, and then form its negative.)

The third basic property of operations that will be needed in what follows (in addition to associativity and the existence of neutral elements) is the possibility of inversion. Suppose that $*$ is an operation (a temporary impartial symbol whose role in applications could be played by either addition or multiplication), and suppose that the domain of $*$ contains a neutral element ε, so that $\varepsilon * \alpha = \alpha * \varepsilon = \alpha$ for all x. Under these circumstances an element β is called an **inverse** of x ($*$ inverse) if

$$\alpha * \beta = \beta * \alpha = \varepsilon.$$

Obvious example: every real number α has a $+$ inverse, namely $-\alpha$. Worrisome example: not every real number has a $\times$ inverse. The exception is 0; there is no real number β such that $0 \times \beta = 1$. That is the only exception: if $\alpha \neq 0$, then the reciprocal $\alpha^{-1} (= \frac{1}{\alpha})$ is a $\times$ inverse. These examples are typical. The use of additive notation is usually intended to suggest the existence of inverses ($+$ inverses, negatives) for every element, whereas for multiplicatively written operations some elements can fail to be **invertible**, that is, can fail to possess inverses ($\times$ inverses, reciprocals).

The definition of $*$ inverse makes sense in complete generality, but it is useful only in case $*$ is associative. The point is that for associative operations an important mini-theorem holds: an element can have at most one inverse. That is: if both β and γ are $*$ inverses of α, so that

$$\alpha * \beta = \beta * \alpha = \varepsilon \qquad \text{and} \qquad \alpha * \gamma = \gamma * \alpha = \varepsilon,$$

then $\beta = \gamma$. Proof: combine all three, γ, and α, and β, in that order, and use the associative law. Looked at one way the answer is

$$\gamma * (\alpha * \beta) = \gamma * \varepsilon = \gamma,$$

whereas the other way it is

$$(\gamma * \alpha) * \beta = \varepsilon * \beta = \beta.$$

The conclusion is that the triple combination $\gamma * \alpha * \beta$ is equal to both γ and β, and hence that γ and β are equal to each other.

Problem 10.   *For complex multiplication (defined in Problem 4), which ordered pairs $\langle \alpha, \beta \rangle$ are invertible? Is there an explicit formula for the inverses of the ones that are?*

11.   11. Affine inverses

Problem 11.   *For the multiplication of affine transformations (defined in Problem 5), which ordered pairs $\langle \alpha, \beta \rangle$ are invertible? Is there an explicit formula for the inverses of the ones that are?*

12.   12. Matrix inverses

Problem 12.   *Which of the 2×2 matrices*

$$\begin{pmatrix} \alpha & \beta \\ \gamma & \delta \end{pmatrix}$$

(for which multiplication was defined in Problem 6) are invertible? Is there an explicit formula for the inverses of the ones that are?

13.   13. Abelian groups

Numbers can be added, subtracted, multiplied, and (with one infamous exception) divided. Linear algebra is about concepts called scalars and vectors. Scalars are usually numbers; to understand linear algebra it is necessary first of all to understand numbers, and, in particular, it is necessary to understand what it means to add and subtract them. The general concept that lies at the heart of such an understanding is that of **abelian groups**.

Consider, as an example, the set $\mathbb{Z}$ of all integers (positive, negative, or zero) together with the operation of addition. The sum of two integers is an integer such that:

addition is **commutative**, meaning that the sum of two integers is independent of the order in which they are added,

$$x + y = y + x;$$

addition is **associative**, meaning that the sum of three integers, presented in a fixed order, is independent of the order in which the two additions between them are performed,

$$(x + y) + z = x + (y + z);$$

the integer **0** plays a special role in that it does not change any integer that it is added to,

$$x + 0 = 0 + x = x;$$

and every addition can be "undone" by another one, namely the addition of the **negative** of what was just added,

$$x + (-x) = (-x) + x = 0.$$

This example is typical. The statements just made about $\mathbb{Z}$ and $+$ are in effect the definition of the concept of abelian group. Almost exactly the same statements can be made about every abelian group; the only differences are terminological (the words "integer" and "addition" may be replaced by others) and notational (the symbols 0 and $+$ may be replaced by others).

Another example is the set $\mathbb{Z}_{12}$ consisting of the integers between 0 and 11 inclusive, with an operation of addition (temporarily denoted by $*$) defined this way: if the sum of two elements of $\mathbb{Z}_{12}$ (in the ordinary meaning of sum) is less than 12, then the new sum is equal to that ordinary sum,

$$x * y = x + y,$$

but if their ordinary sum is 12 or more, then the new sum is the ordinary sum with 12 subtracted,

$$x * y = x + y - 12.$$

The operation $*$ is called addition **modulo 12**, and is usually denoted by just plain $+$, or, if desired, by $+$ followed soon by an explanatory "mod 12". The verification that the four typical sentences stated above for $\mathbb{Z}$ are true for $\mathbb{Z}_{12}$ is a small nuisance, but it's painless and leads to no surprises. (The closest it comes to a surprise is that the role of the negative of x is played by $12 - x$.)

Here is another example of an abelian group: the set $\mathbb{R}_+$ of **positive** real numbers, with an operation, temporarily denoted by $*$, defined as ordinary numerical **multiplication**:

$$x * y = xy.$$

Everybody believes commutativity and associativity; the role of zero is played this time by the real number 1,

$$x * 1 = 1 * x = x,$$

and the role of the negative of x is played by the reciprocal of x

$$x * \left(\frac{1}{x}\right) = \left(\frac{1}{x}\right) * x = 1$$

The general definition of an abelian group should be obvious by now: it is a set $\mathbb{G}$ with an operation of "addition" defined in it (so that whenever x and y are in $\mathbb{G}$, then $x + y$ is again an element of $\mathbb{G}$), satisfying the four conditions discussed above. (They are: commutativity; associativity; the existence of a zero; and, corresponding to each element, the existence of a negative of that element.)

The word "abelian" means exactly the same as "commutative". If an operation in a set satisfies the last three conditions but not necessarily the first, then it is called a **group**. Non-commutative groups also enter the study of linear algebra, but not till later, and not as basically as the commutative ones.

Problem 13. (a) *If a new operation $*$ is defined in the set $\mathbb{R}_+$ of positive real numbers by*

$$x * y = \min(x, y),$$

does $\mathbb{R}_+$ become an abelian group?

(b) *If an operation $*$ is defined in the set $\{1, 2, 3, 4, 5\}$ of positive integers by*

$$x * y = \max(x, y),$$

does that set become an abelian group?

(c) *If x and y are elements of an abelian group such that $x + y = y$, does it follow that $x = 0$?*

Comment. The abbreviations "min" and "max" are for minimum and maximum; $\min(2, 3) = 2$, $\max(-2, -3) = -2$, and $\min(5, 5) = 5$.

Parts (a) and (b) of the problem test the understanding of the definition of abelian group. The beginning of a systematic development of group theory (abelian or not) is usually a sequence of axiom splitting delicacies, which are fussy but can be fun. A sample is the mini-theorem discussed in Problem 9, the one that says that there can never be more than one element that acts the way 0 does. Part (c) of this problem is another sample of the same kind of thing. It is easy, but it's here because it is useful and, incidentally, because it shows how the defining axioms of groups can be useful. What was proved in Problem 9 is that if an element acts the way 0 does for *every* element, then it must be 0; part (c) here asks about elements that act the way 0 does for only *one* element.

14. Groups   14

According to the definition in Problem 13 a set endowed with an operation that has all the defining properties of an abelian group except possibly the first, namely commutativity, is called just simply a **group**. (Recall that the word "abelian" is a synonym for "commutative".) Emphasis: the operation is an essential part of the definition; if two different operations on the same set both satisfy the defining conditions, the results are regarded as two different groups.

Probably the most familiar example of an abelian group is the set $\mathbb{Z}$ of all integers (positive, negative, and zero), or, better said, the group is the pair $(\mathbb{Z}, +)$, the set $\mathbb{Z}$ together with, endowed with, the operation of addition. It is sometimes possible to throw away some of the integers and still have a group left; thus, for instance, the set of all even integers is a group. Throwing things away can, however, be dangerous: the set of positive integers is not an additive group (there is no identity element: 0 is missing), and neither is the set of non-negative integers (once 0 is put back in it makes sense to demand inverses, but the demand can be fulfilled only by putting all the negative integers back in too).

The set of real numbers with addition, in symbols $(\mathbb{R}, +)$, is a group, but $(\mathbb{R}, \times)$, the set of real numbers with multiplication is not—the number 0 has no inverse. The set of non-zero real numbers on the other hand, is a multiplicative group. The same comments apply to the set $\mathbb{C}$ of complex numbers. The set of positive real numbers is a group with respect to multiplication, but the set of negative real numbers is not—the product of two of them is not negative.

Group theory is deep and pervasive: no part of mathematics is free of its influence. At the beginning of linear algebra not much of it is needed, but even here it is a big help to be able to recognize a group when one enters the room.

Problem 14. *Is the set of all affine transformations $\xi \mapsto \alpha\xi + \beta$ (with the operation of functional composition) a group? What about the set of all 2×2 matrices*

$$\begin{pmatrix} \alpha & \beta \\ \gamma & \delta \end{pmatrix},$$

(with matrix multiplication)? Is the set of non-zero integers modulo 6 (that is: the set of numbers 1, 2, 3, 4, 5) a group with respect to multiplication modulo 6? What if 6 is replaced by 7: does the conclusion remain true or does it change?

Comment. The symbol $\mapsto$, called the **barred arrow**, is commonly used for functions; it serves the purpose of indicating the "variable" that a function depends on. To speak of "the function $2x + 3$" is bad form; what the expression $2x + 3$ denotes is not a function but the value of a function at the number x. Correct language speaks of the function

$$x \mapsto 2x + 3,$$

which is an abbreviation for "the function whose value at each x is $2x+3$".

15. Independent group axioms

A group is a set with an operation that has three good properties, namely associativity, the existence of an identity element, and the existence of inverses. Are those properties totally independent of one another, or is it the case that some of them imply some of the others? So, for example, must an associative operation always have an identity element? The answer is no, and that negative answer is one of the first things that most children learn about arithmetic. We all learn early in life that we can add two positive integers and get a third one, and we quickly recognize that that third one is definitely different from both of the numbers that we started with—the discovery of zero came to humanity long after the discovery of addition, and it comes similarly late to each one of us. Very well then, if we have an associative operation that does possess an identity element, does it follow that every element has an inverse? The negative answer to that question reaches most of us not long after our first arithmetic disappointment (see Problem 13): in the set $\{0, 1, 2, 3, \ldots\}$ we can add just fine, and 0 is an identity element for addition, but + inverses are hard to come by—subtraction cannot always be done. After these superficial comments there is really only one sensible question left to ask.

Problem 15. *Can there exist a non-associative operation with an identity element, such that every element has an inverse?*

16. Fields

16

If, temporarily, "number" is interpreted to mean "integer", then numbers can be added and subtracted and multiplied, but, except accidentally as it were, they cannot be divided. If we insist on dividing them anyway, we leave the domain of integers and get the set $\mathbb{Q}$ of all quotients of integers— in other words the set $\mathbb{Q}$ of rational numbers, which is a "field".

(Does everyone know about rational numbers? A real number is called **rational** if it is the ratio of two integers. In other words, x is rational just in case there exist integers m and n such that $x = \frac{m}{n}$. Examples: $\frac{2}{3}$, $-\frac{2}{3}$, 0, 1, 10, -10, 5^5. Note: $\frac{4}{6}$ and $\frac{-12}{-18}$ are additional representations of the rational number $\frac{2}{3}$ already mentioned; there are many others. Celebrated counterexamples: $\sqrt{2}$ and π. A proof that $\sqrt{2}$ is not rational was known to humanity well over 2000 years ago; the news about π is only a little more than 200 years old. That $(\mathbb{Q}, +)$ is a group needs to be checked, of course, but the check is easy, and the same is true for $(\mathbb{Q} - \{0\}, \times)$.)

Probably the best known example of a field is the set $\mathbb{R}$ of real numbers endowed with the operations of addition and multiplication. As far as addition goes, $\mathbb{R}$ is an abelian group, and so is $\mathbb{Q}$. The corresponding statement for multiplication is not true; zero causes trouble. Since $0x = 0$ for every real number x, it follows that there is no x such that $0x = 1$; the number 0 does not have a multiplicative inverse. If, however, $\mathbb{R}$ is replaced by the set $\mathbb{R}^*$ of real numbers different from 0 (and similarly $\mathbb{Q}$ is replaced by the set $\mathbb{Q}^*$ of rational numbers different from 0), then everything is all right again: $\mathbb{R}^*$ with multiplication is an abelian group, and so is $\mathbb{Q}^*$.

It surely does not come as a surprise that the same statements are true about the set $\mathbb{C}$ of all complex numbers with addition and multiplication, and, indeed, $\mathbb{C}$ is another example of a field. The properties of $\mathbb{Q}$ and $\mathbb{R}$ and $\mathbb{C}$ that have been mentioned so far, are, however, not quite enough to guess the definition of a field from. What the examples suggest is that a field has two operations; what they leave out is a connection between them. It is mathematical malpractice to endow a set with two different structures that have nothing to do with each other. In the examples already mentioned addition and multiplication together form a pleasant and useful conspiracy (in fact two conspiracies), called the **distributive law** (or laws):

$$\alpha(x + y) = \alpha x + \alpha y \quad \text{and} \quad (\alpha + \beta)x = \alpha x + \beta x,$$

and once that is observed the correct definition of fields becomes guessable. A **field** is a set $\mathbb{F}$ with two operations $+$ and $\times$ such that with $+$ the entire set $\mathbb{F}$ is an abelian group, with $\times$ the diminished set $\mathbb{F}^*$ (omit 0) is an abelian group, and such that the distributive laws are true.

(Is it clear what "0" means here? It is intended to mean the identity element of the additive group $\mathbb{F}$. The notational conventions for real numbers are accepted for all fields: 0 is always the additive neutral element, 1 is the multiplicative one, and, except at rare times of emphasis, multiplication is indicated simply by juxtaposition.)

Some examples of fields, less obvious than $\mathbb{R}$, $\mathbb{Q}$, and $\mathbb{C}$, deserve mention. One good example is called $\mathbb{Q}(\sqrt{2})$; it consists of all numbers of the form $\alpha + \beta\sqrt{2}$ where α and β are rational; the operations are the usual addition and multiplication of real numbers. All parts of the definition, except perhaps one, are obvious. What may not be obvious is that every non-zero element of $\mathbb{Q}(\sqrt{2})$ has a reciprocal. The proof that it is true anyway, obvious or no, is the process of rationalizing the denominator (used before in Solution 10). That is: to determine $\dfrac{1}{\alpha + \beta\sqrt{2}}$, multiply both numerator and denominator by $\alpha - \beta\sqrt{2}$ and get

$$\frac{\alpha - \beta\sqrt{2}}{\alpha^2 - 2\beta^2} = \frac{\alpha}{\alpha^2 - 2\beta^2} - \frac{\beta}{\alpha^2 - 2\beta^2}\sqrt{2}.$$

The only thing that could possibly go wrong with this procedure is that the denominator $\alpha^2 - 2\beta^2$ is zero, and that cannot happen (unless both α and β are zero to begin with)—the reason it cannot happen is that $\sqrt{2}$ is not rational.

Could it happen in a field that the additive identity element is equal to the multiplicative one, that is that $0 = 1$? That is surely not the intention of the definition. A legalistic loophole that avoids such degeneracy is to recall that a group is never the empty set (because, by assumption, it contains an identity element). It follows that if $\mathbb{F}$ is a field, then $\mathbb{F} - \{0\}$ is not empty. That does it: 1 is an element of $\mathbb{F} - \{0\}$, and therefore $1 \neq 0$.

If $\mathbb{F}$ is a field, then both $\mathbb{F}$ with $+$ and $\mathbb{F}^*$ with $\times$ are abelian groups, but neither of these facts has anything to do with the *multiplicative* properties of the *additive* inverse 0. As far as they are concerned, do fields in general behave the way $\mathbb{Q}$, $\mathbb{R}$, and $\mathbb{C}$ do, or does the generality permit some unusual behavior?

Problem 16.   *Must multiplication in a field be commutative?*

17. Addition and multiplication in fields   17

If a question about the behavior of the elements of a field concerns only one of the two operations, it is likely to be easy and uninteresting. Example: is it true that if

$$\alpha + \gamma = \beta + \gamma,$$

then $\alpha = \beta$ (additive cancellation law)? Answer: yes—just add $-\gamma$ to both sides. A pleasant consequence (that sometimes bothers mathematical beginners): since

$$\alpha + (-\alpha) = 0$$

and

$$(-\alpha) + \big(-(-\alpha)\big) = 0,$$

the commutativity of addition and the additive cancellation law imply that

$$-(-\alpha) = \alpha.$$

The tricky and useful questions about fields concern addition and multiplication simultaneously.

Problem 17.   *Suppose that $\mathbb{F}$ is a field and α and β are in $\mathbb{F}$. Which of the following plausible relations are necessarily true?*
   (a) $0 \times \alpha = 0$.
   (b) $(-1)\alpha = -\alpha$.
   (c) $(-\alpha)(-\beta) = \alpha\beta$.
   (d) $1 + 1 \neq 0$.
   (e) *If $\alpha \neq 0$ and $\beta \neq 0$, then $\alpha\beta \neq 0$.*

Comment.   Observe that both operations enter into each of the five relations. (a) What is the multiplicative behavior of the additive unit? (b) What is the multiplicative behavior of the additive inverse of the multiplicative unit? (c) What is the multiplicative behavior of additive inverses in general? (d) What is the additive behavior of the multiplicative unit? (e) What is the relation of multiplication to the additive unit?

18   ## 18. Distributive failure

Problem 18. *Is there a set $\mathbb{F}$ with three commutative operations $+$, $\times_1$, and $\times_2$ such that $(\mathbb{F}, +)$ is a group, both $(\mathbb{F} - \{0\}, \times_1)$ and $(\mathbb{F} - \{0\}, \times_2)$ are groups, but only one of $(\mathbb{F}, +, \times_1)$ and $(\mathbb{F}, +, \times_2)$ is a field?*

19   ## 19. Finite fields

Could it happen that a field has only finitely many elements? Yes, it could; one very easy example is the set $\mathbb{Z}_2$ consisting of only the two integers 0 and 1 with addition and multiplication defined modulo 2. (Compare the non-field constructed in Problem 16.)

The same sort of construction (add and multiply modulo something) yields the field $\mathbb{Z}_3 = \{0, 1, 2\}$ with addition and multiplication modulo 3. If, however, 4 is used instead of 2 or 3, something goes wrong; the set $\mathbb{Z}_4 = \{0, 1, 2, 3\}$ with addition and multiplication modulo 4 is *not* a field. What goes wrong is not that $2 + 2 = 0$—there is nothing wrong with that— but that $2 \times 2 = 0$—that's bad. The reason it's bad is that it stops 2 from having a multiplicative inverse; the set $\mathbb{Z}_4^* (= \mathbb{Z}_4 - \{0\})$ is not an abelian group.

Further experimentation along these lines reveals that $\mathbb{Z}_5$ is a field, but $\mathbb{Z}_6$ is not; $\mathbb{Z}_7$ is a field, but $\mathbb{Z}_8$, $\mathbb{Z}_9$, and $\mathbb{Z}_{10}$ are not; $\mathbb{Z}_{11}$ is a field, but $\mathbb{Z}_{12}$ is not. (In $\mathbb{Z}_8$, $2 \times 4 = 0$; in $\mathbb{Z}_9$, $3 \times 3 = 0$; etc.) General fact (not hard to prove): $\mathbb{Z}_n$ is a field if and only if the modulus n is a prime.

The fact that $\mathbb{Z}_4$ is not a field shows that a certain way of defining addition and multiplication for four elements does not result in a field. Is it possible that different definitions would lead to a different result?

Problem 19. *Is there a field with four elements?*

CHAPTER **2**

VECTORS

20. Vector spaces

Real numbers can be added, and so can pairs of real numbers. If $\mathbb{R}^2$ is the set of all ordered pairs (α, β) of real numbers, then it is natural to define the sum of two elements of $\mathbb{R}^2$ by writing

$$(\alpha, \beta) + (\gamma, \delta) = (\alpha + \gamma, \beta + \delta)$$

and the result is that $\mathbb{R}^2$ becomes an abelian group. There is also a kind of partial multiplication that makes sense and is useful, namely the process of multiplying an element of $\mathbb{R}^2$ by a real number and thus getting another element of $\mathbb{R}^2$:

$$\alpha(\beta, \gamma) = (\alpha\beta, \alpha\gamma).$$

The end result of these comments is a structure consisting of three parts: an abelian group, namely $\mathbb{R}^2$, a field, namely $\mathbb{R}$, and a way of multiplying the elements of the group by the elements of the field.

For another example of the kind of structure that linear algebra studies, consider the set $\mathbb{P}$ of all polynomials with real coefficients. The set $\mathbb{P}$, endowed with the usual notion of addition of polynomials, is an abelian group. Just as in the case of $\mathbb{R}^2$ there is a multiplication that makes useful sense, namely the process of multiplying a polynomial p by a real number:

$$(\alpha p)(x) = \alpha \cdot p(x).$$

The result is, as before, a triple structure: an abelian group $\mathbb{P}$, a field $\mathbb{R}$, and a way of multiplying elements of $\mathbb{P}$ by elements of $\mathbb{R}$.

The modification of replacing the set $\mathbb{P}$ of all real polynomials (**real polynomial** is a handy abbreviation for "polynomial with real coefficients")

17

by the set $\mathbb{P}_3$ of all real polynomials of degree less than or equal to 3 is sometimes more natural to use than the unmodified version. The sum of two elements of $\mathbb{P}_3$ is again an element of $\mathbb{P}_3$, and so is the product of an element of $\mathbb{P}_3$ by a real number, and that's all there is to it.

One more example, and that will be enough for now. This time let $\mathbb{V}$ be the set of all ordered triples (α, β, γ) of real numbers such that

$$\alpha + \beta + \gamma = 0.$$

Define addition by

$$(\alpha, \beta, \gamma) + (\alpha', \beta', \gamma') = (\alpha + \alpha', \beta + \beta', \gamma + \gamma'),$$

define multiplication by

$$\alpha(\beta, \gamma, \delta) = (\alpha\beta, \alpha\gamma, \alpha\delta),$$

and observe that the result is always an ordered triple with sum zero. The set of all such triples is, once again, an abelian group (namely $\mathbb{V}$), a field, and a sensible way of multiplying group elements by field elements.

The general concept of a vector space is an abstraction of examples such as the ones just seen: it is a triple consisting of an abelian group, a field, and a multiplication between them. Recall, however, that it is immoral, illegal, and unprofitable to endow a set with two or more mathematical structures without tightly connecting them, so that each of them is restricted in an essential way. (The best known instance where that somewhat vague commandment is religiously obeyed is the definition of a field in Problem 16: there is an addition, there is a multiplication, and there is the essential connection between them, namely the distributive law.)

A **vector space over a field** $\mathbb{F}$ (of elements called **scalars**) is an additive (commutative) group $\mathbb{V}$ (of elements called **vectors**), together with an operation that assigns to each scalar α and each vector x a product αx that is again a vector. For such a definition to make good mathematical sense, the operation (called **scalar multiplication**) should be suitably related to the three given operations (addition in $\mathbb{F}$, addition in $\mathbb{V}$, and multiplication in $\mathbb{F}$). The conditions that present themselves most naturally are these.

The vector distributive law:

$$(\alpha + \beta)x = \alpha x + \beta x$$

whenever α and β are scalars and x is a vector. (In other words, multiplication by a vector distributes over scalar addition.)

The scalar distributive law:

$$\alpha(x + y) = \alpha x + \alpha y$$

whenever α is a scalar and x and y are vectors. (In other words, multiplication by a scalar distributes over vector addition).

The associative law:

$$(\alpha\beta)x = \alpha(\beta x)$$

whenever α and β are scalars and x is a vector.

The scalar identity law:

$$1x = x$$

for every vector x. (In other words, the scalar 1 acts as the identity transformation on vectors).

(The reader has no doubt noticed that in scalar multiplication the scalar is always on the left and the vector on the right—since the other kind of multiplication is not even defined, it makes no sense to speak of a commutative law. Nothing is lost by this convention, and something is gained: the very symbol for a product indicates which factor is the scalar and which the vector.)

Many questions can and should be asked about the conditions that define vector spaces: one worrisome question has to do with multiplication, and another one, easier, has to do with zero.

Why, it is natural to ask, is a multiplicative structure not imposed on vector spaces? Wouldn't it be natural and useful to define $(\alpha, \beta) \cdot (\gamma, \delta) = (\alpha\gamma, \beta\delta)$ (similarly to how addition is defined in $\mathbb{R}^2$)? The answer is no. The trouble is that even after the zero element (that is, the element $(0,0)$) of $\mathbb{R}^2$ is discarded, the remainder does not constitute a group; a pair that has one of its coordinates equal to 0, such, for instance, as $(1,0)$, does not have an inverse. The same question for $\mathbb{P}$ is tempting: the elements of $\mathbb{P}$ can be multiplied as well as added. Once again, however, the result does not convert $\mathbb{P}$ into a field; the multiplicative inverse of a polynomial is very unlikely to be a polynomial. Examples such as $\mathbb{P}_3$ add to the discouragement: the product of two elements of $\mathbb{P}_3$ might not be an element of $\mathbb{P}_3$ (the degree might be too large). The example of triples with sum zero is perhaps the most discouraging: the attempt to define the product of two elements of $\mathbb{V}$ collapses almost before it is begun. Even if both $\alpha + \beta + \gamma = 0$ and $\alpha' + \beta' + \gamma' = 0$, it is a rare coincidence if also $\alpha\alpha' + \beta\beta' + \gamma\gamma' = 0$. It is best, at this stage, to resist the temptation to endow vector spaces with a multiplication.

Both the scalar 0 and the vector 0 have to do with addition; how do they behave with respect to multiplication? It is possible that they misbehave, in a sense something like the one for vector multiplication discussed in the preceding paragraph, and it is possible that they are perfectly well behaved—which is true?

Problem 20.   *Do the scalar zero law,*

$$0x = 0,$$

and the vector zero law,

$$\alpha 0 = 0,$$

follow from the conditions in the definition of vector spaces, or could they be false?

Comment.   Note that in the scalar zero law the symbol 0 denotes a scalar on the left and a vector on the right; in the vector zero law it denotes a vector on both sides.

21. Examples

It is always important, in studying a mathematical structure, to be able to recognize an example as a proper one, and to recognize a pretender as one that fails in some respects. Here are a half dozen candidates that may be vector spaces or may be pretenders.

(1) Let $\mathbb{F}$ be $\mathbb{C}$, and let $\mathbb{V}$ also be the set $\mathbb{C}$ of complex numbers. Define addition in $\mathbb{C}$ the usual way, and let scalar multiplication (denoted by $*$) be defined as follows:

$$\alpha * x = \alpha^2 \cdot x.$$

(2) Let $\mathbb{F}$ be a field, let $\mathbb{V}$ be $\mathbb{F}^2$ (the set of all ordered pairs of elements of $\mathbb{F}$), let addition in $\mathbb{V}$ be the usual one (coordinatewise), and define a new scalar multiplication by writing

$$\alpha * (\beta, \gamma) = (\alpha\beta, 0)$$

(for all α, β, and γ).

(3) Let $\mathbb{F}$ be the field of four elements discussed in Problem 19, let $\mathbb{V}$ be $\mathbb{F}^2$ with the usual addition, and define scalar multiplication by

$$\alpha * (\beta, \gamma) = (\alpha\beta, \alpha\gamma) \qquad \text{if } \gamma \neq 0$$

and

$$\alpha * (\beta, 0) = (\alpha^2\beta, 0).$$

(4) Let $\mathbb{F}$ be $\mathbb{R}$ and let $\mathbb{V}$ be the set $\mathbb{R}_+$ of all positive real numbers. Define the "sum" denoted by $\alpha \boxed{+} \beta$ of any two positive real numbers α and β, and define the "scalar product" denoted by $\alpha \boxed{\cdot} \beta$ of any positive real number α by an arbitrary (not necessarily positive) real number β as follows:

$$\alpha \boxed{+} \beta = \alpha\beta$$

and

$$\alpha \boxed{\cdot} \beta = \beta^\alpha$$

(5) Let $\mathbb{F}$ be $\mathbb{C}$, and let $\mathbb{V}$ be $\mathbb{C}$ also. Vector addition is to be defined as the ordinary addition of complex numbers, but the product of a scalar α (in $\mathbb{C}$) and a vector x (in $\mathbb{C}$) is to be defined by forming the real part of α first. That is:

$$\alpha \cdot x = (\operatorname{Re} \alpha)x.$$

(6) Let $\mathbb{F}$ be the field $\mathbb{Q}$ of rational numbers, let $\mathbb{V}$ be the field $\mathbb{R}$ of real numbers and define scalar multiplication by writing

$$\alpha * x = \alpha x$$

for all α in $\mathbb{Q}$ and all x in $\mathbb{R}$.

Problem 21. *Which of the defining conditions of vector spaces are missing in the examples (1), (2), (3), (4), (5), and (6)?*

22. Linear combinations   22

The best known example of a vector space is the space $\mathbb{R}^2$ of all ordered pairs of real numbers, such as

$$\begin{cases} (1,1), \\ (0, \pi^2), \\ \left(\frac{1}{2}, \sqrt{2}\right), \\ (0, -200), \\ \left(\frac{1}{\sqrt{5}}, -\sqrt{5}\right). \end{cases} \tag{$*$}$$

An example of a vector space, different from $\mathbb{R}^2$ but very near to it in spirit, consists not of all ordered pairs of real numbers, but only some of them.

That is, throw away most of the pairs in $\mathbb{R}^2$; typical among the ones to be kept are

$$
\begin{cases}
(0,0), \\
\left(-\dfrac{1}{2}, 1\right), \\
(\sqrt{5}, -2\sqrt{5}), \\
\left(\dfrac{1}{\sqrt{2}}, -\sqrt{2}\right).
\end{cases}
\tag{$**$}
$$

Are these four pairs in $\mathbb{R}^2$ enough to indicate a pattern?—is it clear which pairs are to be thrown away and which are to be kept? The answer is: keep only the pairs in which the second entry is -2 times the first. Right? Indeed: $0 = (-2) \cdot 0$, and $1 = (-2)(-\frac{1}{2})$, and so on. Use $\mathbb{R}_0^2$ as a temporary symbol to denote this new vector space.

Spaces such as $\mathbb{R}^2$ and $\mathbb{R}_0^2$ are familiar from analytic geometry; $\mathbb{R}^2$ is the Euclidean plane equipped with a Cartesian coordinate system, and $\mathbb{R}_0^2$ is a line in that plane, the line with the equation $2x + y = 0$. It is often good to use geometric language in linear algebra; it is comfortable and it suggests the right way to look at things.

A vector was defined as an element of a vector space—any vector space. Caution: the word "vector" is therefore a relative word—it changes its meaning depending on which vector space is under study. (It's like the word "citizen", which changes its meaning depending on which nation is being talked about.) Vectors in the particular vector spaces $\mathbb{R}^2$ and $\mathbb{R}_0^2$ happen to be ordered pairs of real numbers, and the two real numbers that make up a vector are called its **coordinates**. Each of the five pairs in the list $(*)$ is a vector in $\mathbb{R}^2$ (but none of them belongs to $\mathbb{R}_0^2$), and each of the four pairs in the list $(**)$ is a vector in $\mathbb{R}_0^2$.

The most important aspect of vectors is not what they look like but what one can do with them, namely add them, and multiply them by scalars. More generally, if $x = (\alpha_1, \alpha_2)$ and $y = (\beta_1, \beta_2)$ are vectors (either both in $\mathbb{R}^2$ or else both in $\mathbb{R}_0^2$) and if ξ and η are real numbers, then it is possible to form

$$
\xi x + \eta y = (\xi \alpha_1 + \eta \beta_1, \xi \alpha_2 + \eta \beta_2),
$$

which is a vector in the same space called a **linear combination** of the given vectors x and y, and that's what a lot of the theory of vector spaces has to do with. Example: since

$$
3(4,0) - 2(0,5) = (12, -10),
$$

the vector $(12, -10)$ is a linear combination of the vectors $(4,0)$ and $(0,5)$. Even easier example: the vector $(7, \pi)$ is a linear combination of the vectors $(1,0)$ and $(0,1)$; indeed

$$(7, \pi) = 7(1,0) + \pi(0,1).$$

This easy example has a very broad and completely obvious generalization: every vector (α, β) is a linear combination of $(1,0)$ and $(0,1)$. Proof:

$$(\alpha, \beta) = \alpha(1,0) + \beta(0,1).$$

Problem 22. *Is $(2,1)$ a linear combination of the vectors $(1,1)$ and $(1,2)$ in $\mathbb{R}^2$? Is $(0,1)$? More generally: which vectors in $\mathbb{R}^2$ are linear combinations of $(1,1)$ and $(1,2)$?*

Comment. It is important to remember that 0 is a perfectly respectable scalar, so that, in particular $(1,1)$ is a linear combination of $(1,1)$ and $(1,2)$:

$$(1,1) = 1 \cdot (1,1) + 0 \cdot (1,2),$$

and so is $(0,0)$:

$$(0,0) = 0 \cdot (1,1) + 0 \cdot (1,2).$$

23. Subspaces   23

The discussion of Problem 21 established that the four axioms that define vector spaces (the vector and scalar distributive laws, the associative law, and the scalar identity law) are independent. According to the official definition, therefore, a vector space is an abelian group $\mathbb{V}$ on which a field $\mathbb{F}$ "acts" so that the four independent axioms are satisfied. The set $\mathbb{P}$ of all polynomials with, say, real coefficients, is an example of a real vector space, and so is the subset $\mathbb{P}_3$ of all polynomials of degree less than or equal to 3. The set $\mathbb{R}^3$ of all ordered triples of real numbers is a real vector space, and so is the subset $\mathbb{V}$ consisting of all ordered triples with sum zero. (See Problem 20.) The subset $\mathbb{Q}^3$ of $\mathbb{R}^3$ consisting of all triples with rational coordinates is not a vector space over $\mathbb{R}$ (an irrational number times an element of $\mathbb{Q}^3$ might not belong to $\mathbb{Q}^3$), and the subset $\mathbb{X}$ of $\mathbb{R}^3$ consisting of all triples with at least one coordinate equal to 0 is not a vector space (the sum of two elements of $\mathbb{X}$ does not necessarily belong to $\mathbb{X}$).

These examples illustrate and motivate an important definition: a nonempty subset $\mathbb{M}$ of a vector space $\mathbb{V}$ is a **subspace** of $\mathbb{V}$ if the sum of two vectors in M is always in M and if the product of a vector in M with every scalar is always in M. An equivalent way of phrasing the definition is this: a non-empty set M is a subspace if and only if $\alpha x + \beta y$ belongs to M whenever x and y are vectors in M and α and β are arbitrary scalars, or, in other words, subspaces are just the non-empty subsets closed under the formation of linear combinations.

The set $\mathbb{O}$ consisting of the zero vector alone is a subspace of every vector space $\mathbb{V}$ (it is usually referred to as the **trivial** subspace—the others are called non-trivial), and so is the entire space $\mathbb{V}$. The way the words "subset" and "subspace" are used is intended to allow these extremes. (A subspace of $\mathbb{V}$ different from $\mathbb{V}$ is called a **proper** subspace—in this language $\mathbb{V}$ itself is called the improper subspace.) To get more interesting examples of subspaces, it's a good idea to enlarge the stock of examples of vector spaces.

It has already been noted (see Solution 20) that every field is a vector space over itself. In particular, $\mathbb{R}$ is a vector space over $\mathbb{R}$, but, and this is more interesting, $\mathbb{R}$ is a vector space over $\mathbb{Q}$ also—just forget how to multiply real numbers by anything except rational numbers. In this situation, where $\mathbb{R}$ is regarded as a rational vector space, the subset $\mathbb{Q}$ of $\mathbb{R}$ is a new example of a subspace, and so is the larger subset $\mathbb{Q}(\sqrt{2})$ (see Problem 16). In the same spirit, $\mathbb{C}$ (with the operation of addition) is a vector space over $\mathbb{C}$, and it is also a vector space over R; from the latter point of view, the set $\mathbb{R}$ is a subspace (a real subspace of $\mathbb{C}$).

Usually when vector spaces are discussed a field $\mathbb{F}$ has been fixed once and for all, and it is clear that all vector spaces under consideration are over $\mathbb{F}$. If, however, there is some chance that the underlying field may have to be changed during the discussion, then it is necessary to specify the field each time. One way to do that is to speak of an $\mathbb{F}$ vector space; this is the general form of speaking of rational, real, and complex vector spaces.

If $\mathbb{F}$ is a field and n is a positive integer, then the set of all n-tuples $(\xi_1, \xi_2, \ldots, \xi_n)$ of elements of $\mathbb{F}$, is an $\mathbb{F}$ vector space (the addition of the n-tuples that play the role of vectors is coordinate by coordinate, and so is multiplication by an element of $\mathbb{F}$); this space is denoted by $\mathbb{F}^n$. The set of those n-tuples whose first coordinate is equal to 0 is a subspace.

Here is an important non-trivial example: the set of all real-valued functions defined on, say, a closed interval is a vector space over $\mathbb{R}$ if vector addition and scalar multiplication are defined in the obvious pointwise fashion. The set of all continuous functions is an example of a subspace of that space. A different generalization of $\mathbb{R}^n$ is the set of all infinite sequences $\{\xi_1, \xi_2, \xi_3, \ldots\}$, of real numbers; an example of a subspace is the subset consisting of all those sequences for which the series $\sum_{n=1}^{\infty} \xi_n$ is convergent, and a subspace of that subspace is the subset of all those sequences for which the series is absolutely convergent.

For examples with a more geometric flavor, consider the real vector space $\mathbb{R}^2$ and in it the subset M of all vectors of the form $(\alpha, 2\alpha)$, where α is an arbitrary real number. Equivalently: M consists of the vectors whose second coordinate is equal to twice the first; in the usual language of analytic geometry the elements of M are the points on the line through the origin with slope 2; the line described by the equation $y = 2x$. (Examples like this are the reason why linear algebra is called linear: the expression refers to the algebra of lines and their natural higher-dimensional generalizations.) The example is typical: every straight line through the origin is a subspace of $\mathbb{R}^2$, and every non-trivial proper subspace of $\mathbb{R}^2$ is like that. The generalization of these examples to $\mathbb{R}^3$ is straightforward: the non-trivial proper subspaces of $\mathbb{R}^3$ are the lines and planes through the origin.

What's special about the origin? Answer: it necessarily belongs to every subspace. Proof: if M is a subspace, and if x is an arbitrary element of M, then $0 \cdot x$ belongs to M (scalar multiples), and since it is already known that $0 \cdot x = 0$, it follows that $0 \in M$ for all M. The definition of subspaces could have been formulated this way: a subset M of V is a subspace if M itself is a vector space with respect to the same linear operations (vector addition and scalar multiplication, or, in one phrase, linear combination) as are given in V. Since every vector space contains its zero vector, the presence of 0 in M should not come as a surprise.

Problem 23. (a) *Consider the complex vector space $\mathbb{C}^3$ and the subsets M of $\mathbb{C}^3$ consisting of those vectors (α, β, γ) for which*
    (1) $\alpha = 0$,
    (2) $\beta = 0$,
    (3) $\alpha + \beta = 1$,
    (4) $\alpha + \beta = 0$,
    (5) $\alpha + \beta \geqq 0$,
    (6) α *is real.*
In which of these cases is M a subspace of $\mathbb{C}^3$?

(b) *Consider the complex vector space $\mathbb{P}$ and the subsets M of all those vectors (polynomials) p for which*
    (1) p *has degree 3*,
    (2) $2p(0) = p(1)$,
    (3) $p(t) \geqq 0$ *whenever* $0 \leqq t \leqq 1$,
    (4) $p(t) = p(1 - t)$ *for all t.*
In which of these cases is M a subspace of P?

24   **24. Unions of subspaces**

What set-theoretic operations on subspaces produce further examples of
subspaces? One that surely does not is set-theoretic complementation: the
vectors that do not belong to a specified subspace never form a subspace.
To become convinced of that, think of a picture, in the plane for instance:
the complement of a line is not a line. To give a brisk proof, just think of 0:
the complement of a subspace never contains it.

> **Problem 24.**   (a) *Under what conditions is the set-theoretic inter-*
> *section of two subspaces a subspace? What about the intersection of*
> *more than two subspaces (perhaps even infinitely many)—when is*
> *that a subspace?*
>   (b) *Under what conditions is the set-theoretic union of two sub-*
> *spaces a subspace? What about the union of more than two sub-*
> *spaces?*

25   **25. Spans**

Do linear combinations of more than two vectors make sense? Sure. If, for
instance, x, y, and z are three vectors in $\mathbb{R}^3$, or, for that matter, in $\mathbb{R}^2$ or in
$\mathbb{R}_0^2$ (see Problem 22) and if α, β, and γ are scalars, then the vector

$$\alpha x + \beta y + \gamma z$$

is a linear combination of the set $\{x, y, z\}$. Linear combinations of sets of
four vectors, such as $\{x_1, x_2, x_3, x_4\}$, are defined similarly as vectors of the
form

$$\alpha_1 x_1 + \alpha_2 x_2 + \alpha_3 x_3 + \alpha_4 x_4$$

(where α_1, α_2, α_3, and α_4 are scalars, of course), and the same sort of
definition is used for linear combinations of any finite set of vectors. Since

$$\alpha x + \beta y + \gamma z = 1 \cdot (\alpha x + \beta y) + \gamma \cdot z,$$

it is clear that a linear combination of three vectors can be obtained in two
steps by forming linear combinations of two vectors: the first step yields
$\alpha x + \beta y$ and the second step forms a linear combination of that and z. The
same thing is true in complete generality: every finite linear combination
can be obtained in a finite number of steps by forming linear combinations
of two vectors at a time.

A vector space of interest is the set $\mathbb{P}_5$ of all real polynomials p in one
variable x, of degree less than or equal to 5 (an obvious relative of the

space $\mathbb{P}_3$ considered in Problem 20). Examples of such polynomials are

$$p(x) = x + x^5,$$
$$p(x) = -\pi + x^3,$$
$$p(x) = 7 + x + (\sqrt{2} + e^\pi)x^5,$$
$$p(x) = 7,$$
$$p(x) = 0,$$
$$p(x) = x^4,$$

and it is clear that to get more, richer, examples of vectors in $\mathbb{P}_5$ "long" linear combinations of examples such as these need to be formed.

Objects that naturally arise in this connection are the large sets of vectors that can be obtained from small sets by forming all possible linear combinations. Problem 22, for instance, asked which vectors in $\mathbb{R}^2$ are linear combinations of $(1, 1)$ and $(1, 2)$, and the answer turned out to be that every vector in $\mathbb{R}^2$ is such a linear combination. A similar question is this: which vectors in $\mathbb{R}^3$ are linear combinations of $(1, 1, 0)$ and $(1, 2, 0)$? The solution of Problem 22 makes the answer to this question obvious: the answer is all vectors of the form $(x, y, 0)$. The technical word for "set of all linear combinations" is **span**. So, for example, the span of the vectors $(1, 1, 0)$ and $(1, 2, 0)$ is the set of all vectors of the form $(\xi, \eta, 0)$, or, to say the same thing in different words, the set $\{(1, 1, 0), (1, 2, 0)\}$ **spans** the set of all vectors of the form $(\xi, \eta, 0)$.

In geometrical language $\mathbb{R}^3$ is 3-dimensional Euclidean space. In that space the set of all those vectors (ξ, η, ζ) for which $\eta = \zeta = 0$, or, in other words, the set of all $(\xi, 0, 0)$, is called the ξ-axis, and, similarly, the η-axis is the set of all $(0, \eta, 0)$, and the ζ-axis is the set of all $(0, 0, \zeta)$. These coordinate axes are lines. The coordinate planes are the (ξ, η)-plane, which is the set of all (ξ, η, ζ) with $\zeta = 0$, or, in other words, the set of all $(\xi, \eta, 0)$, and, similarly, the (η, ζ)-plane, which is the set of all $(0, \eta, \zeta)$, and the (ξ, ζ)-plane, which is the set of all $(\xi, 0, \zeta)$. In this language: the ξ-axis and the ζ-axis span the (ξ, ζ)-plane, and the set $\{(1, 1, 0), (1, 2, 0)\}$ spans the (ξ, η)-plane.

What is the span of the set $\{(1, 1, 1), (0, 0, 0)\}$? Answer: it is the set of all vectors of the form (ξ, ξ, ξ), or, geometrically, it is the line through the origin that makes an angle of $45°$ with each of the three coordinate axes.

How about this: does the vector $(1, 4, 9)$ in $\mathbb{R}^3$ belong to the span of $\{(1, 1, 1), (0, 1, 1), (0, 0, 1)\}$? The answer is probably not obvious, but it is not difficult to get. If $(1, 4, 9)$ did belong to the span of

$$\{(1, 1, 1), (0, 1, 1), (0, 0, 1)\},$$

then scalars α, β, and γ could be found so that

$$\alpha(1, 1, 1) + \beta(0, 1, 1) + \gamma(0, 0, 1) = (1, 4, 9),$$

and then it would follow that

$$\alpha = 1,$$
$$\alpha + \beta = 4,$$
$$\alpha + \beta + \gamma = 9.$$

This in turn implies that

$$\alpha = 1,$$
$$\beta = 4 - \alpha = 4 - 1 = 3,$$
$$\gamma = 9 - \alpha - \beta = 9 - 1 - 3 = 5.$$

Check: $1 \cdot (1, 1, 1) + 3 \cdot (0, 1, 1) + 5 \cdot (0, 0, 1) = (1, 4, 9)$.

Among the simplest of the polynomials (vectors) in the vector space $\mathbb{P}_5$ are 1, x^2, and x^4. What is their span? Answer: it is the set of all polynomials of the form

$$\alpha + \beta x^2 + \gamma x^4.$$

These polynomials happen to have a pleasant property that characterizes them: the replacement of x by $-x$ does not change them. Polynomials with this property are called **even**. Symbolically said: a polynomial p is even if it satisfies the identity $p(-x) = p(x)$. A polynomial p is called **odd** if it satisfies the identity $p(-x) = -p(x)$. What do the odd polynomials in $\mathbb{P}_5$ look like?

Problem 25. (a) *Can two disjoint subsets of $\mathbb{R}^2$, each containing two vectors, have the same span?* (b) *What is the span in $\mathbb{R}^3$ of*

$$\{(1, 1, 1), (0, 1, 1), (0, 0, 1)\}?$$

26   26. Equalities of spans

Span is a set-theoretic operation that converts sets (of vectors) into other sets (subspaces). (In other words, "span" applies to *sets* of vectors, not to vectors themselves, and an expression such as "the span of the vectors x and y" is not really a proper one.) What can be said about the relation between a set of vectors and its span? There are three easy statements on the most abstract (and therefore most shallow) level, namely that

(1) every set is a subset of its span,
(2) if a set $\mathbb{E}$ is a subset of a set $\mathbb{F}$, then the span of $\mathbb{E}$ is a subset of the span of $\mathbb{F}$,

and

(3) the span of the span of a set is the same as the span of the set.

It is often convenient to have a symbol to denote "span", and one possible symbol is $\bigvee$ (which is intended to be reminiscent of the ordinary set-theoretic symbol for union). In terms of that symbol the statements just made can be expressed as follows:

$$\mathbb{E} \subset \bigvee \mathbb{E}, \tag{1}$$

$$\text{if } \mathbb{E} \subset \mathbb{F}, \text{ then } \bigvee \mathbb{E} \subset \bigvee \mathbb{F}, \tag{2}$$

and

$$\bigvee \bigvee \mathbb{E} = \bigvee \mathbb{E}. \tag{3}$$

In technical language (which is not especially useful here) (1) says that the span operation is increasing, (2) says that it is monotone, and (3) says that it is idempotent.

Knowledge about the span of a set of vectors provides geometric insight about the set, and, for instance, the knowledge that two sets have the same span (compare Problem 25 (a)) provides geometric insight about the relations between them. Here is an example of the kind of question about some spans that might arise: if we know about three vectors x, y, and z that $x \in \bigvee\{y, z\}$, are we allowed to infer that $\bigvee\{x, z\} = \bigvee\{y, z\}$? The answer is no. If, for instance, x is a scalar multiple of z but y is not, then x obviously belongs to $\bigvee\{x, z\}$, but y does not.

A related question is this: if $\mathbb{M}$ is a subspace and x and y are vectors such that

$$x \in \bigvee\{\mathbb{M}, y\},$$

does it follow that

$$\bigvee\{\mathbb{M}, x\} = \bigvee\{\mathbb{M}, y\}?$$

(Here $\{\mathbb{M}, x\}$ is an abbreviation for $\mathbb{M} \cup \{x\}$.) The answer is trivially no: it could, for instance, happen that $\mathbb{M}$ is the subspace spanned by x, in which case the assumption is obviously true, but the questioned conclusion can be true only if y belongs to that subspace, which it may fail to do. Is that the only thing that can go wrong?

> **Problem 26.** *If x and y are vectors and $\mathbb{M}$ is a subspace such that $x \notin \mathbb{M}$ but $x \in \bigvee\{\mathbb{M}, y\}$, does it follow that*
>
> $$\bigvee\{\mathbb{M}, x\} = \bigvee\{\mathbb{M}, y\}?$$

27. Some special spans

If x is a vector in $\mathbb{V}$, what is the intersection of all the subspaces of $\mathbb{V}$ that contain x? (Caution: are there any?) In view of Problem 24, one thing is for sure: that intersection, call it $\mathbb{M}$, is a subspace. Since, moreover, $\mathbb{M}$ is the intersection of sets (subspaces) each of which contains x, it too contains x. What else does it have to contain? Answer: since a subspace containing x must contain all scalar multiples of x, it follows that αx is in $\mathbb{M}$ for every α. The set $\mathbb{M}_0$ of all scalar multiples of x is itself a subspace, and the preceding sentence says exactly that $\mathbb{M}_0 \subset \mathbb{M}$. Since, moreover, the subspace $\mathbb{M}_0$ contains x, so that it is a member of the collection that was intersected to form $\mathbb{M}$, it follows that $\mathbb{M} \subset \mathbb{M}_0$. Consequence: $\mathbb{M} = \mathbb{M}_0$.

The same argument can be applied to two vectors as easily as to one. If x and y are in $\mathbb{V}$, there surely exist subspaces that contain them both ($\mathbb{V}$ is one), and the intersection of all those subspaces is a subspace $\mathbb{M}$ that contains both. Being a subspace, it contains all linear combinations $\alpha x + \beta y$ also. The set $\mathbb{M}_0$ of all such linear combinations is itself a subspace, the one that was called the span of $\{x, y\}$ in Problem 25, and, obviously $\mathbb{M}_0 \subset \mathbb{M}$. Argue as above and conclude that $\mathbb{M} = \mathbb{M}_0$.

These examples are special cases of a general concept that applies to arbitrary subsets of a vector space. If $\mathbb{E} \subset \mathbb{V}$ (the set $\mathbb{E}$ can be a singleton $\{x\}$, a pair $\{x, y\}$, or, for that matter, an arbitrary finite or infinite set), the intersection, call it $\mathbb{M}$, of all subspaces that include $\mathbb{E}$ is a subspace. (Recall that there always exists at least one subspace that includes $\mathbb{E}$, namely $\mathbb{V}$.) The argument given in the preceding paragraphs can be given again and it proves that $\mathbb{M} = \bigvee \mathbb{E}$.

Is the last sentence correct? There is a curious degenerate case to be considered: what is the span of the empty set of vectors? In view of the definition, the question is this: which vectors can be obtained as linear combinations of no vectors at all? This formulation calls attention to a blemish of the definition; it doesn't apply to the conceptually trivial but technically very important empty set. The cure is to rephrase the definition: the span of a set of vectors is the intersection of all the subspaces that include the set. It is a non-profound exercise to show that for non-empty sets the rephrasing is equivalent to the original definition; its virtue is that it applies with no change to *every* set—including, in particular, the empty set. Since every subspace includes the empty set, and, in particular, the trivial subspace $\mathbb{O}$ includes the empty set, it follows that the span of the empty set is $\mathbb{O}$. The result is worth mentioning, if only to show that the concept of span works smoothly in all cases, with no troublesome exceptions.

The span of a set (finite or infinite) consists, by definition, of the set of all linear combinations of elements of the set. There is a way of saying the same thing that uses exactly one more word and that word seems to come naturally to some people: they say that the span of a set consists of the set of all **finite** linear combinations of elements of the set. The added word "finite" is harmless, in the sense that it doesn't change the meaning of the sentence (no other kind of linear combination has been defined), but at the same time it might be harmful because it suggests that "infinite linear combinations" could have been considered but were deliberately excluded. That is not true, and that way confusion lies.

To get acquainted with the notion of span it is a good idea to look at a handful of special cases, various more or less randomly selected (small or large) subsets of $\mathbb{R}^2$ or $\mathbb{R}^3$ or $\mathbb{P}$, and examine their spans.

Problem 27. (a) *Is there a vector that spans* $\mathbb{R}^2$? (b) *Are there two vectors that span* $\mathbb{R}^2$? (c) *Are there two vectors that span* $\mathbb{R}^3$? (d) *Is there any finite set of vectors that spans* $\mathbb{P}$?

28. Sums of subspaces   28

Which vectors in $\mathbb{R}^2$ can be obtained from the two subspaces (lines) $\mathbb{M}$ and $\mathbb{N}$, where $\mathbb{M}$ is the line through the origin with slope 2 and $\mathbb{N}$ is the line through the origin with slope 3, by adding a vector in $\mathbb{M}$ to a vector in $\mathbb{N}$? The vectors in $\mathbb{M}$ are those of the form $(\alpha, 2\alpha)$ and the vectors in $\mathbb{N}$ are those of the form $(\alpha, 3\alpha)$. The question is: which vectors can be represented in the form $(\alpha + \beta, 2\alpha + 3\beta)$ as α and β are allowed to vary over all real numbers? The answer is easy enough to figure out (all vectors in $\mathbb{R}^2$), but there is a general concept here, waiting to be recognized, that's much more useful than the special answer.

If $\mathbb{M}$ and $\mathbb{N}$ are subspaces of a vector space $\mathbb{V}$, which vectors can be obtained by adding a vector in $\mathbb{M}$ to a vector in $\mathbb{N}$? That is: choose x in $\mathbb{M}$ and y in $\mathbb{N}$, form $x + y$, and ask which vectors can be so represented as x and y vary over all vectors in $\mathbb{M}$ and $\mathbb{N}$ respectively. Whatever the answer, the symbol for it is

$$M + N.$$

This is a new operation on subspaces, a kind of addition. Its main use comes from its relation to spans.

Since 0 belongs to every subspace, it follows that both M and N are included in $M + N$, and hence that $M \cup N \subset M + N$. Since $M + N$ is a subspace (that is easy to check), it follows that

$$\bigvee(M \cup N) \subset M + N.$$

(Right? If a subspace includes a set E, then it includes $\bigvee E$.) But both M and N are included in $\bigvee(M \cup N)$, and $\bigvee(M \cup N)$ is closed under vector addition. It follows that $M + N \subset \bigvee(M \cup N)$, and therefore, finally, that

$$\bigvee(M \cup N) = M + N.$$

Summary: to form the span of two subspaces is the same as forming their sum.

Addition of subspaces is a curious operation. It is commutative and associative—that's easy. It has an identity, namely the trivial subspace $\mathbb{O}$. It does not, however, make the set of subspaces of a vector space into a group—inverses do not exist. Indeed, since $M \subset M + N$ whenever M and N are subspaces, it follows that $M + N = \mathbb{O}$ is out of the question unless $M = N = \mathbb{O}$. Note also that $M + M = M$—not the sort of behavior that groups permit. A related unorthodox property of subspace addition is that $M + N = N$ can happen quite easily even when $M \neq \mathbb{O}$. (Under what conditions does it happen? Answer: if and only if $M \subset N$.)

Subspaces have a kind of multiplicative structure too, namely intersection. Intersection is commutative and associative—that's easy—and it has an identity, namely the improper subspace V. That's as far as the good properties go. Inverses do not exist. Indeed: $M \supset M \cap N$ whenever M and N are subspaces, so that $M \cap N = V$ is out of the question unless $M = N = V$. Note also that $M \cap M = M$, and that $M \cap N = N$ can happen quite easily, namely just when $N \subset M$.

Related to the additive and multiplicative structure of subspaces and symmetrically connected with both there is a geometrically important possibility that has some of the properties of set-theoretic complementation. It's best to begin its study by looking at some examples.

Consider two distinct non-trivial proper subspaces M and N of $\mathbb{R}^2$ (or, if geometric language is preferred, consider two distinct lines in the ordinary Euclidean plane). It cannot be true that $M + N = \mathbb{O}$ (that is, M

and $\mathbb{N}$ are not additive inverses of one another), and it cannot be true that $\mathbb{M} \cap \mathbb{N} = \mathbb{R}^2$ (that is, $\mathbb{M}$ and $\mathbb{N}$ are not multiplicative inverses of one another). The extreme opposite is true: $\mathbb{M} + \mathbb{N} = \mathbb{R}^2$ and $\mathbb{M} \cap \mathbb{N} = \mathbb{O}$. (Look at the picture.)

Another example. Let $\mathbb{M}$ be the set of all even polynomials (with, say, real coefficients) and $\mathbb{N}$ the set of all odd ones; the definitions of these terms appear in Problem 25. Can a polynomial be both even and odd? A moment's thought should reveal the answer: if and only if it is identically zero. Note that it follows that both $\mathbb{M}$ and $\mathbb{N}$ are subspaces of the vector space $\mathbb{P}$ of all polynomials. (Caution: if the underlying field is such that $1 + 1 = 0$, the two definitions of evenness and oddness are not equivalent. Over the field of integers modulo 2, for instance, every polynomial is both even and odd in the second sense.)

Can every polynomial be written as a sum of an even one and an odd one? Sure: given p, define q and r by

$$q(x) = \frac{1}{2}\big(p(x) + p(-x)\big) \qquad \text{and} \qquad r(x) = \frac{1}{2}\big(p(x) - p(-x)\big),$$

and verify that q is even, r is odd, and $p = q + r$. Conclusion:

$$\mathbb{M} \cap \mathbb{N} = \mathbb{O} \qquad \text{and} \qquad \mathbb{M} + \mathbb{N} = \mathbb{P}.$$

In a general vector space $\mathbb{V}$ two subspaces $\mathbb{M}$ and $\mathbb{N}$ are called **complements** of one another if

$$\mathbb{M} \cap \mathbb{N} = \mathbb{O} \qquad \text{and} \qquad \mathbb{M} + \mathbb{N} = \mathbb{V}.$$

The concept is illustrated by the examples: every line in the plane is a complement of every other line (no hope of uniqueness), and the subspaces of even and odd polynomials are complements in $\mathbb{P}$. A trivial example can be given in any $\mathbb{V}$, namely the trivial subspace $\mathbb{O}$ and the improper subspace $\mathbb{V}$.

Does every subspace in every vector space have at least one complement? Does every non-trivial proper subspace in every vector space have many complements? (What does "many" mean? It is a relative notion; it depends on the coefficient field. For finite fields "many" might just mean "more than one".) The answers are yes both times, but the proofs depend on set-theoretic techniques (such as Zorn's lemma) foreign to the spirit of introductory linear algebra. For the vector spaces that will presently start occupying the center of the stage the answer will be obtained by more easily accessible methods.

The set of subspaces with addition ($+$) and intersection ($\cap$) misses being a field because neither operation admits inverses. What about the connection between the two operations: how well behaved is it? A frontal attack on the question would try to prove or disprove the distributive law. It is advisable to approach the question more modestly by asking about easier algebraic properties of subspace addition. One well-known possible property is called the **modular identity**; that's what this problem is about.

Problem 28.   *Is it true that if* $\mathbb{L}$, $\mathbb{M}$, *and* $\mathbb{N}$ *are subspaces of a vector space, then*

$$\mathbb{L} \cap (\mathbb{M} + (\mathbb{L} \cap \mathbb{N})) = (\mathbb{L} \cap \mathbb{M}) + (\mathbb{L} \cap \mathbb{N})?$$

29   29. Distributive subspaces

Problem 29.   *For which vector spaces* $\mathbb{V}$ *is it true that if* $\mathbb{L}$, $\mathbb{M}$, *and* $\mathbb{N}$ *are subspaces of* $\mathbb{V}$, *then*

$$\mathbb{L} \cap (\mathbb{M} + \mathbb{N}) = (\mathbb{L} \cap \mathbb{M}) + (\mathbb{L} \cap \mathbb{N})?$$

30   30. Total sets

Is there a subset $\mathbb{E}$ of a vector space $\mathbb{V}$ such that the only subspace of $\mathbb{V}$ that includes $\mathbb{E}$ is $\mathbb{V}$ itself? Sure: several such examples have already been seen. An example in $\mathbb{R}^1$ is the singleton $\{x\}$ of any non-zero x; an example in $\mathbb{R}^2$ is $\mathbb{E} = \{(1,0), (0,1)\}$.

If the only subspace of $\mathbb{V}$ that includes $\mathbb{E}$ is $\mathbb{V}$, then, of course, the intersection of all the subspaces that include $\mathbb{E}$ is just $\mathbb{V}$, so that

$$\bigvee \mathbb{E} = \mathbb{V}.$$

A set $\mathbb{E}$ with this property, a set whose span is the entire vector space, is called a **total** set. By a slight extension of the language, a set $\mathbb{E}$ that spans a subspace $\mathbb{M}$ of $\mathbb{V}$ is called total for $\mathbb{M}$. In the vector space $\mathbb{P}_2$ of all polynomials of degree less than or equal to 2 the set

$$\mathbb{E} = \{1, 1 + x, 1 + x + x^2\}$$

is a total set, and in the larger vector space $\mathbb{P}$ of all polynomials the infinite set

$$\mathbb{E} = \{1, x, x^2, x^3, \ldots\}$$

of monomials is a total set. For the space $\mathbb{O}$, the empty set is total.

The good vector spaces in linear algebra, the easiest ones to work with and the ones that the subject is rooted in, are the ones that have a finite total set; vector spaces like that are called **finite-dimensional**. The space $\mathbb{P}_2$ is finite-dimensional, but (see Problem 27) the space $\mathbb{P}$ is not.

The first natural question about finite-dimensional vector spaces sounds deceptively simple: is every subspace of a finite-dimensional vector space finite-dimensional? That sounds like asking whether every subset of a finite set is finite, but it is not. The question is surprisingly delicate; it is not the kind for which all that's necessary is to feed the definitions into a machine and turn the crank. Here is a step toward acquiring the necessary insight.

Problem 30. *If $\mathbb{E}$ is a total subset of a vector space $\mathbb{V}$, and if $\mathbb{M}$ is a subspace of $\mathbb{V}$, does it follow that some subset of $\mathbb{E}$ is total for $\mathbb{M}$?*

31. Dependence   31

The three vectors

$$x = (1,0), \qquad y = (0,1), \qquad \text{and} \qquad z = (1,1)$$

form a total set for $\mathbb{R}^2$; in fact the first two are enough and the third one is superfluous. There is a simple doctrine at work here: adjoining extra vectors to a total set leaves it total. The new vectors do no harm, but they give no new information.

The vector z in this example is the sum of x and y, and that makes it obvious that every linear combination of x, y, and z is already a linear combination of x and y. The presence of superfluous vectors in a total set is not always so clearly visible. For a look at a more hidden kind of superfluity, let x, y, and z this time be defined by

$$x = (1,7), \qquad y = (2,8), \qquad \text{and} \qquad z = (3,6).$$

It doesn't jump to the eye that x, y, and z form a total set, but they do—and it doesn't jump to the eye that one of x, y, and z is superfluous, but it is true. One way to become convinced of totality is to verify that

$$x - \frac{7}{4}y + \frac{7}{6}z = (1,0) \qquad \text{and} \qquad \frac{1}{7}x + \frac{7}{6}y - \frac{2}{21}z = (0,1).$$

Once that is granted, totality does jump to the eye: since every vector in $\mathbb{R}^2$ is a linear combination of $(1,0)$ and $(0,1)$, it follows that every vector in $\mathbb{R}^2$ is a linear combination of x, y, and z. As for superfluity: since

$$4x - 5y + 2z = 0,$$

it follows that z is "superfluous" in the sense that z is a linear combination of x and y ($z = \frac{5}{2}y - 2x$). Similarly, of course, x is also superfluous, and so is y. If *any one* of x, y, and z is omitted from $\{x, y, z\}$, what's left is still a total set.

The way x, y, and z in these examples depend on one another ("depend" is the crucial word) is an instance of the basic general concept called dependence. No matter what x, y, and z are, it is always possible to find scalars α, β, and γ so that the linear combination $\alpha x + \beta y + \gamma z$ becomes 0; just choose $\alpha = 0$, $\beta = 0$, and $\gamma = 0$. That's a trivial statement, and the linear combination $0 \cdot x + 0 \cdot y + 0 \cdot z$ is justly called the trivial linear combination. The same language is used for any finite number of vectors. With that settled, the ground is prepared for the appropriate general definition: a finite set of vectors is called **dependent** (usually the longer expression **linearly dependent** is used) if some non-trivial linear combination of them vanishes. If the set is $\{x_1, \ldots, x_n\}$, then dependence means that there exist scalars $\alpha_1, \ldots, \alpha_n$, not *all* zero, such that

$$\alpha_1 x_1 + \cdots + \alpha_n x_n = 0.$$

Example: no matter what vector x is, the set $\{0, x\}$ is dependent. Reason: $1 \cdot 0 + 0 \cdot x = 0$. (Note: the scalar coefficients are 1 and 0, and they are not *all* zero.) A trivial example of a dependent set is the set consisting of the vector 0 alone. Reason: $1 \cdot 0 = 0$. Here is a more nearly typical example: if x and y are arbitrary vectors, the set $\{x, y, x + y\}$ is dependent. Reason:

$$1 \cdot x + 1 \cdot y + (-1) \cdot (x + y) = 0.$$

Still another: if x, y, u, and v are arbitrary vectors, then the set

$$\{x, y, x + y, u, v\}$$

is dependent. Reason:

$$1 \cdot x + 1 \cdot y + (-1) \cdot (x + y) + 0 \cdot u + 0 \cdot v = 0.$$

This last example illustrates that in at least one respect dependence behaves the way totality does: adjoining extra vectors doesn't change the property. A set larger than a dependent set is still dependent.

Here is a final easy example of dependence in the concrete vector space $\mathbb{R}^1$: if x and y are any two vectors in that space (that is any two real numbers), then the set $\{x, y\}$ is dependent. Reason: if both x and y are 0, the assertion is trivial; if at least one of them is different from 0, then $yx + (-x)y$ is a non-trivial linear combination that vanishes.

The concept of dependence was introduced via a discussion of "superfluous" vectors in total sets. Does the same connection between dependence and superfluity hold in general?

Problem 31. *If a vector x_0 is a linear combination of $\{x_1, \ldots, x_n\}$, does it follow that the set $\{x_0, x_1, \ldots, x_n\}$ is dependent? If, conversely, a finite set $\{x_0, x_1, \ldots, x_n\}$ of vectors is dependent, does it follow that at least one of them is a linear combination of the others?*

32. Independence   32

If x is a non-zero vector, then the only linear combination of the set $\{x\}$ that can vanish is the trivial one—or, in plain English, the only time αx can be 0 is when $\alpha = 0$. In other words, with one exception the singleton $\{x\}$ is not dependent.

In the vector space $\mathbb{R}^2$ if $x = (1, 0)$ and $y = (0, 1)$, then the only linear combination of the set $\{x, y\}$ that can vanish is the trivial one—that is, the only time $\alpha x + \beta y$ can be zero is when $\alpha = \beta = 0$. In other words, the pair $\{x, y\}$ is not dependent.

General definition: a set that is not dependent is called **independent** (usually **linearly independent**).

Dependence and independence are properties of sets (of vectors), but most people find it comfortable to speak a little loosely and apply the adjectives to vectors. Instead of speaking of an "independent set" [of vectors], they speak of [a set of] "independent vectors". The slightly less sharp usage isn't really dangerous.

It is often convenient to extend the use of the language to two extreme cases, very large sets and very small sets. Very large: infinite. Very small: empty.

An infinite set is called independent if every finite subset of it is independent. Example: the monomials $1, x, x^2, x^3, \ldots$ form an infinite independent set in $\mathbb{P}$. Reason: the only time a linear combination of powers is the zero polynomial is when every coefficient is zero. (This, by the way, is not a statement about the algebra of polynomials: it is merely a reminder of what "zero polynomial" means in contexts such as this.)

Is the empty set dependent or independent? The question is not intrinsically important, but it would be inconvenient to proceed without examining it. The point is that the empty set is quite likely to occur in the middle of a deduction (when, for instance, the intersection of two sets has to be formed), and it would be awkward to have to keep making case distinctions.

The best way (the only convincing way?) to answer questions about the empty set (questions of "vacuous implication") is to ask how they can be false. The present question is this: if a linear combination of the empty set is 0, does it follow that every coefficient must be 0? How could that be false? It is false only if some non-trivial linear combination of the empty set turns out to have the value 0. To say that a linear combination is non-trivial means that it has at least one coefficient different from 0—and that cannot happen. Reason: there are no coefficients at all, and hence, in particular, there is no coefficient different from 0. Conclusion: the assertion that the empty set is independent cannot be false. A consistent use of language demands that the empty set be declared independent. Note, by the way, that this conclusion is in harmony with the assertion that every subset of an independent set is independent.

"Independent" is the accepted dignified way to say of a set of vectors that it has no "superfluous" elements. An independent total set in a vector space $\mathbb{V}$ is called a **basis** of $\mathbb{V}$. Examples: if x is any real number different from 0, then $\{x\}$ is a basis for $\mathbb{R}^1$; if

$$x = (1,0) \qquad \text{and} \qquad y = (0,1),$$

then $\{x, y\}$ is a basis for $\mathbb{R}^2$, and so is $\{x - y, x + y\}$; the monomials

$$1, x, x^2, x^3, \ldots$$

constitute a basis for $\mathbb{P}$; the monomials 1, x, x^2, x^3, x^4, x^5 constitute a basis for $\mathbb{P}_5$.

Does the vector space $\mathbb{O}$ (consisting of the vector 0 only) have a basis? Since the only possible element in a basis for $\mathbb{O}$ is 0, and since the set $\{0\}$ is dependent, it looks as if the answer must be no—but that's wrong. The answer is yes; the space $\mathbb{O}$ does have a basis, namely the empty set. Indeed: the empty set is independent and its span contains 0. This sort of thing looks strange on first encounter, but it's easy to get used to, and it works smoothly—there is nothing wrong, either logically or linguistically.

Do vector spaces always have bases? That's a surprisingly difficult question; the techniques for the general answer are necessarily transfinite. For finite-dimensional vector spaces, however, the tools already available are adequate.

Problem 32. *Does every finite-dimensional vector space have a finite basis?*

CHAPTER 3

BASES

33. Exchanging bases

The most useful questions about total sets, and, in particular, about bases, are not so much how to make them, but how to change them. Which vectors can be used to replace some element of a prescribed total set and have it remain total? Which sets of vectors can be used to replace some subset of a prescribed total set and have it remain total? What restriction is imposed by the relation between the prescribed set and the prescribed total set?

> **Problem 33.** *Under what conditions on a total set* $\mathbb{T}$ *of a vector space* $\mathbb{V}$ *and a finite subset* $\mathbb{E}$ *of* $\mathbb{V}$ *does there exist a subset* $\mathbb{F}$ *of* $\mathbb{T}$ *such that* $(\mathbb{T} - \mathbb{F}) \cup \mathbb{E}$ *is total for* $\mathbb{V}$?

Does that sound awkward? In less stilted language the question is this: under what conditions can one replace a part of a total set by a prescribed set without ruining totality?

Comment. The way the problem is stated the answer is "always": just take $\mathbb{F} = \varnothing$. Consequence: it is necessary to think about the problem before beginning to solve it. Under what conditions on $\mathbb{T}$ and $\mathbb{E}$ and $\mathbb{F}$ does the question make good sense?

34. Simultaneous complements

If $\mathbb{M}$ is a subspace of a vector space $\mathbb{V}$, a **complement** of $\mathbb{M}$ was defined in Problem 28 as a subspace $\mathbb{N}$ of $\mathbb{V}$ such that $\mathbb{M} \cap \mathbb{N} = \{0\}$ and $\mathbb{M} + \mathbb{N} = \mathbb{V}$. (Recall that $\mathbb{M} + \mathbb{N}$ denotes the set of all vectors of the form $x + y$

with $x \in M$ and $y \in N$, or, which for subspaces comes to the same thing, it denotes the span of the set $M \cup N$.) It is easy for a subspace to have more than one complement, or, to put the same thing another way, it is easy for several subspaces to have a "simultaneous" complement, meaning a complement in common. It's easy enough, but that doesn't mean that it always happens. Sample question (which will cause even the experts to think for a nanosecond): if two subspaces are complements, can they have a simultaneous complement? Must they always have one?

Problem 34.   *Under what conditions does a finite collection of subspaces of a finite-dimensional vector space have a simultaneous complement?*

35   35. Examples of independence

Linear independence is one of the most important concepts of linear algebra. A good way to acquire it in one's bloodstream is to look at many examples, and this problem, and several of the ones that follow, are intended to provide some practice in the use of the concept. Most such problems require very little thought—just a little work will solve them.

Problem 35.   (a) *For which real numbers x is it true that the vectors x and 1 are linearly independent in the vector space $\mathbb{R}$ of real numbers (over the field $\mathbb{Q}$ of rational numbers)?*

(b) *Under what conditions on the scalar ξ are the vectors $(1 + \xi, 1 - \xi)$ and $(1 - \xi, 1 + \xi)$ in $\mathbb{R}^2$ (over the field $\mathbb{Q}$ of rational numbers) linearly independent?*

36   36. Independence over $\mathbb{R}$ and $\mathbb{Q}$

Problem 36.   *Is there a subset of $\mathbb{R}^3$ that is independent over $\mathbb{Q}$ but dependent over $\mathbb{R}$?*

37   37. Independence in $\mathbb{C}^2$

Problem 37.   (a) *Under what conditions on the scalars α and β are the vectors $(1, \alpha)$ and $(1, \beta)$ in $\mathbb{C}^2$ linearly independent?*

(b) *Is there a set of three linearly independent vectors in $\mathbb{C}^2$?*

38. Vectors common to different bases

Problem 38. (a) *Do there exist two bases in* $\mathbb{C}^4$ *such that the only vectors common to them are* $(0,0,1,1)$ *and* $(1,1,0,0)$?

(b) *Do there exist two bases in* $\mathbb{C}^4$ *that have no vectors in common so that one of them contains the vectors* $(1,0,0,0)$ *and* $(1,1,0,0)$ *and the other one contains the vectors* $(1,1,1,0)$ *and* $(1,1,1,1)$?

39. Bases in $\mathbb{C}^3$

Problem 39. (a) *Under what conditions on the scalar* x *do the vectors* $(1,1,1)$ *and* $(1,x,x^2)$ *form a basis of* $\mathbb{C}^3$?

(b) *Under what conditions on the scalar* x *do the vectors* $(0,1,x)$, $(x,0,1)$, *and* $(x,1,1+x)$ *form a basis of* $\mathbb{C}^3$?

40. Maximal independent sets

Problem 40. *If* $\mathbb{X}$ *is the set consisting of the six vectors in* $\mathbb{R}^4$,

$$(1,1,0,0), \quad (1,0,1,0), \quad (1,0,0,1),$$
$$(0,1,1,0), \quad (0,1,0,1), \quad (0,0,1,1),$$

do there exist two different maximal linearly independent subsets of $\mathbb{X}$?

(A **maximal linearly independent** subset of $\mathbb{X}$ is a subset $\mathbb{Y}$ of $\mathbb{X}$ that becomes linearly dependent every time that a vector of $\mathbb{X}$ that is not already in $\mathbb{Y}$ is adjoined to $\mathbb{Y}$.)

41. Complex as real

A vector space is not only a set of vectors; it is a set of vectors together with a coefficient field that acts on it. It follows that one and the same set of vectors can well be a vector space in several different ways, depending on what scalars are admitted. So, for instance, the set $\mathbb{C}$ of complex numbers is a vector space over the field $\mathbb{C}$, but that's not especially thrilling; what is more interesting is that $\mathbb{C}$ is a vector space over the field $\mathbb{R}$ of real numbers: just forget that multiplication by non-real scalars is possible. The following question is a generalization of the one just hinted at.

Problem 41. *Every complex vector space* $\mathbb{V}$ *is intimately associated with a real vector space* $\mathbb{V}^{\text{real}}$; *the space* $\mathbb{V}^{\text{real}}$ *is obtained from* $\mathbb{V}$ *by refusing to multiply vectors in* $\mathbb{V}$ *by anything other than real scalars. If the dimension of the complex vector space* $\mathbb{V}$ *is* n, *what is the dimension of the real vector space* $\mathbb{V}^{\text{real}}$?

42   42. Subspaces of full dimension

Problem 42. *Can a proper subspace of a finite-dimensional vector space have the same dimension as the whole space?*

43   43. Extended bases

Which vectors are fit to belong to a basis of a vector space? The vector 0 is not; is that the only exception? Which sets of vectors are fit to be subsets of a basis? A dependent set is not; is that the only exception?

Consider a special example. The vectors

$$x_1 = (1, 0, 0, 0),$$
$$x_2 = (0, 1, 0, 0),$$
$$x_3 = (0, 0, 1, 0),$$
$$x_4 = (0, 0, 0, 1),$$

form a basis for $\mathbb{C}^4$—that's easy. Suppose, however, that for some application a different basis is needed, one that contains the vectors

$$u = (1, 1, 1, 1)$$

and

$$v = (1, 2, 3, 4).$$

Is there such a basis? What is easy to check is that u and v are independent, so that they might be fit to be part of a basis, but the question is whether independence by itself is a sufficient condition.

Problem 43. *Can every (finite) independent set in a finite-dimensional vector space be extended to a basis?*

44. Finite-dimensional subspaces   44

In Problem 30 the question arose whether every subspace of a finite- dimensional vector space is finite-dimensional, but it was not answered there. Now that the technique of making independent sets larger is at hand, that question can be raised with more profit.

> **Problem 44.** *Is every subspace of a finite-dimensional vector space finite-dimensional?*

45. Minimal total sets   45

Total sets are "large" in some rough sense, and, in particular, it is obviously true that if a set is total, then any larger set is necessarily total also. This obvious remark calls attention to the fact that sometimes it is possible to omit some of the elements of a total set and still end up with a total set. When that is not possible, it is natural to call the total set **minimal**. Some total sets are minimal, and some (for example bases) are independent. Is there an implication relation between these two possible properties of total sets, either way?

> **Problem 45.** *Is every minimal total set independent? Is every independent total set minimal?*

46. Existence of minimal total sets   46

Minimal total sets exist all right (any basis is one), but how easy are they to come by in prescribed contexts?

> **Problem 46.** *Does every total set have a minimal total subset?*

47. Infinitely total sets   47

Do there exist total sets that remain total when any one of their elements is discarded, but cease being total if an appropriately chosen set of two elements is discarded? Caution: the question is about *sets* not *sequences*; duplication of elements is not appropriate in this context. If, for instance, $\mathbb{V}$ is a 2-dimensional vector space, and $\{x, y\}$ is a basis for $\mathbb{V}$, then the "set" $\{x, x, y, y\}$ is not an acceptable answer. A small modification of this unacceptable construction does, however, yield an answer, namely the set $\{x, 2x, y, 2y\}$ (provided that the underlying scalar field does not have characteristic 2). An obvious extension of the technique gives for each positive integer n a total set that remains total when any n of its elements are discarded, but ceases being total if an appropriately chosen set of $n + 1$ elements is discarded. It is even possible for a set to be **infinitely total** in the sense that it remains total when any finite subset of it is discarded. A specific example for a 2-dimensional vector space with basis $\{x, y\}$ over $\mathbb{R}$, say, is the set

$$\{x, y, 2x, 2y, 3x, 3y, \ldots\}.$$

In this example even infinite omissions can be allowed, if they are carefully made. If every term after the second one is omitted (that's being very careful), the result is still total, but if all the terms in even positions are omitted (not careful enough), the remainder is not total. Is this a very special case, or are its properties shared by all infinitely total sets?

> **Problem 47.** *Does every infinitely total set $\mathbb{E}$ have an infinite subset $\mathbb{F}$ such that the relative complement $\mathbb{E} - \mathbb{F}$ is total?*

48 48. Relatively independent sets

Every set of $n + 1$ vectors in an n-dimensional vector space is dependent. It is, however, trivial to find three vectors in $\mathbb{R}^2$ such that no two of them are dependent (or, in geometric language, such that no two of them are collinear with the origin), and it is equally trivial to find four vectors in $\mathbb{R}^3$ such that no three of them are coplanar with the origin, and, generally, it is easy to find $n + 1$ vectors in $\mathbb{R}^n$ such that every n of them constitute an independent set. It is temporarily convenient to call a subset $\mathbb{E}$ of $\mathbb{R}^n$ with this property **relatively independent**, the property being that every n vectors in $\mathbb{E}$ are independent. A relatively independent set in $\mathbb{R}^n$ can have $n + 1$ vectors; can the number $n + 1$ be improved?

> **Problem 48.** *What is the largest possible number of vectors in a relatively independent subset of $\mathbb{R}^n$?*

49 49. Number of bases in a finite vector space

Properties of the coefficient field of a vector space obviously have an effect on the linear algebraic properties of the space. Finite fields are especially important in some applications, and the subject as a whole is not properly understood without at least a little insight into how they work.

The best known examples of finite fields are the ones of the form $\mathbb{Z}_p$, that is, the integers modulo p, where p is a prime. These examples are not the only ones; see Problem 19. Granted that they exist, linear algebra can be used to prove a little theorem about them, namely that the number of elements is always a power of a prime. The necessary tools from field theory are these: a finite field must have prime characteristic; every field of characteristic p has a subfield isomorphic to $\mathbb{Z}_p$; every field is a vector space over any subfield. (These statements belong to the part of field theory that is right next to the definitions; they are easy to prove.) An additional tool that is needed is from linear algebra, and it will be discussed later; the proof of the "little theorem" will be postponed till then.

It is true that if q is a power of a prime, then there does indeed exist a field with q elements (and, to within a change of notation, there is only one such field); the case $q = 2^2$ discussed in Problem 19 is more or less typical. A typical vector space over a field $\mathbb{F}$ with $q \ (= p^k)$ elements, is $\mathbb{F}^n$. How many vectors does that vector space contain? The answer is q^n and that's easy. The following considerably trickier counting question asks not for the number of elements but for the number of subsets of a certain kind.

Problem 49. *If $\mathbb{F}$ is a field with q elements, how many bases are there in $\mathbb{F}^n$?*

50. Direct sums

The Euclidean plane can be viewed as the result of a construction that starts from two lines (the x-axis and the y-axis) and puts them together to form a new vector space. That construction is an instance of a general one; other instances of it occur throughout linear algebra (or, for that matter, throughout mathematics). If $\mathbb{U}$ and $\mathbb{V}$ are vector spaces, then their **direct sum**, denoted by

$$\mathbb{U} \oplus \mathbb{V}$$

is the set of all ordered pairs (x, y), with x in $\mathbb{U}$ and y in $\mathbb{V}$, and with vector addition and scalar multiplication defined by the natural equations

$$(x_1, y_1) + (x_2, y_2) = (x_1 + x_2, y_1 + y_2)$$

and

$$\alpha(x, y) = (\alpha x, \alpha y).$$

The vectors x (in $\mathbb{U}$) and y (in $\mathbb{V}$) are called the **coordinates** of (x, y) (in $\mathbb{U} \oplus \mathbb{V}$). In that language, the definitions just described can be expressed by saying that the linear operations in $\mathbb{U} \oplus \mathbb{V}$ are defined **coordinatewise**. It must, of course, be checked that the definitions are correct, meaning that they do indeed define a vector space. That check is painless, and requires no new techniques; it's no different from the proof that if $\mathbb{F}$ is a field, then $\mathbb{F}^2$ is a vector space. That familiar assertion is, in fact, a special case of what is now being asserted; in the present language $\mathbb{F}^2$ is the direct sum of $\mathbb{F}^1$ and $\mathbb{F}^1$.

If $\mathbb{U}$ and $\mathbb{V}$ are well-behaved vector spaces, how well-behaved is $\mathbb{U} \oplus \mathbb{V}$?

Problem 50. *If $\mathbb{U}$ and $\mathbb{V}$ are finite-dimensional vector spaces, of dimensions n and m respectively, what is the dimension of $\mathbb{U} \oplus \mathbb{V}$?*

Reminder. The dimension of a (finite-dimensional) vector space was defined in Solution 33 as the number of elements in a basis of $\mathbb{V}$ immediately after the statement that all bases have the same number of elements.

51. Quotient spaces

If $\mathbb{M}$ is a subspace of a vector space $\mathbb{V}$, then there are, usually, many subspaces $\mathbb{N}$ such that

$$\mathbb{M} \cap \mathbb{N} = \mathbb{O}$$

and

$$\mathbb{M} + \mathbb{N} = \mathbb{V},$$

or, in other words, $\mathbb{M}$ can have many complements, and there is no natural way of choosing one from among them. There is, however, a natural construction that associates with $\mathbb{M}$ and $\mathbb{V}$ a new vector space that plays, for all practical purposes, the role of a complement of $\mathbb{M}$. The theoretical advantage that the construction has over the formation of an arbitrary complement inside $\mathbb{V}$ is precisely its "natural" character, that is, it does not depend on choosing a basis, or, for that matter, on choosing anything at all.

To understand the construction it is a good idea to keep a picture in mind. Suppose, for instance, that $\mathbb{V} = \mathbb{R}^2$ and that $\mathbb{M}$ consists of all those vectors (x_1, x_2) for which $x_2 = 0$ (the horizontal axis). Each complement of $\mathbb{M}$ is a line (other than the horizontal axis) through the origin. Observe that each such complement has the property that it intersects every horizontal line in exactly one point. The idea of the construction to be described now is to make a vector space out of the set of all horizontal lines.

Begin (back in the general case) by using $\mathbb{M}$ to single out certain subsets of $\mathbb{V}$. If x is an arbitrary vector in $\mathbb{V}$, the set $x + \mathbb{M}$ consisting of all the vectors of the form $x + y$ with y in $\mathbb{M}$ is called a **coset** of $\mathbb{M}$, and sets like that are the ones that are of interest now. As for the notation: it is consistent with that used before for vector sums (see Problem 28). In the case of the plane-line example, the cosets are the horizontal lines. Note that one and the same coset can arise from many different vectors: it is quite possible that $x + M = y + M$ even when $x \neq y$. It makes good sense, just the same, to speak of a coset, say $\mathbb{K}$, of M, without specifying which element (or elements) $\mathbb{K}$ comes from; to say that $\mathbb{K}$ is a coset (of M) means simply that there is at least one x such that $\mathbb{K} = x + M$.

If $\mathbb{H}$ and $\mathbb{K}$ are cosets (of M), the vector sum $\mathbb{H} + \mathbb{K}$ is also a coset of M. Indeed, if

$$\mathbb{H} = x + M$$

and

$$\mathbb{K} = y + M,$$

then every element of $\mathbb{H} + \mathbb{K}$ belongs to the coset $(x + y) + M$ (note that $M + M = M$), and, conversely, every element of $(x + y) + M$ is in $\mathbb{H} + \mathbb{K}$. (If, for instance, z is in M, then $(x + y) + z = (x + z) + (y + 0)$.) In other words,

$$\mathbb{H} + \mathbb{K} = (x + y) + M,$$

so that $\mathbb{H} + \mathbb{K}$ is a coset, as asserted.

It is easy to verify that coset addition is commutative and associative. The coset M (that is, $0 + M$) is such that

$$\mathbb{K} + M = \mathbb{K}$$

for every coset $\mathbb{K}$, and, moreover, M is the only coset with this property. (If $(x+M)+(y+M) = x+M$, then $x+M$ contains $x+y$, so that $x+y = x+u$ for some u in M; this implies that y is in M, and hence that $y + M = M$.) If $\mathbb{K}$ is a coset, then the set consisting of all the vectors $-u$, with u in $\mathbb{K}$, is itself a coset, which is denoted by

$$-\mathbb{K}.$$

The coset $-\mathbb{K}$ is such that $\mathbb{K} + (-\mathbb{K}) = M$, and, moreover, $-\mathbb{K}$ is the only coset with this property. To sum up: with the operation of vector sum, the cosets of M form an abelian group.

If $\mathbb{K}$ is a coset and if α is a non-zero scalar, write $\alpha \cdot \mathbb{K}$ for the set consisting of all the vectors αu with u in $\mathbb{K}$; the coset $0 \cdot \mathbb{K}$ is defined to be M. A simple verification shows that with scalar multiplication so defined the cosets of M form a vector space. This vector space is called the **quotient space** of V modulo M; it is denoted by V/M.

The quotient space V/M could have been defined differently. According to an alternative definition, the elements of V/M are the same as the elements (vectors) of V, but the concept of equality is redefined: two vectors are to be regarded as the same if they differ from one another only by a vector in M. In other words, if x and y are vectors in V, say that $x = y$ modulo M, best pronounced as "x is **congruent** to y modulo M", and perhaps more honestly written as

$$x \equiv y \quad (\text{mod } M)$$

when $x - y \in M$. This alternative formulation is intended to be reminiscent of the discussion of polynomials modulo (the multiples of) a fixed polynomial (see Solution 19). There are two approaches to the study of "quotients": in one the new elements are sets, with the necessary operations, such as addition, suitably defined, and in the other the new elements are the same as the old ones, and so are the operations, but equality is suitably re-defined. The coset approach could have been used in Solution 19 and the congruence approach could have been used in the definition of quotient spaces—and, in what follows, the latter will in fact be used whenever it seems convenient to do so.

There are three constructions that are **universal** in the sense that they occur in every kind of mathematical structure and are important whenever they occur: they are usually referred to by expressions such as **substructures, direct sum structures,** and **quotient structures.** Such constructions appear, in particular, in group theory, and in topology, and, of course, in linear algebra. It is a good idea to acquire some facility in handling them, and, in the particular case of linear algebra, the most obvious questions concern dimensions.

Problem 51.   (a) *Is there an example of a vector space V and a subspace M such that neither M nor V/M is finite-dimensional?*

(b) *Is there an example of a vector space V and subspaces M and N such that V/M is finite-dimensional but V/N is not?*

Comment.   The quotient language and the quotient notation (V/M) might strike some people as inappropriate—shouldn't the language and the notation indicate subtraction rather than division? Yes and no. In many parts of mathematics sets of ordered pairs (such as a direct sum $U \oplus V$), are called **Cartesian products**, and in such cases it is natural to look at the reverse as a kind of division. The trouble is that different parts of linear algebra come, historically, from different sources, and the terminological clash is unchangeable by now. It's not hard to learn to live with it.

52. Dimension of a quotient space 52

Problem 52. *If $\mathbb{V}$ is an n-dimensional vector space and $\mathbb{M}$ is an m-dimensional subspace, what is the dimension of $\mathbb{V}/\mathbb{M}$?*

53. Additivity of dimension 53

If $\mathbb{M}$ and $\mathbb{N}$ are subsets of a set, there is a natural third set associated with them, namely their union, $\mathbb{M} \cup \mathbb{N}$. If $\mathbb{M}$ and $\mathbb{N}$ are finite, and if **card** (for cardinal number) is used to denote "the number of elements in", then *sometimes*

$$\text{card}(\mathbb{M} \cup \mathbb{N}) = \text{card}\,\mathbb{M} + \text{card}\,\mathbb{N}.$$

More precisely, the equation is true when $\mathbb{M}$ and $\mathbb{N}$ are **disjoint** ($\mathbb{M} \cap \mathbb{N} = \varnothing$). If they are not disjoint, then the right side counts the elements common to $\mathbb{M}$ and $\mathbb{N}$ twice, and the equation is false.

If $\mathbb{M}$ and $\mathbb{N}$ are subspaces of a vector space, there is a natural third subspace associated with them, namely their span, $\mathbb{M} + \mathbb{N}$. If $\mathbb{M}$ and $\mathbb{N}$ are finite-dimensional, then sometimes

$$\dim(\mathbb{M} + \mathbb{N}) = \dim\mathbb{M} + \dim\mathbb{N}.$$

More precisely , the equation is true when $\mathbb{M}$ and $\mathbb{N}$ are disjoint (which means that $\mathbb{M} \cap \mathbb{N} = \mathbb{O}$); otherwise it's false. What is always true?

Problem 53. *If $\mathbb{M}$ and $\mathbb{N}$ are finite-dimensional subspaces of a vector space, what relation, if any, is always true among the numbers $\dim\mathbb{M}$, $\dim\mathbb{N}$, $\dim(\mathbb{M} + \mathbb{N})$, and $\dim(\mathbb{M} \cap \mathbb{N})$?*

Terminological caution. For subspaces "disjoint" means that their intersection is $\mathbb{O}$, not $\varnothing$. Since the zero vector belongs to every subspace, the latter is impossible.

CHAPTER 4

TRANSFORMATIONS

54. Linear transformations

Here is where the action starts. Till now the vectors in a vector space just sat there; the action begins when they move, when they change into other vectors. A typical example of a change can be seen in the vector space $\mathbb{P}_5$ (all polynomials of degree less than or equal to 5): replace each polynomial p by its derivative Dp.

What is visible here? If $p_1(x) = 3x$ and $p_2(x) = 5x^2$, then

$$Dp_1(x) = 3 \quad \text{and} \quad Dp_2(x) = 10x;$$

if moreover s is the sum $p_1 + p_2$,

$$s(x) = 3x + 5x^2,$$

then

$$Ds(x) = 3 + 10x.$$

This simple property of differentiation is from the present point of view its most important one: the derivative of a sum is the sum of the derivatives. An almost equally important property is illustrated by

$$D\big(7p_2(x)\big) = 70x;$$

the general assertion is that

$$D\big(\alpha p(x)\big) = \alpha Dp(x)$$

for any polynomial p and for any scalar α. In words: the derivative of a scalar multiple is the same scalar multiple of the derivative.

51

These two features of the change that differentiation effects can be described by just one statement: whether you form a linear combination first and then change, or change first and then form the same linear combination—the result will be the same. That's what makes differentiation important in linear algebra; the property is described by saying that D is a **linear transformation**.

Here is another example: consider the vector space $\mathbb{R}^3$, and stretch each vector (x, y, z) by a factor of 7. Let S be the symbol for this stretch, so that S changes $(1, 0, 2)$ (call it u) into $(7, 0, 14)$ ($= Su$), and S changes $(3, -1, 5)$ (call it v) into $(21, -7, 35)$ ($= Sv$), and, generally,

$$S(x, y, z) = (7x, 7y, 7z).$$

Look at a linear combination such as

$$3u - 2v,$$

which is equal to

$$(3, 0, 6) - (6, -2, 10)$$

and therefore to

$$(-3, 2, -4),$$

and then stretch to get

$$(-21, 14, -28).$$

Other possibility: stretch u and v separately and then form the linear combination to get

$$Su = (7, 0, 14), \quad Sv = (21, -7, 35)$$

and

$$3Su - 2Sv = (21, 0, 42) - (42, -14, 70) = (-21, 14, -28)$$

—the same final answer.

It works every time. Given two vectors (or, for that matter, any finite number), if you form a linear combination of them and then stretch, or if you stretch each vector first and then form the same linear combination, the results will always be the same, and, for that reason, the act of stretching is called a linear transformation. (Symbols such as Sv and $S(1, 0, 2)$ are pronounced the way we are all taught when we learn the language of functions: they are "S of v" and "S of $(1, 0, 2)$".)

To understand a mathematical phenomenon it is essential to see several places where it occurs and several where it does not. In accordance with that principle, what follows now is a description of each of five transformations ("changes") on a vector space; some of them are linear transformations and some of them are not.

(1) The vector space is $\mathbb{R}^2$; the transformation T changes each vector by interchanging its coordinates:

$$T(x, y) = (y, x).$$

(2) The vector space is $\mathbb{R}^2$; the transformation T replaces each coordinate by its square:

$$T(x, y) = (x^2, y^2).$$

(3) The vector space is $\mathbb{R}^2$; the transformation T replaces each coordinate by its exponential:

$$T(x, y) = (e^x, e^y).$$

(4) The vector space is $\mathbb{P}$; the transformation T integrates:

$$Tp(x) = \int_2^x p(t)\, dt.$$

(5) The vector space is $\mathbb{R}^2$; the transformation T replaces each coordinate by a certain specific linear combination of the two coordinates:

$$T(x, y) = (2x + 3y, 7x - 5y).$$

The result is that (1), (4), and (5) define linear transformations and (2) and (3) do not. The verification for (1) and (5) is easy. In each case, just replace (x, y) by an arbitrary linear combination

$$\alpha_1(x_1, y_1) + \alpha_2(x_2, y_2),$$

apply T, and compare the result with the result of doing things in the other order. (Is "other order" clear? It means apply T to each of (x_1, y_1) and (x_2, y_2) and then form the linear combination.) In other words, the verification consists of applying the very definition of linear transformation, and that yields what is wanted. The truth of (4) depends on known facts about integration: the integral of a sum is the sum of the integrals, and the integral of a scalar multiple is the scalar multiple of the integral.

As for (2): everything goes wrong. If it were true that

$$(s + t)^2 = s^2 + t^2$$

for all real numbers s and t (an identity that a few beginning students of mathematics are in fact tempted to believe), then it would follow that the transformation T satisfies one of the necessary conditions for being linear. Namely, it would then be true that

$$\left((x_1 + x_2)^2, (y_1 + y_2)^2\right) = (x_1^2 + x_2^2, y_1^2 + y_2^2),$$

and hence that

$$T\big((x_1, y_1) + (x_2, y_2)\big) = T(x_1, y_1) + T(x_2, y_2).$$

But even if that were right, scalar multiples would still misbehave. For linearity it is necessary that $T(\alpha x, \alpha y)$ should equal $\alpha T(x, y)$, but

$$T(\alpha x, \alpha y) = (\alpha^2 x^2, \alpha^2 y^2)$$

and

$$\alpha T(x, y) = \alpha(x^2, y^2) = (\alpha x^2, \alpha y^2),$$

and except in the rare cases when $\alpha = \alpha^2$ the two right sides of these equations are not eager to be the same.

Warning: this argument would be regarded with disapproval by many professional mathematicians. The trouble is that the argument does not prove that T fails to be a linear transformation; it just points out that the natural way to try to prove that T is a linear transformation doesn't succeed. The only convincing way to prove that T does not satisfy the identity that linearity requires is to exhibit *explicitly*, with concrete scalars and vectors, a linear combination that T does not cooperate with. That's easy enough to do: if, for instance, $(x, y) = (1, 1)$ and $\alpha = 2$, then

$$T(\alpha x, \alpha y) = (2^2, 2^2) = (4, 4)$$

and

$$\alpha T(x, y) = 2 \cdot (1, 1) = (2, 2).$$

The negative assertion that the transformation T described in (3) is not a linear transformation either is proved similarly. An explicit counterexample is given by $(x, y) = (0, 0)$ and $\alpha = 2$; in that case

$$T(\alpha x, \alpha y) = T(0, 0) = (1, 1)$$

and

$$\alpha T(x, y) = 2 \cdot (1, 1) = (2, 2).$$

Problem 54. (a) *Which of the following three definitions of transformations on $\mathbb{R}^2$ give linear transformations? (The equations are
intended to hold for arbitrary real scalars α, β, γ, δ.)*
   (1) $T(\xi, \eta) = (\alpha\xi + \beta\eta, \gamma\xi + \delta\eta)$.
   (2) $T(\xi, \eta) = (\alpha\xi^2 + \beta\eta^2, \gamma\xi^2 + \delta\eta^2)$.
   (3) $T(\xi, \eta) = (\alpha^2\xi + \beta^2\eta, \gamma^2\xi + \delta^2\eta)$.

   (b) *Which of the following three definitions of transformations
on $\mathbb{P}$ give linear transformations? (The equations are intended to hold
for arbitrary polynomials p.)*
   (1) $Tp(x) = p(x^2)$.
   (2) $Tp(x) = \big(p(x)\big)^2$.
   (3) $Tp(x) = x^2 p(x)$.

55. Domain and range   55

Integration on the vector space $\mathbb{P}$ is a perfectly good linear transformation
(see Problem 54), but the same equation

$$Tp(x) = \int_2^x p(t)\, dt \qquad (*)$$

as the one that worked there does not work on the space $\mathbb{P}_5$; the trouble is
that the degree of the polynomial that it gives may be too large. Right? If,
for instance, $p(x) = x^5$ then

$$\int_2^x p(t)\, dt = \left[\frac{1}{6}t^6\right]_2^x = \frac{x^6 - 2^6}{6}.$$

Differentiation on the vector space $\mathbb{P}_5$ might seem to run into similar
trouble—it *lowers* degrees instead of raising them—but, in fact, there is
nothing wrong with it. Sure, it's true that D applied to a polynomial of
degree less than or equal to 5 always yields a polynomial of degree less
than or equal to 4, but 4 is less than 5, and vectors in $\mathbb{P}_5$ stay in $\mathbb{P}_5$.

These two examples prepare the ground for a small but useful generalization of the concept of linear transformation and for introducing two
important constructs associated with each linear transformation. The generalization is to a transformation (= change, function, mapping, map, operator, etc.) that changes each vector v in one vector space into a vector
Tv in a possibly different vector space, and does it in such a way that it "co-
operates" with linear combinations. The technical word is "commutes"; it
means, of course, that the result of forming a linear combination and then
transforming is always the same as the result of transforming first and then

forming the linear combination. Symbolically:

$$T(\alpha u + \beta v) = \alpha T u + \beta T v.$$

Example: if T is the integration defined by the equation $(*)$ above, then T is a linear transformation **from** $\mathbb{P}_5$ **to** $\mathbb{P}_6$ (where the meaning of $\mathbb{P}_6$ is surely guessable: it is the vector space of all real polynomials of degree not more than 6). The same equation can also be regarded as defining a linear transformation from $\mathbb{P}_5$ to $\mathbb{P}$, and many others in similar contexts. A linear transformation from a vector space $\mathbb{V}$ to itself is called a linear transformation **on** $\mathbb{V}$; that's the kind that was introduced in Problem 54.

The set of vectors where a linear transformation starts is called the **domain** of T, and the set of vectors that result from applying T to them is called the **range** of T; the abbreviations

$$\operatorname{dom} T \quad \text{and} \quad \operatorname{ran} T$$

are quite commonly accepted. So, for example, if T is differentiation on $\mathbb{P}_5$, then

$$\operatorname{dom} T = \mathbb{P}_5 \quad \text{and} \quad \operatorname{ran} T = \mathbb{P}_4;$$

if T is differentiation from $\mathbb{P}_5$ to $\mathbb{P}$, then also $\operatorname{dom} T = \mathbb{P}_5$ and $\operatorname{ran} T = \mathbb{P}_4$. Some confusion is possible here and should be avoided. Integration can be considered to define a linear transformation from $\mathbb{P}_5$ to $\mathbb{P}_6$ or from $\mathbb{P}_5$ to $\mathbb{P}_7$ or from $\mathbb{P}_5$ to $\mathbb{P}_{200}$; in each of these cases the domain is $\mathbb{P}_5$ and the range is a part of $\mathbb{P}_6$. (Which part? The question deserves a moment's thought.) The vector space that follows the specification "to" plays a much smaller role than the range, a subsidiary role, and it does not have a commonly accepted name; the word "codomain" is sometimes used for it.

Important observation: the domain of a linear transformation is always a vector space, *and so is the range.*

Integration is not the only useful linear transformation from one vector space to a different one. The change of variables example in Problem 54,

$$Tp(x) = p(x^2),$$

can be regarded as a linear transformation on $\mathbb{P}$, or, alternatively, the same equation can be used to define a linear transformation from $\mathbb{P}_5$ to $\mathbb{P}_{10}$. Right? If $p(x) = x^4$, then $Tp(x) = x^8$. Similarly the multiplication example in Problem 54,

$$Tp(x) = x^2 p(x),$$

can be regarded as a linear transformation on $\mathbb{P}$, or, alternatively, the same equation can be used to define a linear transformation from $\mathbb{P}_5$ to $\mathbb{P}_7$. (If $p(x) = x^4$, then $Tp(x) = x^6$.)

The domains and ranges of the linear transformations given as examples in Problem 54 are not difficult to find. Challenge: check that for the stretching example on $\mathbb{R}^3$ both the domain and the range are equal to $\mathbb{R}^3$, and for the interchange example ((1) in the discussion of Problem 54), and for the linear combination example ((5) in the discussion of Problem 54), both the domain and the range are equal to $\mathbb{R}^3$.

When trying to understand domains and ranges for linear transformations between possibly different vector spaces, it is a good idea to study at least a few new examples. The problems that follow describe some.

Problem 55. (1) *The set $\mathbb{R}$ of all real numbers is a real vector space, which in that capacity is denoted by $\mathbb{R}^1$. The sum of two real numbers x and y, considered as vectors, is just the ordinary sum obtained by considering them as the real numbers they are and adding them; the multiple of a "vector" (real number) by a "scalar" (real number) is just the product obtained by forming the product of the two real numbers. The equation*

$$F(x, y) = x + 2y$$

*defines a linear transformation from $\mathbb{R}^2$ to $\mathbb{R}^1$. This example is a special case of an important class of linear transformations: a linear transformation from any real vector space $\mathbb{V}$ to the special vector space $\mathbb{R}^1$ is called a **linear functional** on $\mathbb{V}$. (The use of "on" here is in slight collision with the use explained before, but that's life—with a little care confusion can be avoided.) What are the domain and the range of the particular linear functional here defined? What, in general, can be said about the range of a linear functional?*

(2) *Does the equation*

$$Tp(x) = p(x + 2)$$

define a linear transformation from $\mathbb{P}_5$ to $\mathbb{P}_{10}$? If so, what are its domain and its range?

(3) *Does the equation*

$$T(x, y, z) = (0, 0)$$

define a linear transformation from $\mathbb{R}^3$ to $\mathbb{R}^2$? If so, what are its domain and its range?

(4) *Does the equation*

$$T(x, y, z) = (x + 2, y + 2)$$

define a linear transformation from $\mathbb{R}^3$ to $\mathbb{R}^2$? If so, what are its domain and its range?

   (5) *Let $\mathbb{R}_+$ be the set of all positive real numbers, and try to make it into a real vector space. To do that, it is necessary to define an "addition" for any two positive real numbers and to define the "scalar multiple" of any positive real number by an arbitrary (not necessarily positive) real number. In trying to do that it would be dangerous to use the ordinary symbols for addition and multiplication—that way confusion lies. To avoid that confusion, the sum about to be defined will be denoted by $\boxed{+}$ and the product by $\boxed{\cdot}$, and the actual definitions are as follows: if s and t are positive real numbers, then*

$$s \boxed{+} t = st$$

(that is, the new sum of s and t is the plain old product of s and t), and if s is a positive real number and x is an arbitrary real number, then

$$x \boxed{\cdot} s = s^x$$

(that is, the new product of s by x is s to the power x in the usual sense). This is a weird procedure, but it works; it actually defines a real vector space. Does the equation

$$T(s) = \log s$$

define a linear functional on that vector space? If so, what is its range? Caution: what does "log" mean—does it mean $\log_{10}$, or $\log_e$, or $\log_2$?

56   56. Kernel

There is a real number 0 and (in every vector space) there is a vector 0, and no confusion will ever arise between the number 0 and the vector 0; on the rare occasions when one threatens, a few cautionary words will dispel it.

   The symbol 0 has more than just two uses in mathematics, and even in linear algebra; here, for instance, is a third. Consider any two vector spaces $\mathbb{V}$ and $\mathbb{W}$, and define a transformation T from $\mathbb{V}$ to $\mathbb{W}$ by writing

$$Tv = 0$$

for all v in $\mathbb{V}$. Since the description of T warned that its range will be in $\mathbb{W}$, it is clear that the symbol 0 on the right side of this equation must stand for the zero vector in $\mathbb{W}$. Is T a linear transformation? Sure—obviously. Form all the linear combinations you like in $\mathbb{V}$ and then apply T to them. Since T sends everything to 0 in $\mathbb{W}$, the result obtained by applying T first and forming linear combinations later will always be 0. That's just fine:

$$T(\alpha u + \beta v)$$

is indeed the same as

$$\alpha Tu + \beta Tv$$

because $T(\alpha u + \beta v)$ is 0 and $\alpha Tu + \beta Tv$ is $\alpha \cdot 0 + \beta \cdot 0$, which is also 0. This very special linear transformation (that can be used between any two vector spaces) has a special name, namely 0.

Linguistic interruption: if T is a linear transformation and v is a vector in the domain of T, the corresponding vector Tv in the range is often called the **image** of v, or the **transform** of v, under the action of T. (Caution: here "transform" is the right word, not "transformation".) So: the zero linear transformation from $\mathbb{V}$ to $\mathbb{W}$ is the one that maps (= sends) every vector v to 0, or, in other words, it is the one for which every vector in $\mathbb{V}$ has the same image, namely 0.

Zero plays an important role in linear algebra. So, for instance, every linear transformation sends 0 to 0. (A precise proof of that comment is easy, but it belongs to the hairsplitting axiomatics of the subject, which will be treated later.) There could perfectly well be many other vectors, different from 0, that a linear transformation T sends to 0 also. The collection (set) of all those vectors gives vital information about T; it is called the **kernel** of T, abbreviated

$$\ker T.$$

For a first example, consider the zero transformation from, say, $\mathbb{R}^3$ to $\mathbb{R}^2$: what is its kernel? In other words: what is the set of all vectors in $\mathbb{R}^3$ that the transformation 0 sends to the vector 0 in $\mathbb{R}^2$? Answer: 0 sends *every* vector in $\mathbb{R}^3$ to 0 in $\mathbb{R}^2$; the kernel of 0 is the whole space $\mathbb{R}^3$.

Consider next the linear transformation T on $\mathbb{P}_5$ defined by

$$Tp = p$$

for all p. Is it really a linear transformation? Sure—that's very easy. What is its kernel? In other words, what is the set of all polynomials in $\mathbb{P}_5$ whose image under this T is the zero polynomial? Answer: T can send *no* polynomial to 0, except only the polynomial 0; the kernel of this T is the set consisting of the polynomial 0 only. The cautious notation for that set is {0}. The braces are needed: it is important to realize and to emphasize that the kernel of a linear transformation is always a *set*. The set might, to be sure, consist of just one object, but it is a set just the same. (Analogy: a hatbox with just one hat in it is not the same as a hat.)

By the way: the equation $Tp = p$ defines a linear transformation on every vector space, and that linear transformation is an important one with a special name. (Recall that "on" in this context describes a linear transformation *from* the given vector space *to* the same space.) It is always called the **identity** transformation, but it is not always denoted by the same symbol. Some people call it

$$1$$

(so that the number "one" and the identity transformation have the same symbol), others use the letter I, and still others indicate the vector space under consideration by using the symbol I_V. In this book the first of these possibilities, the numerical symbol, is the one that will be used most of the time; "I" will be used when that practice threatens to lead to confusion.

Problem 56. *What are the kernels of the linear transformations named below?*

(1) The linear transformation T defined by integration, say, for instance,

$$Tp(x) = \int_{-3}^{x+9} p(t)\, dt,$$

from $\mathbb{P}_6$ to $\mathbb{P}_7$.

(2) The linear transformation D of differentiation on $\mathbb{P}_5$.

(3) The linear transformation T on $\mathbb{R}^2$ defined by

$$T(x, y) = (2x + 3y, 7x - 5y);$$

see example (5) in Problem 48.

(4) The linear transformation T from $\mathbb{P}_5$ to $\mathbb{P}_{10}$ defined by the change of variables

$$Tp(x) = p(x^2);$$

see part (4) of Problem 48.

(5) The linear transformation T on $\mathbb{R}^2$ defined by

$$T(x, y) = (x, 0).$$

(6) *The linear transformation F from $\mathbb{R}^2$ to $\mathbb{R}^1$ defined by*

$$F(x, y) = x + 2y;$$

see part (1) of Problem 55.

57. Composition

Differentiation (denoted by D) is a linear transformation on the vector space $\mathbb{P}$ of all polynomials, and so is the transformation M (multiplication by the variable) defined by

$$Mp(x) = xp(x).$$

What happens to a polynomial if both those transformations act on it, one after another? Suppose, to be specific, that D sends p to q,

$$q = Dp,$$

and then M sends q to r,

$$Mq = r;$$

what can be said about the passage from p to r? Write

$$r = Tp,$$

and, just to see what happens in a special example, let $p(x)$ be

$$2 + 3x + 4x^2.$$

In that case

$$q(x) = 3 + 8x,$$

and therefore

$$r(x) = 3x + 8x^2.$$

Suppose now that the same thing is done not for one p but for two, and then a linear combination is formed, so that the result looks like

$$\alpha_1 T p_1(x) + \alpha_2 T p_2(x).$$

What if the linear combination had been formed before the two-step transformation T? Would the result

$$\big(T(a_1 p_1 + a_2 p_2)\big)(x)$$

be the same? In other words, is T a linear transformation? Isn't it clear that the answer must be yes? To say that D is linear means that D **distributes** over vector addition and scalar multiplication—that is, that D converts a linear combination of vectors into the same linear combination of their D images. If M is allowed to act on the linear combination so obtained, then the linearity of M means that M distributes over it—that is, that M converts it into the same linear combination of the M images of the vectors that enter. These two sentences together say that T distributes over linear combinations, or, in official language, that T is linear.

The reasoning just described is quite general: it proves that the **composition** of two linear transformations is a linear transformation. The concept of composition so introduced is just as often called **product**. The official definition is easy to state: if S and T are linear transformations on the same vector space, then the composition of S and T, denoted by

$$ST,$$

is the transformation that sends each vector v in the space to the vector obtained by applying T to v and then applying S to the result.

Caution: the order of events is important. What would have happened to the polynomial $2 + 3x + 4x^2$ if it had been multiplied by x first and then differentiated? Answer: the multiplication would have produced

$$2x + 3x^2 + 4x^3,$$

and then the differentiation would have produced

$$2 + 6x + 12x^2,$$

which is not at all the same as the

$$3x + 8x^2$$

obtained before. In other words:

$$MD \neq DM.$$

For a different enlightening example, consider the multiplication transformation N defined on $\mathbb{P}$ by

$$Np(x) = (1 - 3x^2)p(x).$$

In that case

$$N(2 + 3x + 4x^2) = (1 - 3x^2)(2 + 3x + 4x^2)$$
$$= 2 + 3x - 2x^2 - 9x^3 - 12x^4,$$

and the result of applying M to that is

$$2x + 3x^2 - 2x^3 - 9x^4 - 12x^5.$$

On the other hand

$$M(2 + 3x + 4x^2) = 2x + 3x^2 + 4x^3,$$

and the result of applying N to that is

$$(1 - 3x^2)(2x + 3x^2 + 4x^3) = 2x + 3x^2 - 2x^3 - 9x^4 - 12x^5$$

—the same thing. This could have been obvious without any calculation: the first result was obtained by two multiplications ($1 - 3x^2$ followed by x), as was the second (x followed by $1 - 3x^2$). Since the order in which multiplications are performed doesn't matter, it is no surprise that

$$MN = NM.$$

The result is described by saying that the linear transformations M and N **commute** (or they are **commutative**). The example of M and D shows that linear transformations may fail to commute—they may be **non-commutative**.

Can transformation multiplication be defined for linear transformations that go from one vector space to a different one? Yes—sometimes. If $\mathbb{U}$ and $\mathbb{V}$ are vector spaces, and if T is a linear transformation from $\mathbb{U}$ to $\mathbb{V}$, then it makes sense to follow an action of T by another linear transformation, say S, but only if S starts where T left off, or, in more dignified language, if the domain of S includes the range of T. If, in other words, for each vector u in $\mathbb{U}$, the image Tu belongs to the domain of S, then STu makes sense, and if, to be specific, the range of S is included in a vector space $\mathbb{W}$, then the product ST is defined and is a linear transformation from $\mathbb{U}$ to $\mathbb{W}$.

Here is an example. Suppose that $\mathbb{U}$ is $\mathbb{R}^3$, $\mathbb{V}$ is $\mathbb{R}^2$, and $\mathbb{W}$ is $\mathbb{R}^1$; let T be the linear transformation from $\mathbb{R}^3$ to $\mathbb{R}^2$ defined by

$$T(x, y, z) = (x, y),$$

and let S be the linear transformation from $\mathbb{R}^2$ to $\mathbb{R}^1$ defined by

$$S(x, y) = x + y.$$

In that case

$$ST(x, y, z) = S(x, y) = x + y.$$

Note: the product TS cannot be defined. The only way a symbol such as TS could be interpreted is as a transformation that starts where S starts, that is, as a transformation with domain $\mathbb{R}^2$. But then the result of applying S yields a vector in $\mathbb{R}$, and it does not make sense to apply T to it —T can only work on vectors in $\mathbb{R}^3$. If the product of two linear transformations S and T is defined, is it clear how to read a symbol such as ST? A possible source of confusion, which should be avoided, is that the symbol must be read "backward", from right to left. To see how ST acts on a vector, let T act on it first and then, second, let S act on the result. The reason is that transformations are, after all, functions, and the usual functional notation (as in $f(x)$) puts the name of the function next to the name of the variable *and to the left of it.* Students of mathematics realize early that if

$$f(x) = x^2 \quad \text{and} \quad g(x) = x + 2,$$

then

$$f\big(g(x)\big) = (x+2)^2;$$

the first function to act is the one next to x—the one on the right. (The other order would yield $g\big(f(x)\big)$, which is $x^2 + 2$—not at all the same thing. Non-commutativity raises its head again.)

Problem 57.   (1) *If S is the stretching transformation on $\mathbb{R}^2$,*

$$S(x, y) = (7x, 7y),$$

(see Problem 54) and T is the transformation on $\mathbb{R}^2$ defined by

$$T(x, y) = (2x + 3y, 7x - 5y),$$

do S and T commute?
    (2) *If S is the stretching transformation on $\mathbb{R}^3$,*

$$S(x, y, z) = (7x, 7y, 7z),$$

and T is the "projection" transformation from $\mathbb{R}^3$ to $\mathbb{R}^2$ defined by

$$T(x, y, z) = (x, y),$$

do S and T commute?
    (3) *If S is the change of variables on $\mathbb{P}$ defined by*

$$Sp(x) = p(x^2),$$

and T is the multiplication transformation defined by

$$Tp(x) = x^2 p(x),$$

do S and T commute?

(4) *If S is the transformation from $\mathbb{R}^2$ to $\mathbb{R}^1$ defined by*

$$S(x, y) = x + 2y,$$

and T is the transformation from $\mathbb{R}^1$ to $\mathbb{R}^2$ defined by

$$T(x) = (x, x),$$

do S and T commute?

(5) *If S is the change of variables defined on $\mathbb{P}_3$ by*

$$Sp(x) = p(x + 2),$$

and T is the transformation defined by

$$T(\alpha + \beta x + \gamma x^2 + \delta x^3) = \alpha + \gamma x^2,$$

(for all α, β, γ, δ) do S and T commute?

(6) *In each of the preceding cases, what are the domains, ranges, and kernels of ST and TS (when they make sense)?*

58. Range inclusion and factorization   58

When is one linear transformation divisible by another? In view of the difference between right and left, the question doesn't quite make sense, but once divisibility is interpreted one way or the other then it does.

Suppose, for instance, that A is called **left divisible** by B in case there exists a linear transformation T such that $A = BT$. One relation between A and B is an immediate consequence, namely that

$$\operatorname{ran} A \subset \operatorname{ran} B.$$

Is that necessary condition for left divisibility sufficient also?

Problem 58. *If two linear transformations A and B on a vector space are such that $\operatorname{ran} A \subset \operatorname{ran} B$, does it follow that A is left divisible by B? What is the analogous necessary condition for right divisibility? Is it sufficient?*

59. Transformations as vectors   59

Two transformations A and B on a vector space can form a conspiracy: for each vector v the results of the actions of A and B on v can be added to yield a new vector. The result of this conspiracy assigns a vector, call it Sv, to each starting vector v—in other words, the passage S from v to $Av + Bv$

is a transformation defined on the underlying vector space V. It is natural to call the transformation S the **sum** of A and B. The commutativity and associativity of addition in V imply immediately that the addition of linear transformations is commutative and associative.

Much more than that is true. The sum of any linear transformation A and the linear transformation 0 (see Problem 56) is A. If for each linear transformation A, the symbol $-A$ is used to denote the transformation defined by

$$(-A)v = -(Av),$$

then

$$A + (-A) = 0,$$

and the linear (!) transformation $-A$ is uniquely characterized by that property. To sum up: the set of all linear transformations on V is an abelian group with respect to the operation of addition.

If, moreover, for any scalar α and any linear transformation A a product αA is defined by

$$(\alpha A)v = \alpha(Av),$$

it follows that the set of all linear transformations on a vector space V is itself a vector space; a usable symbol for it might be $\mathbb{L}(V)$.

The set $\mathbb{L}(V)$ has a structural property that not every vector space has, namely it has a natural multiplication defined on it, the composition of linear transformations. If A and B are linear transformations, then not only are αA and $A + B$ linear transformations, but so also is AB; it is possible to form not only linear combinations of linear transformations, but also linear combinations of powers of a single linear transformation. "Linear combinations of powers" is a long way to say **polynomials**; if, that is,

$$p(x) = \alpha_0 + \alpha_1 x + \cdots + \alpha_n x^n$$

is a polynomial, and A is a linear transformation, then $p(A)$ makes sense:

$$p(A) = \alpha_0 + \alpha_1 A + \cdots + \alpha_n A^n.$$

How are the linear and multiplicative properties of $\mathbb{L}(V)$ related to the vector space properties of V?

Problem 59. (1) *What can be said about the dimension of* $\mathbb{L}(V)$ *in terms of the dimension of* V*?*

(2) *If A is a linear transformation on a finite-dimensional vector space, does there always exist a non-zero polynomial p such that $p(A) = 0$?*

(3) *If x_0 is a vector in a vector space $\mathbb{V}$, and y_0 is a linear functional on $\mathbb{V}$, and if Ax is defined for every x in $\mathbb{V}$ by*

$$Ax = y_0(x)x_0,$$

then A is a linear transformation on $\mathbb{V}$; what is the smallest possible degree of a polynomial p such that $p(A) = 0$?

60. Invertibility   60

Addition can be undone—reversed—by subtraction; can the multiplication of linear transformations also be undone somehow? Is there a process like division for linear transformations? Central special case: can the identity transformation 1 be "divided" by any other linear transformation T—or, in other words, does a linear transformation always have a "reciprocal"?

Suppose that T is a transformation on a vector space $\mathbb{V}$, so that T assigns to each vector in $\mathbb{V}$ another vector (or possibly the same one) in $\mathbb{V}$ again. A candidate for a reciprocal should presumably be a transformation that goes backward, in the sense that it assigns to each vector in $\mathbb{V}$ the vector that it comes from. That may sound plausible, but there is a catch in it—two catches in fact. To be able to go backward, *each vector* in $\mathbb{V}$ must be the image under T of something (in other words, $\operatorname{ran} T$ must be $\mathbb{V}$), and it must make sense to speak of *the* vector that a vector comes from (in other words, T must never send two different vectors onto the same vector). Equivalent language: the transformation T must map $\mathbb{V}$ *onto* $\mathbb{V}$ (technical word: T is **surjective**), and T must do so in a *one-to-one* manner (technical word: T is **injective**).

A typical violent counterexample (in case the vector space $\mathbb{V}$ is not the trivial space $\mathbb{O}$) is the linear transformation 0: it fails to satisfy the conditions just described in the worst possible way. Not only is it false that T is surjective, but in fact $\operatorname{ran} 0$ is as small as can be—it consists of one vector only—and not only is it false that 0 is injective, but in fact 0 is as **many-to-one** as can be—it sends every vector onto the same one. Here are a couple of other examples that are bad enough but not quite that bad: the linear transformations defined for every vector $\begin{pmatrix} \xi \\ \eta \end{pmatrix}$ in $\mathbb{R}^2$ by

$$A \begin{pmatrix} \xi \\ \eta \end{pmatrix} = \begin{pmatrix} \xi + \eta \\ \xi + \eta \end{pmatrix}$$

and

$$B \begin{pmatrix} \xi \\ \eta \end{pmatrix} = \begin{pmatrix} 0 \\ \eta \end{pmatrix}.$$

(Vectors are written vertically, as $\begin{pmatrix} \xi \\ \eta \end{pmatrix}$, or horizontally, as (ξ, η) ad lib. The vertical symbol will seem more appropriate when matrices enter the picture, but the horizontal one is typographically more convenient.) The range of A consists only of vectors whose two coordinates are equal— and that's nowhere near all of $\mathbb{R}^2$. As for B, the vector $(0, 0)$ comes from infinitely many vectors—the transformation is nowhere near injective. (Question: do A and B fail for just one reason each, or for two? That is: granted that A is not surjective, is it at least injective? And: granted that B is not injective, is it at least surjective?)

A linear transformation T on a vector space $\mathbb{V}$ is called **invertible** if it is both injective and surjective. If T is invertible, then the transformation that assigns to each vector v in $\mathbb{V}$ the unique vector u that it comes from (that is, the unique vector u such that $Tu = v$) is called the **inverse** of T; it is denoted by T^{-1} (pronounced "T inverse").

If an invertible T acts on a vector v, what happens when the inverse T^{-1} acts on the result? The answer is obvious: T^{-1} sends Tv back to v, so that $T^{-1}(Tv) = v$. Does it work the other way? In other words, what happens when T is made to go backward before it goes forward? That may be colorful language, but it's sloppy enough to be dangerous. What the question really asks is: what is $T(T^{-1}v)$? To find the answer, write $u = T^{-1}v$, which says the same as $Tu = v$, and then unscramble the notation:

$$T(T^{-1}v) = Tu = v.$$

Another way of expressing the same question (and its answer) is to note that the inverse of a transformation, as here defined, is really a **left inverse**, and to ask whether it works from the **right** also. The question makes sense (since both T and T^{-1} map $\mathbb{V}$ into $\mathbb{V}$, the product can be formed in either order), and, as the discussion above shows, the answer is yes.

The simplest example of an invertible transformation is 1, the identity; it is obviously both injective and surjective, and it is its own inverse.

Is the linear transformation T defined for all $\begin{pmatrix} \xi \\ \eta \end{pmatrix}$ in $\mathbb{R}^2$ by

$$T \begin{pmatrix} \xi \\ \eta \end{pmatrix} = \begin{pmatrix} 2\xi + \eta \\ \xi + \eta \end{pmatrix}$$

invertible? That's two questions: is T surjective?, and is T injective?—but, as it turns out an examination of surjectivity alone yields the full answer.

The surjective question for T reduces to the solvability of the equations

$$2\xi + \eta = \alpha$$
$$\xi + \eta = \beta$$

for all α and β. That's a routine matter: ξ must be $\alpha - \beta$ and η must be $-\alpha + 2\beta$. That is: the candidate for the inverse of T is the linear transformation S defined by

$$S\begin{pmatrix} \xi \\ \eta \end{pmatrix} = \begin{pmatrix} \xi - \eta \\ -\xi + 2\eta \end{pmatrix}.$$

To check whether the candidate works, form the product ST. That is: find ST, at each $\begin{pmatrix} \xi \\ \eta \end{pmatrix}$, by forming $S\begin{pmatrix} \gamma \\ \delta \end{pmatrix}$, where

$$\begin{pmatrix} \gamma \\ \delta \end{pmatrix} = \begin{pmatrix} 2\xi + \eta \\ \xi + \eta \end{pmatrix}.$$

The result of the substitution is

$$ST\begin{pmatrix} \xi \\ \eta \end{pmatrix} = \begin{pmatrix} (2\xi + \eta) - (\xi + \eta) \\ -(2\xi + \eta) + 2(\xi + \eta) \end{pmatrix} = \begin{pmatrix} \xi \\ \eta \end{pmatrix},$$

and that does it—indeed, $ST = 1$.

These examples suggest a question to which the answer must be known before the theory can proceed.

Problem 60. *Must the inverse of an invertible linear transformation be a linear transformation?*

61. Invertibility examples 61

Problem 61. (1) *Is the linear transformation defined by*

$$T\begin{pmatrix} \xi \\ \eta \end{pmatrix} = \begin{pmatrix} 2\xi + \eta \\ 2\xi + \eta \end{pmatrix}$$

invertible?

(2) *What about*

$$T\begin{pmatrix} \xi \\ \eta \end{pmatrix} = \begin{pmatrix} \eta \\ \xi \end{pmatrix}?$$

(3) *Is the differentiation transformation D on the vector space $\mathbb{P}_5$ invertible?*

62   **62. Determinants: 2×2**

If α and ξ are known numbers, can the equation

$$\alpha x = \xi$$

be solved for x? Maybe. If α is not 0, then all that's needed is to divide by it. If $\alpha = 0$, there is trouble: if $\xi \neq 0$, the equation has no solutions, and if $\xi = 0$, it has too many—every x works.

The question could have been asked this way: if α is a known scalar, and if a linear transformation T is defined on $\mathbb{R}^1$ by

$$Tx = \alpha x,$$

is T invertible? The answer is yes if and only if $\alpha \neq 0$.

That answer is in fact the answer to a very special case of a very general question. The general question is this: how do the entries of an $n \times n$ matrix determine whether or not it is invertible? When $n = 1$, there is only one entry and the answer is that the matrix is invertible if and only if that entry is different from 0.

What happens when $n = 2$? In other words, under what conditions on the numbers α, β, γ, δ is the 2×2 matrix

$$M = \begin{pmatrix} \alpha & \beta \\ \gamma & \delta \end{pmatrix}$$

invertible, or, equivalently, when can the equations

$$\alpha x + \beta y = \xi$$
$$\gamma x + \delta y = \eta$$

be solved for x and y?

To find x, eliminate y, which means multiply the first equation by δ, the second by β, and subtract the second from the first to get

$$(\alpha\delta - \beta\gamma)x = \delta\xi - \beta\eta.$$

To find y, multiply the first equation by γ, the second by α, and subtract the first from the second to get

$$(\alpha\delta - \beta\gamma)y = \alpha\eta - \beta\xi.$$

If $\alpha\delta - \beta\gamma \neq 0$, then all that's needed is to divide by it, but if $\alpha\delta - \beta\gamma = 0$, there is trouble: the results obtained give no information about x and y. If either one of $\delta\xi - \beta\eta$ or $\alpha\eta - \beta\xi$ is not 0, the equations have no solutions, and if both are 0, they have too many—every pair (x, y) works.

The expression $\alpha\delta - \beta\gamma$ is called the **determinant** of M, abbreviated

$$\det M,$$

and the answer to the 2×2 question is that M is invertible if and only if $\det M \neq 0$. The function $\det$ has some obvious properties and some non-obvious ones. An obvious property is that

$$\det 1 = 1,$$

and, more generally, that if M is diagonal (that is, $\beta = \gamma = 0$), then $\det M$ is the product of the diagonal entries. Another easy property is the behavior of $\det$ with respect to scalar multiplication:

$$\det(\lambda M) = \lambda^2 \det M$$

for every real scalar λ.

A non-obvious and possibly surprising property of $\det$ is that it is multiplicative, which means that if M_1 and M_2 are two 2×2 matrices, then

$$\det(M_1 \cdot M_2) = (\det M_1) \cdot (\det M_2).$$

If, in other words,

$$M_1 = \begin{pmatrix} \alpha_1 & \beta_1 \\ \gamma_1 & \delta_1 \end{pmatrix} \quad \text{and} \quad M_2 = \begin{pmatrix} \alpha_2 & \beta_2 \\ \gamma_2 & \delta_2 \end{pmatrix},$$

so that

$$M_1 \cdot M_2 = \begin{pmatrix} \alpha_1\alpha_2 + \beta_1\gamma_2 & \alpha_1\beta_2 + \beta_1\delta_2 \\ \gamma_1\alpha_2 + \delta_1\gamma_2 & \gamma_1\beta_2 + \delta_1\delta_2 \end{pmatrix},$$

then it is to be proved that

$$(\alpha_1\alpha_2 + \beta_1\gamma_2)(\gamma_1\beta_2 + \delta_1\delta_2) - (\alpha_1\beta_2 + \beta_1\delta_2)(\gamma_1\alpha_2 + \delta_1\gamma_2)$$
$$= (\alpha_1\delta_1 - \beta_1\gamma_1)(\alpha_2\delta_2 - \beta_2\gamma_2),$$

and to prove that, there is no help for it—compute.

Multiplicativity has a pleasant consequence about inverses. If M happens to be invertible, then

$$(\det M) \cdot (\det M^{-1}) = \det(MM^{-1}) = \det 1 = 1.$$

so that

$$\det M^{-1} = \frac{1}{\det M}.$$

Can the results about the determinants of 2×2 matrices be generalized to bigger sizes? One tempting direction of generalization is this: replace the entries α, β, γ, δ of a 2×2 matrix by 2×2 matrices (α), (β), (γ), (δ), thus getting a 4×4 matrix

$$\begin{pmatrix} (\alpha) & (\beta) \\ (\gamma) & (\delta) \end{pmatrix}.$$

The attempt to generalize the concept of determinant runs into a puzzle before it can get started: since the matrix entries may fail to commute, it is not clear whether the generalized determinant ought to be defined as

$$(\alpha)(\delta) - (\beta)(\gamma), \quad \text{or} \quad (\alpha)(\delta) - (\gamma)(\beta), \quad \text{or}$$
$$(\delta)(\alpha) - (\beta)(\gamma), \quad \text{or} \quad (\delta)(\alpha) - (\gamma)(\beta).$$

These four "formal determinants" seem to play equal roles—no reason to prefer one to the others is apparent. The best thing to do therefore is to be modest: assume as much as possible and conclude as little as possible. Here is one possible question.

Problem 62. *Which (if either) of the following two assertions is true?*

(1) If a 2×2 matrix M of 2×2 matrices is invertible, then at least one of its formal determinants is invertible.

(2) If all the formal determinants of a 2×2 matrix M of 2×2 matrices are invertible, then M is invertible.

63. Determinants: $n \times n$

When linear algebra began it wasn't called "linear algebra"—the central part of the subject was thought to be the concept of determinant. Nowadays determinant theory is considered a very small part of linear algebra. One reason is that this part of linear algebra is not really linear—the subject is an intricately combinatorial one that some people love for that reason, while others insist that the only elegant way to proceed is to avoid it whenever possible. Every one admits, however, that a little of it must be known to every student of the subject, and here comes a little.

It is not obvious how the definition of determinant for 2×2 matrices can be extended to $n \times n$ matrices. Even the basic definition with which most treatments begin is a little frightening. If $M = (\alpha_{ij})$ is an $n \times n$ matrix, the **cofactor** of an entry α_{ij}, call it A_{ij}, is the $(n-1) \times (n-1)$ matrix obtained by removing from M the entire row and the entire column that contain α_{ij}—that is, removing row i and column j. Thus, for instance, the cofactor of α in $\begin{pmatrix} \alpha & \beta \\ \gamma & \delta \end{pmatrix}$ is the 1×1 matrix (δ), and the cofactor of γ is the

1×1 matrix (β). The standard definition of determinants uses cofactors
to proceed inductively. The determinant of a 1×1 matrix (x) is defined to
be the number x, and the determinant of an $n \times n$ matrix, in terms of its
$(n-1) \times (n-1)$ submatrices (written here for $n = 4$), is given by

$$\det M = \alpha_{11} \cdot \det A_{11} - \alpha_{21} \cdot \det A_{21} + \alpha_{31} \cdot \det A_{31} - \alpha_{41} \cdot \det A_{41}.$$

In words: multiply each element of the first column by the determinant of
its cofactor, and then form the sum of the resulting products with alternat-
ing signs. Special case:

$$\det \begin{pmatrix} \alpha & \beta \\ \gamma & \delta \end{pmatrix} = \alpha \cdot \det(\delta) - \gamma \cdot \det(\beta),$$

which agrees with the $\alpha\delta - \beta\gamma$ definition of Problem 62.

Important comment: the definition does not have to be based on the
first column—with a small modification any other column gives the same
result. The modification has to do with signs. Think of the entries of a
matrix as having signs painted on them: the sign painted on α_{ij} is to be
plus if $i + j$ is even and minus if $i + j$ is odd. (Equivalently: the sign of
α_{ij} is the sign of the number $(-1)^{i+j}$.) Thus, α_{11} and α_{31} are plus whereas
α_{21} and α_{41} are minus. The definition of determinant in terms of column
j instead of column 1 (written here for $n = 4$) is either

$$\alpha_{1j} \cdot \det A_{1j} - \alpha_{2j} \cdot \det A_{2j} + \alpha_{3j} \cdot \det A_{3j} - \alpha_{4j} \cdot \det A_{4j}$$

or the negative of that depending on whether the sign of α_{1j} is plus or
minus. Example: if

$$M = \begin{pmatrix} \alpha_{11} & \alpha_{12} & \alpha_{13} \\ \alpha_{21} & \alpha_{22} & \alpha_{23} \\ \alpha_{31} & \alpha_{32} & \alpha_{33} \end{pmatrix},$$

then

$$\det M = \alpha_{11}(\alpha_{22}\alpha_{33} - \alpha_{23}\alpha_{32})$$
$$- \alpha_{21}(\alpha_{12}\alpha_{33} - \alpha_{13}\alpha_{32})$$
$$+ \alpha_{31}(\alpha_{12}\alpha_{23} - \alpha_{13}\alpha_{22}),$$

and also

$$\det M = -\alpha_{12}(\alpha_{21}\alpha_{33} - \alpha_{23}\alpha_{31})$$
$$+ \alpha_{22}(\alpha_{11}\alpha_{33} - \alpha_{13}\alpha_{31})$$
$$- \alpha_{32}(\alpha_{11}\alpha_{23} - \alpha_{13}\alpha_{21}).$$

These formulas are called the **expansion** of det M in terms of the first and
the second columns; there is also, of course, an expansion in terms of the

third column, and there are completely similar expansions in terms of the rows.

The assertion that these various signed sums all give the same answer (in every case, not only for 3×3 matrices) needs proof, but the proof is not likely to elevate the reader's soul and is omitted here. Comment: if all parentheses are removed, so that the expansion of the determinant of a 3×3 matrix is written out in complete detail, then each of the resulting six terms is a product (with its appropriate sign) of exactly one element from each row and each column. A similar statement is true about $n \times n$ matrices for every n: the value of the determinant is the sum, with appropriate signs, of the $n!$ products that contain exactly one factor from each row and each column. A description such as that is sometimes used as an alternative definition of determinants.

Another (definitely non-trivial) assertion is that (for matrices of all sizes) determinant is multiplicative: if M_1 and M_2 are two $n \times n$ matrices, then

$$\det(M_1 \cdot M_2) = (\det M_1) \cdot (\det M_2).$$

An easier property (with its easier proof also omitted) is that the determinant of an upper triangular matrix, such as for instance

$$\begin{pmatrix} * & * & * & * \\ 0 & * & * & * \\ 0 & 0 & * & * \\ 0 & 0 & 0 & * \end{pmatrix},$$

is the product of the diagonal entries. (The picture is intended to convey that the entries below the main diagonal are 0 while the entries on and above are arbitrary. A similar picture and similar comment apply to lower triangular matrices also.) Special case: the determinant of a diagonal matrix is the product of the diagonal entries. Special case of that special case: the determinant of a scalar matrix $\lambda \cdot 1$ is λ^n, and, in particular, the determinant of 1 is 1. For invertible matrices

$$\det M^{-1} = \frac{1}{\det M},$$

and, for any matrix,

$$\det(\lambda M) = \lambda^n \det M$$

(where λ is an arbitrary scalar).

Far and away the most useful property of determinants is that $\det M \neq 0$ if and only if M is invertible. (The proof is just as omitted as the other proofs so far in this section.) That property is probably the one that is most frequently exploited, and it will indeed be exploited at some crucial spots in the sequel.

The actual numerical values of determinants are rarely of any interest; what is important is the theoretical information that the existence of a function of matrices with the properties described above implies. The questions that follow, however, do have to do with the numerical evaluation of some determinants—they are intended to induce at least a small familiarity with (and perhaps a friendly feeling toward) such calculations.

Problem 63.   *If M_1, M_2, and M_3 are the matrices below, what are the values of* $\det M_1$, $\det M_2$, *and* $\det M_3$?

$$M_1 = \begin{pmatrix} 1 & 2 & 0 & 0 \\ 2 & 1 & 0 & 0 \\ 0 & 0 & 3 & 4 \\ 0 & 0 & 4 & 3 \end{pmatrix},$$

$$M_2 = \begin{pmatrix} 2 & 0 & 0 & 0 & 2 \\ 0 & 2 & 0 & 2 & 0 \\ 0 & 0 & 2 & 0 & 0 \\ 0 & 2 & 0 & 2 & 0 \\ 2 & 0 & 0 & 0 & 2 \end{pmatrix},$$

$$M_3 = \begin{pmatrix} 3 & 0 & 0 & 0 & 0 & 2 \\ 0 & 3 & 0 & 0 & 2 & 0 \\ 0 & 0 & 3 & 2 & 0 & 0 \\ 0 & 0 & 2 & 3 & 0 & 0 \\ 0 & 2 & 0 & 0 & 3 & 0 \\ 2 & 0 & 0 & 0 & 0 & 3 \end{pmatrix}.$$

64. Zero-one matrices

Matrices with only a small number of different entries intrigue mathematicians—such matrices have some pleasant, and curious, and sometimes surprising properties. Suppose, for instance, that every entry of an $n \times n$ matrix is either 0 or 1—how likely is such a matrix to be invertible? It is clear that it doesn't have to be. The extreme example is the matrix in which every entry is equal to 0—the zero matrix. Equally clearly a matrix of 0's and 1's (would "01-matrix" be a useful abbreviation?) can succeed in being invertible—the trivial example is the identity matrix. It seems reasonable that the more zeros a matrix has, the more likely it is that its determinant is 0, and hence the less likely the matrix is to be invertible. What in fact is the largest number of 0's that a 01-matrix of size $n \times n$ can have and still be invertible? Equivalently: what is the smallest number of 1's that a 01-matrix of size $n \times n$ can have and still be invertible? That's easy: the answer is n. The number of 1's in the identity matrix is n; if a matrix has fewer 1's, then at least one of its rows must consist of 0's only, which implies that its determinant is 0. (Why?) Consequence: the maximum number of 0's that an invertible 01-matrix can have is $n^2 - n$.

What about 1's? An $n \times n$ matrix has n^2 entries; if they are all equal to 1, then, for sure, it is not invertible; in fact its determinant is 0. What is the largest number of 1's that an $n \times n$ 01-matrix can have and be invertible? Triangular examples such as

$$\begin{pmatrix} 1 & 1 & 1 & 1 \\ 0 & 1 & 1 & 1 \\ 0 & 0 & 1 & 1 \\ 0 & 0 & 0 & 1 \end{pmatrix}$$

suggest the conjecturable answer

$$n^2 - \left(1 + \cdots + (n-1)\right) = \frac{n(n+1)}{2}.$$

Is that right?

Problem 64. *What is the largest number of 1's that an invertible 01-matrix of size $n \times n$ can have?*

Comment. The displayed formula has a not uncommon problem at its lowest value: when $n = 1$, the left side has to be interpreted with some kindness.

65   65. Invertible matrix bases

The set $\mathbb{L}(\mathbb{V})$ of all linear transformations of an n-dimensional real vector space $\mathbb{V}$ to itself, or, equivalently, the set of all $n \times n$ real matrices, is a vector space that has a basis consisting of n^2 elements—see Problem 59. How good can that basis be?

Problem 65. *Does $\mathbb{L}(\mathbb{V})$ have a basis consisting of invertible linear transformations?*

66. Finite-dimensional invertibility

The two conditions needed for invertibility are easy to characterize individually: to say that a linear transformation T on a vector space $\mathbb{V}$ is surjective means just that $\operatorname{ran} T = \mathbb{V}$, and to say that it is injective means that $\ker T = \{0\}$. The first statement is nothing but language: that's what the words mean. The second statement is really two statements: (1) if T is injective, then $\ker T = \{0\}$, and (2) if $\ker T = \{0\}$, then T is injective. Of these (1) is obvious: if T is injective, then, in particular, T can never send a vector other than 0 onto 0. Part (2) is almost equally obvious: if T never sends a non-zero vector onto 0, then T cannot send two different vectors u_1 and u_2 onto the same vector, because if it did, then (by linearity) T would send $u_1 - u_2$ onto 0.

The interesting question along these lines is whether there is any relation between the two conditions. Is it possible for T to be surjective but not injective? What about the other way around? These questions, and their answers, turn out to have a deep (and somewhat complicated) theory in the infinite parts of linear algebra. In the finite-dimensional case the facts are nearer to the surface.

Problem 66. *Can a linear transformation on a finite-dimensional vector space be injective but not surjective? How about the other way around?*

67. Matrices

A matrix is a square array of scalars, such as

$$
\begin{pmatrix}
-3 & 7 & \frac{1}{2} & \pi \\
3 & 0 & 0 & 1 \\
\sqrt[3]{10} & \sqrt{\pi} & e^2 & 19 \\
0 & 1 & 0 & \frac{1}{2\pi}
\end{pmatrix}.
$$

This example is called a 4×4 matrix (pronounced "four by four"); all other sizes (2×2, 11×11) are equally legitimate. An extreme case is a 1×1 matrix, which is hard to tell apart from a scalar. In some contexts (slightly more general than the ones here considered) even rectangular matrices are allowed (instead of only square ones).

Matrices have already occurred in these problems. Solution 59 shows that a basis $\{e_1, e_2, \ldots, e_n\}$ of a finite-dimensional vector space can be used to establish a one-to-one correspondence between all linear transformations A on that space and all matrices $\{\alpha_{ij}\}$. The relation between a linear transformation A and its matrix $\{\alpha_{ij}\}$ is that

$$Ae_j = \sum_{i=1}^{n} \alpha_{ij} e_i$$

for each $j = 1, 2, \ldots, n$. What is already known is that the linear proper-ties of linear transformations correspond just fine to the linear properties of matrices: if two transformations A and B have matrices $\{\alpha_{ij}\}$ and $\{\beta_{ij}\}$, then a transformation such as $\xi A + \eta B$ has the matrix $\{\xi \alpha_{ij} + \eta \beta_{ij}\}$. What is not known is how matrices behave under multiplication: what, to be def-inite, is the matrix of AB? The only way to find the answer is to indulge in a bit of not especially fascinating computation, like this:

$$(AB)e_j = A(Be_j) = A\left(\sum_k \beta_{kj} e_k\right) = \sum_k \beta_{kj} A e_k$$

$$= \sum_k \beta_{kj} \left(\sum_i \alpha_{ik} e_i\right) = \sum_i \left(\sum_k \alpha_{ik} \beta_{kj}\right) e_i.$$

Conclusion: the $\langle i, j \rangle$ entry of the matrix of AB is

$$\sum_k \alpha_{ik} \beta_{kj}.$$

It may look complicated to someone who has never seen it before, but all it takes is a little getting used to—it is really quite easy to work with.

Here is an easy example: the set $\mathbb{C}$ of complex numbers is a real (!) vec-tor space, the set $\{1, i\}$ is a basis for that space, and the action C of complex conjugation is a linear transformation on that space—what is the corre-sponding matrix? That's two questions: what is the expansion, in terms of $\{1, i\}$ of $C1$, and what is the expansion in terms of $\{1, i\}$ of Ci? Since the answers are obvious: $C1 = 1$ and $Ci = -i$, the matrix is

$$\begin{pmatrix} 1 & 0 \\ 0 & -1 \end{pmatrix}.$$

That's too easy. For a little more revealing example, consider the vec-tor space $\mathbb{P}_4$ of all polynomials of degree 4 or less, with basis

$$e_0 = 1, \; e_1 = t, \; e_2 = t^2, \; e_3 = t^3, \; e_4 = t^4,$$

and the linear transformation A on $\mathbb{P}_4$ defined by

$$Ax(t) = x(t+1);$$

what is the matrix? Since

$$Ae_0 = 1,$$

$$Ae_1 = t + 1,$$

$$Ae_2 = (t + 1)^2 = t^2 + 2t + 1, \text{ etc.,}$$

it follows that the matrix (look at its columns) is

$$\begin{pmatrix} 1 & 1 & 1 & 1 & 1 \\ 0 & 1 & 2 & 3 & 4 \\ 0 & 0 & 1 & 3 & 6 \\ 0 & 0 & 0 & 1 & 4 \\ 0 & 0 & 0 & 0 & 1 \end{pmatrix}.$$

Problem 67. *What happens to the matrix of a linear transformation on a finite-dimensional vector space when the elements of the basis with respect to which the matrix is computed are permuted among themselves?*

Comment. The considerations of invertibility introduced in Problem 60 can be formulated more simply and more naturally in terms of matrices. Thus, for instance, the question about the transformation T of that problem could have been asked this way: is the matrix

$$\begin{pmatrix} 2 & 1 \\ 1 & 1 \end{pmatrix}$$

invertible?

68. Diagonal matrices   68

Some matrices are easier to work with than others, and the easiest, usually, are the diagonal ones—they are the ones, such as

$$\begin{pmatrix} 5 & 0 & 0 & 0 \\ 0 & -e & 0 & 0 \\ 0 & 0 & \frac{2}{3} & 0 \\ 0 & 0 & 0 & 100 \end{pmatrix},$$

in which every entry not on the main diagonal is 0. (The main diagonal of a matrix $\{\alpha_{ij}\}$ is the set of entries of the form α_{ii}, that is the entries whose two subscripts are equal.) The sum of two diagonal matrices is a diagonal matrix, and, extra pleasant, even the multiplication formula for diagonal matrices is extra simple. The product of two diagonal matrices is the diagonal matrix obtained by multiplying each entry of one of them by the corresponding entry in the other. It follows in particular that within the territory of diagonal matrices multiplication is commutative. The product of two matrices usually depends on the order in which they are multiplied, but if A and B are diagonal matrices, then always $AB = BA$.

The simplest among the diagonal matrices are the scalar matrices, which means the matrices that are scalar multiples of the identity: matrices in which all the diagonal entries are equal. They are the simplest to look at, but it happens that sometimes the extreme opposite, the diagonal matrices in which all diagonal entries are different, are the easiest ones to work with.

Problem 68. *If A is a diagonal matrix whose diagonal entries are all different, and if B is a matrix that commutes with A, must B also be diagonal?*

69   69. Universal commutativity

Which matrices commute with all matrices? To keep the question from being nonsense, it must be assumed of course that a size is fixed once and for all and only square matrices of that size are admitted to the competition. One example is the identity matrix; another is the zero matrix. Which are the non-trivial examples?

Problem 69. *Which $n \times n$ matrices B have the property that $AB = BA$ for all $n \times n$ matrices A?*

70   70. Invariance

The set $\mathbb{M}$ of all vectors of the form

$$(0, \alpha, \beta, \beta)$$

in $\mathbb{R}^4$ (that is, the set of all vectors whose first coordinate is 0 and whose last two coordinates are equal) is a subspace of $\mathbb{R}^4$, and the matrix

$$A = \begin{pmatrix} 1 & 0 & 1 & -1 \\ 2 & 1 & 2 & 1 \\ 0 & 1 & 6 & 1 \\ 3 & 1 & 3 & 4 \end{pmatrix}$$

defines a linear transformation on $\mathbb{R}^4$. If u is a vector in $\mathbb{M}$, then the vector Au has its first coordinate equal to 0 and its last two coordinates equal to one another (true?)—in other words, it too is in $\mathbb{M}$ (along with u). This phenomenon is described by saying that the subspace $\mathbb{M}$ is **invariant** under the linear transformation A. (The facts are often stated in this language: $\mathbb{M}$ is an **invariant subspace** of A.)

If B is the scalar matrix

$$\begin{pmatrix} 2 & 0 & 0 & 0 \\ 0 & 2 & 0 & 0 \\ 0 & 0 & 2 & 0 \\ 0 & 0 & 0 & 2 \end{pmatrix},$$

and if u is in $\mathbb{M}$, then Bu is in $\mathbb{M}$—the subspace $\mathbb{M}$ is invariant under B also.

If C is the diagonal matrix

$$\begin{pmatrix} 1 & 0 & 0 & 0 \\ 0 & 2 & 0 & 0 \\ 0 & 0 & 1 & 0 \\ 0 & 0 & 0 & 2 \end{pmatrix},$$

and if $u = (0, 1, 2, 2)$, then u is in $\mathbb{M}$, but Cu is not; the subspace $\mathbb{M}$ is not invariant under C.

The concept of invariance plays a central role in linear algebra (or, for that matter, in all mathematics). The most primitive questions about invariance are those of counting: how many invariant subspaces are there? If, for instance, the underlying vector space $\mathbb{V}$ is the trivial one, that is the space $\mathbb{O}$, then every linear transformation has exactly one invariant subspace (namely $\mathbb{O}$ itself). If $\mathbb{V}$ is $\mathbb{R}^1$ and if A is the identity transformation on $\mathbb{R}^1$, then there are just two invariant subspaces (namely $\mathbb{O}$ and $\mathbb{R}^1$.)

Problem 70. *Is there a vector space $\mathbb{V}$ and a linear transformation A on $\mathbb{V}$, such that A has exactly three invariant subspaces?*

71. Invariant complements

Invariant subspaces can be used to simplify the appearance of matrices. If A is a linear transformation on a vector space $\mathbb{V}$ of dimension n, say, and if $\mathbb{M}$ is a subspace of $\mathbb{V}$ invariant under A, of dimension m, say, then there exists a basis $\{e_1, e_2, \ldots, e_n\}$ of $\mathbb{V}$ such that the vectors e_j $(j = 1, \ldots, m)$ belong to $\mathbb{M}$. With respect to such a basis, the matrix corresponding to A has the form

$$\begin{pmatrix} P & Q \\ 0 & R \end{pmatrix},$$

where P is an $m \times m$ matrix, R is an $(n - m) \times (n - m)$ matrix (square), and Q is an array (a rectangular matrix) with m rows and $n - m$ columns.

The entry 0 in the lower left corner deserves mention too: it is an array (a rectangular matrix) with $n - m$ rows and m columns.

The more zeroes a matrix has, the easier it is to compute with. For each linear transformation it is natural to look for a basis so as to make the corresponding matrix have many zeroes. What the preceding comment about matrices shows is that if the basis begins with vectors in an invariant subspace, then the matrix has an array of zeroes in the lower left corner. Under what circumstances will the other corners also consist of zeroes?

When, for instance, will it happen that $P = 0$? That's easy: $P = 0$ means that $Ae_j = 0$ for $j = 1, \ldots, m$, so that the span $\mathbb{M}$ of $\{e_1, e_2, \ldots, e_m\}$ is in the kernel of A—that's necessary and sufficient.

When is $Q = 0$? Answer: if and only if the coefficients of the vectors $e_1, e_2, \ldots, e_m$ in the expansion of Ae_j $(j = m + 1, \ldots, n)$ are all 0. Better said: if and only if the image under A of each e_j $(j = m+1, \ldots, n)$ belongs to the span of $\{e_{m+1}, \ldots, e_n\}$. Reformulation: the span of $\{e_{m+1}, \ldots, e_n\}$ is invariant under A. In other words, a necessary and sufficient condition that the matrix of A have the form

$$\begin{pmatrix} P & 0 \\ 0 & R \end{pmatrix}$$

with respect to some basis $\{e_1, e_2, \ldots, e_n\}$ is that for some m both the span of $\{e_1, \ldots, e_m\}$ and the span of $\{e_{m+1}, \ldots, e_n\}$ be invariant under A. Best answer: the matrix of A has the form $\begin{pmatrix} P & 0 \\ 0 & R \end{pmatrix}$ if and only if there exist two complementary subspaces each of which is invariant under A.

How likely is that to happen?

Problem 71. *If A is a linear transformation on a finite-dimensional vector space, does every invariant subspace of A have an invariant complement?*

72. Projections

If $\mathbb{M}$ and $\mathbb{N}$ are complementary subspaces of a finite-dimensional vector space $\mathbb{V}$, then (Problem 28) corresponding to every vector z in $\mathbb{V}$ there are two vectors x and y, with x in $\mathbb{M}$ and y in $\mathbb{N}$, such that $z = x + y$, and, moreover, x and y are uniquely determined by z. The vector x is called the **projection** of z into (or just plain "to") $\mathbb{M}$ along (or **parallel** to) $\mathbb{N}$, and, similarly, y is the projection of z to $\mathbb{N}$ along $\mathbb{M}$.

A picture for the case in which $\mathbb{M}$ and $\mathbb{N}$ are distinct lines through the origin in the plane helps to see what's going on. If z is a typical point not on either $\mathbb{M}$ or $\mathbb{N}$, then draw a line through z parallel to $\mathbb{M}$; the point where that parallel intersects N is the projection of z to N. (Similarly, of course, the intersection with M of the line through z parallel to N is the projection of z to M.) What happens if z does belong to M or N?

Consider the correspondence (transformation) that assigns to each z the vector x—call it E—and, similarly, let F be the transformation that assigns to each z the vector y. The verification that E and F are linear transformations is dull routine (but necessary). Since $x + y = z$, it follows that $E + F = 1$. The word **projection** is used for the transformations E and F (as well as for the vectors x and y in their ranges): E is called the projection on M (along N) and F the projection on N (along M). Warning: E emphasizes M, but its definition depends crucially on both M and N, and, similarly, F emphasizes N, but depends on both subspaces also. In other words if N_1 and N_2 are two different complements of M, then the projections E_1 and E_2 on M along N_1 and N_2 are different transformations. If M $=$ V (in which case N is uniquely determined as $\mathbb{O}$), then $E = 1$; if M $= \mathbb{O}$, then $E = 0$. These are trivial examples of projections; what are some non-trivial ones? Could it be, for instance, that every linear transformation is the projection to some subspace along one of its complements?

Problem 72. (a) *Which linear transformations are projections?*

(b) *If E is the projection on M along N, what are* ran E *and* ker E?

Comment. Question (a) asks for a characterization of some algebraic kind (as opposed to the geometric definition) that puts all projections into one bag and all other linear transformations into an another.

73. Sums of projections

73

Is the sum of two projections always a projection? Of course not—the identity transformation 1, for instance, is a projection, but the sum of the identity and itself, the transformation 2, is certainly not a projection. (In fact the double of a projection different from 0 is never a projection. Proof?)

On the other hand the sum of two projections can often be a projection. A trivial example is given by any projection E and the projection 0. An only slightly less trivial example is

$$\begin{pmatrix} 1 & 0 \\ 0 & 0 \end{pmatrix} \quad \text{and} \quad \begin{pmatrix} 0 & 0 \\ 0 & 1 \end{pmatrix}.$$

A numerically somewhat cumbersome example, in, however, the same spirit, is given by

$$\frac{1}{25}\begin{pmatrix} 9 & 12 \\ 12 & 16 \end{pmatrix} \quad \text{and} \quad \frac{1}{25}\begin{pmatrix} 16 & -12 \\ -12 & 9 \end{pmatrix}$$

What's going on?

Problem 73.   *When is the sum of two projections a projection?*

74   ## 74. Not quite idempotence

A linear transformation E is a projection if and only if it is idempotent, and that condition is equivalent to the equation $E(1-E) = 0$. If that equation is satisfied then $E = E^2$ and $1 - E = (1 - E)^2$. It follows that if E is idempotent, then the slightly weaker equations

$$E^2(1 - E) = 0 \quad \text{and} \quad E(1 - E)^2 = 0$$

are satisfied also. Are these necessary conditions sufficient? Caution: does the question mean together or separately?

Problem 74.   *If E is a linear transformation such that*

$$E^2(1 - E) = 0,$$

does it follow that E is idempotent? Is that condition equivalent to

$$E(1 - E)^2 = 0?$$

CHAPTER 5

DUALITY

75. Linear functionals

The most useful functions on vector spaces are the linear functionals. Recall their definition (Problem 55): a linear functional is a scalar-valued function ξ such that

$$\xi(x + y) = \xi(x) + \xi(y)$$

and

$$\xi(\alpha x) = \alpha \xi(x)$$

whenever x and y are vectors in $\mathbb{V}$ and α is a scalar.

Example on $\mathbb{R}^n$: $\xi(x_1, x_2, \ldots, x_n) = 3x_1$.

Example on $\mathbb{R}^3$: $\xi(x, y, z) = x + 2y + 3z$.

Examples on the space $\mathbb{P}$ of polynomials:

$$\xi(p) = \int_{-1}^{+1} p(t)\, dt, \quad \text{or} \quad \int_{-2}^{+1} t^2 p(t)\, dt,$$

$$\text{or} \quad \int_{0}^{9} t^2 p(t^2)\, dt, \quad \text{or} \quad \frac{d^2 p}{dt^2}\bigg|_{t=1}.$$

The most trivial linear functional is also the most important one, namely 0. That is: if ξ is defined by

$$\xi(x) = 0$$

for all x, then ξ is a linear functional. Except for this uninteresting case, every linear functional takes on every scalar value—that's an easy exercise (see Problem 55(1)). So, for instance, if ξ_0 is a non-zero linear functional on a finite-dimensional vector space V, then there always exists a vector x in V such that $\xi_0(x) = 3$. Does it work the other way? That is: if x_0 is a non-zero vector in a finite-dimensional vector space V, does there always exist a linear functional ξ such that $\xi(x_0) = 3$? That takes a little more thought, but the answer is still yes. To see that, let $\{x_1, x_2, \ldots, x_n\}$ be a basis for V that has the prescribed vector x_0 as one of its elements, say, $x_1 = x_0$; then the first example above does the job.

These statements are mainly about the ranges of linear functionals; can something intelligent be said about their kernels? If, for instance, ξ and η are linear functionals (on the same vector space V) and α is a scalar such that $\eta(x) = \alpha\xi(x)$ for every x in V, then, clearly, $\eta(x) = 0$ whenever $\xi(x) = 0$. Is there a chance that the converse statement is true?

Problem 75. *If ξ and η are linear functionals (on the same vector space) such that $\eta(x) = 0$ whenever $\xi(x) = 0$, must there always exist a scalar α such that $\eta(x) = \alpha\xi(x)$ for all x?*

76 76. Dual spaces

Old vector spaces can sometimes be used to make new ones. A typical example of such a happening is the formation of the subspaces of a vector space. Another example starts from two spaces, a vector space and a subspace, and forms the quotient space. Still another example starts from two arbitrary vector spaces and forms their direct sum.

One of the most important ways to get new vector spaces from old ones is **duality**: corresponding to every real vector space V there is a so-called **dual space** V'. The elements of V' are easy enough to describe: they are the linear functionals on V. Linear functionals can be added: if ξ and η are linear functionals on V, then a linear functional $\sigma = \xi + \eta$ is defined for all x by

$$\sigma(x) = \xi(x) + \eta(x).$$

Linear functionals can be multiplied by scalars: if ξ is a linear functional and α is a scalar, then a linear functional $\tau = \alpha\xi$ is defined for all x by

$$\tau(x) = \alpha\xi(x).$$

With these definitions of addition and scalar multiplication the set V' of all linear functionals becomes a vector space, and that is the dual of V.

Problem 76. *Is the dual of a finite-dimensional vector space finite-dimensional?*

77. Solution of equations

Every vector in a vector space (except the zero vector) is just like every other vector, and every linear functional (except the zero linear functional) is just like every other. If, however, $\mathbb{F}$ is a field, and, for some positive integer n, the particular vector space under consideration is $\mathbb{F}^n$, then the existence of built-in coordinates makes it possible to single out certain special vectors and special linear functionals. Thus, for instance, the vectors that have only one coordinate different from 0 and that one equal to 1 are often made to play an important role, and the same is true of the linear functionals p_j defined by

$$p_j(x_1, \ldots, x_n) = x_j, \qquad j = 1, \ldots, n,$$

called the **coordinate functionals** or **coordinate projections**.

These vectors and functionals are most conspicuous in the procedure called "solving" systems of n linear equations in n unknowns. Here is what that seems to mean to most students: keep forming linear combinations of the given equations till they take the form

$$x_1 = \alpha_1, \ldots, x_n = \alpha_n,$$

and then feel justified in deciding that $(\alpha_1, \ldots, \alpha_n)$ is the sought for solution.

The most fussily honest way of describing what the reasoning shows is to say that IF there is a solution, THEN the procedure leads to one, but an existence proof of some kind is probably called for, and so is a uniqueness proof. Some teachers worry about putting such an incomplete tool into the hands of their students, and feel called upon to justify themselves by an airy wave of the hand and the suggestion "just follow the steps backwards, and you'll be all right". Is that always true?

A linear equation is presumably something of the form

$$y(x) = \alpha,$$

where y is a linear functional (given), α is a scalar (known), and x is a vector (unknown). A system of n linear equations in n unknowns, then, involves n linear functionals $y_1, \ldots, y_n$ and n scalars $\alpha_1, \ldots, \alpha_n$; what is wanted is a vector $x = (x_1, \ldots, x_n)$ such that

$$y_j(x) = \alpha_j, \qquad j = 1, \ldots, n.$$

Another way of saying all this is to define a linear transformation T from $\mathbb{F}^n$ to $\mathbb{F}^n$ by writing

$$T(x) = T(x_1, \ldots, x_n) = \big(y_1(x), \ldots, y_n(x)\big),$$

and then hope that enough is known about T to guarantee that its range is all of $\mathbb{F}^n$ (existence of solutions) and that its kernel is $\{0\}$ (uniqueness). The hope, in other words, is that T is invertible; if it is, then the desired solution $(\alpha_1, \ldots, \alpha_n)$ is given by $T^{-1}(\alpha_1, \ldots, \alpha_n)$.

What hidden assumption is there in the usual solution procedure that might be enough to imply the invertibility of T? The not especially well hidden assumption is that it is possible to

"keep forming linear combinations of the given equations till they take the form $x_1 = \alpha_1, \ldots, x_n = \alpha_n$".

In other words: the span of the linear functionals $y_1, \ldots, y_n$ (their span in the dual space of $\mathbb{F}^n$, of course) contains each of the coordinate projections.

Problem 77. *If $y_1, \ldots, y_n$ are linear functionals on $\mathbb{F}^n$ such that each of the coordinate projections belongs to their span, does it always follow that the linear transformation T from $\mathbb{F}^n$ to $\mathbb{F}^n$ defined by*

$$T(x) = \big(y_1(x), \ldots, y_n(x)\big)$$

is invertible?

78. Reflexivity

Sometimes two different vector spaces can resemble each other so much that it's hard to tell them apart. At first blush the space $\mathbb{P}_3$ (polynomials of degree 3 or less) and the space $\mathbb{R}^4$ (sequences of length 4) may not look much alike, but on second thought maybe they do. What are vector properties of polynomials such as

$$\alpha_0 + \alpha_1 t + \alpha_2 t^2 + \alpha_3 t^3,$$

and how do they compare with the vector properties of sequences such as

$$(\xi_1, \xi_2, \xi_3, \xi_4)?$$

In this question the traditional use of the alphabet hurts a little rather than helps. Change the α's to ξ's and jack up their indices by one, or else change the ξ's to α's and lower their indices by one, and the differences tend to go away: the only difference between the vector

$$\xi_1 + \xi_2 t + \xi_3 t^2 + \xi_4 t^3$$

in $\mathbb{P}_3$ and the vector

$$(\xi_1, \xi_2, \xi_3, \xi_4)$$

in $\mathbb{R}^4$ is that one uses plus signs and t's and the other uses parentheses and commas—notation, not substance. More elegantly said: there is a natural one-to-one correspondence between $\mathbb{R}^4$ and $\mathbb{P}_3$, namely the correspondence T defined by

$$T(\xi_1, \xi_2, \xi_3, \xi_4) = \xi_1 + \xi_2 t + \xi_3 t^2 + \xi_4 t^3,$$

and both that correspondence and its inverse are linear. Is the meaning of that sentence clear? It says that not only is T a one-to-one correspondence, but, moreover, if $x = (\xi_1, \xi_2, \xi_3, \xi_4)$ and $y = (\eta_1, \eta_2, \eta_3, \eta_4)$ are in $\mathbb{R}^4$ and α and β are scalars, then

$$T(\alpha \xi + \beta \eta) = \alpha T(\xi) + \beta T(\eta),$$

and also that if p and q are in $\mathbb{P}_3$, then

$$T^{-1}(\alpha p + \beta q) = \alpha T^{-1}(p) + \beta T^{-1}(q).$$

(See Problem 60.) An informal but not misleading way of saying all this is to say that $\mathbb{P}_3$ and $\mathbb{R}^4$ differ in notation only, or even more informal and perhaps a tiny bit misleading, that $\mathbb{P}_3$ and $\mathbb{R}^4$ are essentially the same.

The dignified word for "essentially the same" is **isomorphic**. An **isomorphism** from a vector space $\mathbb{V}$ to a vector space $\mathbb{W}$ is a one-to-one linear correspondence T from (all of) $\mathbb{V}$ to (all of) $\mathbb{W}$; two spaces are called isomorphic if there exists an isomorphism between them. Trivially $\mathbb{V}$ and $\mathbb{V}$ are always isomorphic, and if $\mathbb{V}$ and $\mathbb{W}$ are isomorphic, then so are $\mathbb{W}$ and $\mathbb{V}$ (the inverse of an isomorphism is an isomorphism), and, finally, if $\mathbb{U}$, $\mathbb{V}$, and $\mathbb{W}$ are vector spaces such that $\mathbb{U}$ is isomorphic to $\mathbb{V}$, and $\mathbb{V}$ is isomorphic to $\mathbb{W}$, then $\mathbb{U}$ is isomorphic to $\mathbb{W}$ (the composition of two isomorphisms is an isomorphism). What was just said is that isomorphism is an equivalence relation—and nothing that has been said on the subject so far is deep at all.

Isomorphisms preserve all important properties of vector spaces. Example: isomorphic vector spaces have the same dimension. Proof: if $\{\xi_1, \ldots, \xi_n\}$ is a basis for $\mathbb{V}$ and if T is an isomorphism from $\mathbb{V}$ to $\mathbb{W}$, then $\{T\xi_1, \ldots, T\xi_n\}$ is a basis for $\mathbb{W}$. Reason: if

$$\alpha_1 T\xi_1 + \cdots + \alpha_n T\xi_n = 0,$$

then

$$T(\alpha_1 \xi_1 + \cdots + \alpha_n \xi_n) = 0,$$

which implies that

$$\alpha_1 \xi_1 + \cdots + \alpha_n \xi_n = 0,$$

and hence that

$$\alpha_1 = \cdots = \alpha_n = 0$$

—the Tx_j's are linearly independent. If, moreover, y is an arbitrary vector in W, then $T^{-1}y$ (in V) is a linear combination of the x_j's—that is $T^{-1}y$ has the form $\alpha_1 \xi_1 + \cdots + \alpha_n \xi_n$ for suitable α's. Apply T to conclude that $y = \alpha_1 T\xi_1 + \cdots + \alpha_n T\xi_n$—the $T\xi_j$'s span W.

The converse of the result just proved is also true: if two finite-dimensional vector spaces V and W have the same dimension, then they are isomorphic. Proof: if $\{\xi_1, \ldots, \xi_n\}$ is a basis for V and $\{\eta_1, \ldots, \eta_n\}$ is a basis for W, then there is one and only one linear transformation T from V to W such that $T\xi_j = \eta_j$ $(j = 1, \ldots, n)$, and that transformation is an isomorphism from V to W.

The proof just sketched is correct, but in the view of some mathematicians it is ugly—it is unnatural. Reason: it depends on two arbitrary (unnatural) choices—a basis for V must be chosen and a basis for W must be chosen, and the argument depends on those two bases. If those bases are changed, the isomorphism T changes. Yes, sure, there exists an isomorphism, and, in fact, there are many isomorphisms from V to W, but there is no reason to prefer any one of them to any other—there is no natural way to make the choice between them.

There is one celebrated circumstance in which a **natural** isomorphism between two vector spaces does spring to the eye, and that is between a finite-dimensional vector space V and the dual space $(V')'$ of its dual space V'. (It is more convenient and customary to denote that **second dual space** by V''.) The elements of V'' are linear functionals of linear functionals. What is an example of such a thing? Here is one: fix a vector ξ_0 in V, and then, for each element η of V'— that is, for each linear functional η on V—write

$$\xi_0'(\eta) = \eta(x_0).$$

Emphasis: ξ_0 is held fixed here (in V) and η is allowed to vary (in V'). Consequence: ξ_0' is a function on V', a function of η, and half a minute's staring at the way $\eta(\xi_0)$ (which equals $\xi_0'(\eta)$) depends on η should convince everyone that ξ_0' is a linear functional on V'.

Does every linear functional on V' arise in this way?

Problem 78. *If V is a vector space, and if for each vector x in V an element x' ($= Tx$) of V'' is defined by*

$$x'(\eta) = \eta(x),$$

then T is always a linear transformation from V to V'' (verification?); is it always an isomorphism?

Comment. The question belongs to one of the subtlest parts of linear algebra; to put it into the right context a couple of comments are advisable.

(1) The mapping T here defined is called the **natural** mapping from V to V''. In case it happens that T maps V onto V'' (so that it is an isomorphism), the vector space V is called **reflexive**. (Warning: for infinite-dimensional, topological, vector spaces the same word is defined with an extra twist—the use of the present definition in those cases can lead to error and confusion.)

(2) If a finite-dimensional vector space V is reflexive—if, that is, not only are V and V'' isomorphic, but, in fact, the natural mapping is an isomorphism—then it is frequently convenient (though mildly sloppy) to identify the isomorphic spaces V and V''. In more detail: if V is reflexive, then each vector x in V is regarded as the **same** as its image x' ($= Tx$) in V''. As a special case, recall the construction (described in Solution 76) of a basis dual to a prescribed one. Start with a basis

$$X = \{x_1, \ldots, x_n\}$$

in V, let

$$U = \{u_1, \ldots, u_n\}$$

be the dual basis in V' (so that $u_i(x_j) = \delta_{ij}$ for all i and j), and do it again: let

$$X' = \{x'_1, \ldots, x'_n\}$$

be the basis in V'' dual to U (so that

$$x'_j(u_i) = \delta_{ij}$$

for all i and j). Since

$$x'_j(u_i) = u_i(x_j)$$

for all i and j, it follows that

$$x'_j(u) = u(x_j)$$

for all u in V' and for all j, so that x'_j is exactly the image of x_j under the natural mapping T from V to V''. The proposed identification declares that for each j the vectors x'_j and x_j are the same, and hence that the bases X' and X are the same.

79. Annihilators

79

The kernel of a linear transformation (see Problem 56), and, in particular, the kernel of a linear functional is the set (subspace) of vectors at which it takes the value 0. A question can be asked in the other direction: if x is a vector in a vector space V, what is the set of all linear functionals u on V such that $u(x) = 0$? Frequently used reformulation: what is the **annihilator** of x? An obviously related more general question is this: if M is a subset (possibly but not necessarily a subspace) of a vector space V, what is the set of all linear functionals u on V such that $u(x) = 0$ for every x in M? The answer, whatever it is, is denoted by M^0 and is referred to as the annihilator of M.

   Trivial examples: $\mathbb{O}^0 = V'$ and $V^0 = \mathbb{O}$ (in V'). If V is finite-dimensional and M contains a non-zero vector, then (see the discussion preceding Problem 63) $M^0 \neq V'$.

   The annihilator of a singleton ($M = \{x\}$) consists of all those linear functionals u for which x is in $\ker u$. In the abbreviated language of double duality (see Problem 77) $\{x\}^0$ is the kernel (in V') of the linear functional x in V'' (originally denoted by x' in Problem 77). Consequence: the annihilator of $\{x\}$ is always a subspace of V'. This consequence could have been derived perfectly easily without the double duality discussion, but that is where the result naturally belongs. If M is an arbitrary subset of V, then M^0 is the intersection of all the annihilators, such as $\{x\}^0$, of the vectors x in M. Consequence: the annihilator of every set (in V) is a subspace (of V').

   Since M^0 is a subspace, it makes sense to speak of $\dim M^0$; can anything intelligent be said about it?

   Problem 79. *If M is an m-dimensional subspace of an n-dimensional vector space V, what is the dimension of M^0 in V'?*

80. Double annihilators

80

It is a good rule in most of mathematics that if you can do something once, you should do it again, and again and again and again: iterate whenever possible. Example: if it is a good thing to start with a subspace M of a vector space V and form its annihilator M^0 in V', then it is a good thing to do it again. Doing it again means to start with M^0 in V' and form its annihilator $(M^0)^0$ in V'' (except that, just as for $(V')'$, it is saner to denote that double annihilator by M^{00}).

Problem 80. *If M is a subspace of a finite-dimensional vector space V, what is the relation between M and M^{00}?*

Comment. Strictly speaking M and M^{00} are incomparable: they are subspaces of the different vector spaces V and V''. Since, however, every finite-dimensional vector space is reflexive, the identification convention of Problem 77 can and should be applied. According to that convention the space V'' is the same as the space V, and both M and M^{00} are subspaces of that space.

81. Adjoints   81

The concept of duality was defined in terms of very special linear transformations, namely linear functionals; does it have anything to do with more general linear transformations? Yes, it does.

Suppose that A is an arbitrary linear transformation on a vector space V and that u is an arbitrary element of the dual space V'. With those two tools at hand, there are two natural things to do to any particular vector x in V: form the vector Ax and form the scalar $u(x)$. And there is also a third thing: the two tools can form a conspiracy by being used one after the other (in the only order in which that is possible). It makes sense, that is, in addition to applying either A or u to x, to apply both by forming $u(Ax)$. If A and u are regarded as temporarily fixed, the expression $u(Ax)$ depends on x alone—it is a function of x. Since both A and u depend linearly on x, so does their composition. If, in other words, a function v on V is defined by

$$v(x) = u(Ax),$$

then v is a linear functional, an element of V'.

A minor miracle just occurred: a linear transformation on V and a linear functional in V' collaborated to produce a new element of V'. That new element v can be viewed as the result of operating on the old element u by a transformation A', so that

$$v = A'u.$$

The transformation A' is called the **adjoint** of A; it sends vectors in $\mathbb{V}'$ to vectors in $\mathbb{V}'$.

How does A' operate on $\mathbb{V}'$? Unsurprising answer: linearly. That is: if

$$u = \alpha_1 u_1 + \alpha_2 u_2,$$

then

$$A'u = \alpha_1 A'u_1 + \alpha_2 A'u_2;$$

the verification of that equation should by now be regarded as dull routine. That answers the question of how $A'u$ depends on u. A more interesting and less commonplace question is this: how does $A'u$ depend on A? Answer: simply and beautifully, with only one small and harmless surprise.

It is, for instance, child's play to verify that if $A = 0$ (on $\mathbb{V}$), then $A' = 0$ (on $\mathbb{V}'$), and, similarly, that if $A = 1$ (on $\mathbb{V}$), then $A' = 1$ (on $\mathbb{V}'$). Simpler put: $0' = 0$ and $1' = 1$. The proof that if A and B are linear transformations on $\mathbb{V}$, then

$$(A + B)' = A' + B'$$

is just as easy—the only difference is that a few more symbols are involved. Since, moreover, if A is a linear transformation on $\mathbb{V}$ and α is a scalar, then

$$(\alpha A)' = \alpha A',$$

a part of the relation between linear transformations and their adjoints can be described by saying that A' depends linearly on A.

What about products? If A and B are linear transformations (on $\mathbb{V}$), and if $C = AB$, what can be said about C' (on $\mathbb{V}'$) in terms of A' and B'? Here comes the small and harmless surprise:

$$(C'u)(x) = u(Cx) = u\big((AB)x\big) = u\big(A(Bx)\big)$$

$$= (A'u)(Bx) \qquad \text{(by the definition of } A')$$

$$= B'\big((A'u)x\big) \qquad \text{(by the definition of } B')$$

$$= \big((B'A')u\big)(x).$$

Conclusion: $(AB)' = B'A'$—the order of the product became reversed.

The last question of this kind concerns inverses: if A is invertible, what can be said about $(A^{-1})'$? Answer: if A is invertible, (on $\mathbb{V}$), then A' is invertible (on $\mathbb{V}'$) and

$$(A^{-1})' = (A')^{-1}.$$

The proof makes obvious use of the multiplicativity equation just proved and of the basic relation $1' = 1$.

At this point another opportunity presents itself to act on the principle of doing something doable again and again: if a linear transformation A on a vector space V yields a linear transformation A' on V', then A' yields a linear transformation $(A')'$ on V''—what is the relation between A and its double adjoint? First comment: notational sanity suggests that $(A')'$ be denoted by A''. Second comment: in order for A'' and A to be comparable, it would be good to have them live on the same domain, and, to achieve that, it is a good idea to assume that V is finite-dimensional and, therefore, reflexive. Once that's done, the answer to the question becomes simple:

$$A'' = A.$$

Proof: if x is in V and u is in V', then

$$u(Ax) = ((A'u)x) \qquad \text{(by the definition of } A')$$

and

$$(A'u)(x) = u(A''x) \qquad \text{(by the definition of } A'').$$

So much for the general properties of adjoints; it is high time that a deeper understanding of them be acquired by studying their more special properties, which means their relations to other concepts. Example: linear transformations are intimately associated with certain subspaces, namely their kernels and ranges. What does the formation of adjoints do to kernels and ranges?

Problem 81. *If A is a linear transformation on a finite-dimensional vector space V and if A' is its adjoint on V', what is the relation between the ranges and the kernels of A and A'?*

Comment. The question can be asked in any case, but the restriction to finite-dimensionality is a familiar and sane precaution.

82. Adjoints of projections   82

Can the adjoint of a concretely presented transformation be exhibited concretely? What, for instance, happens with projections?

Problem 82. *If M and N are subspaces of a finite-dimensional vector space V, what is the adjoint of the projection of V to M along N?*

83   83. Matrices of adjoints

Linear transformations have adjoints; what do matrices have?

Problem 83.   *What is the relation between the matrix, with respect to some basis, of a linear transformation on a finite-dimensional vector space and the matrix, with respect to the dual basis, of its adjoint on the dual space?*

CHAPTER **6**
SIMILARITY

84. Change of basis: vectors 84

If V is an n-dimensional vector space, with a prescribed ordered basis $\{x_1, \ldots, x_n\}$, then each vector x determines an ordered n-tuple of scalars. This is an elementary fact by now: if the expansion of x in terms of the x_j's is

$$x = \alpha_1 x_1 + \cdots + \alpha_n x_n,$$

then the ordered n-tuple determined by x is just the n-tuple

$$(\alpha_1, \ldots, \alpha_n)$$

of coefficients. The game can, of course, be played in the other direction: once a basis is fixed, each ordered n-tuple of scalars determines a vector, namely the vector whose sequence of coefficients it is.

Now change the rules again, or, rather, play the already changed rules but change the emphasis. Given an n-dimensional vector space V and an ordered n-tuple $(\alpha_1, \ldots, \alpha_n)$ of scalars, note that each basis $\{x_1, \ldots, x_n\}$ of V determines a vector in V, namely the vector described by the first equation above. That vector depends, obviously, on the basis. If $\{y_1, \ldots, y_n\}$ is also a basis of V, it would be a surprising coincidence if the determined vector

$$y = \alpha_1 y_1 + \cdots + \alpha_n y_n$$

turned out to be the same as x; as the basis changes, the vector changes. How?

Problem 84. *If $\{x_1, \ldots, x_n\}$ and $\{y_1, \ldots, y_n\}$ are bases of a vector space, and if*

$$x = \alpha_1 x_1 + \cdots + \alpha_n x_n$$

and

$$y = \alpha_1 y_1 + \cdots + \alpha_n y_n,$$

what is the relation between the vectors x and y?

Reformulation.   What happens to vectors under a "change of basis"?

Emphasis. The same coefficients $(\alpha_1, \ldots, \alpha_n)$ appear in the two displayed equations.

85. Change of basis: coordinates

85

If $\{x_1, \ldots, x_n\}$ and $\{y_1, \ldots, y_n\}$ are bases of a vector space, the problem of changing from one to the other can be thought of in two ways.

(1) Given an ordered n-tuple $(\alpha_1, \ldots, \alpha_n)$ of scalars, what is the relation between the vectors

$$x = \alpha_1 x_1 + \cdots + \alpha_n x_n$$

and

$$y = \alpha_1 y_1 + \cdots + \alpha_n y_n?$$

(2) Given a vector z, what is the relation between its coordinates with respect to the x's and the y's? A preceding problem (84) took the first point of view. How does the answer obtained compare with the one demanded by the second point of view?

Problem 85. *If $\{x_1, \ldots, x_n\}$ and $\{y_1, \ldots, y_n\}$ are bases of a vector space, and if*

$$\xi_1 x_1 + \cdots + \xi_n x_n = \eta_1 y_1 + \cdots + \eta_n y_n,$$

what is the relation between the coordinates ξ and η?

86. Similarity: transformations

86

Vectors are in the easy part of linear algebra; the more challenging and more useful part deals with linear transformations. One and the same "coordinate vector" $(\alpha_1, \ldots, \alpha_n)$ can correspond to two different elements of a vector space via two different bases—that's what Problem 83 is about. A parallel statement on the higher level is that one and the same matrix can correspond to two different linear transformations on a vector space— that's what the present discussion is about.

Problem 86. *If* $\{x_1, \ldots, x_n\}$ *and* $\{y_1, \ldots, y_n\}$ *are bases of a vector space, and if*

$$Bx_j = \sum_i \alpha_{ij} x_i$$

and

$$Cy_j = \sum_i \alpha_{ij} y_i,$$

what is the relation between the linear transformations B *and* C*?*

Reformulation. What happens to linear transformations under a change of basis?

Emphasis. The same matrix (α_{ij}) appears in the two displayed equations.

Comment. This problem is only slightly harder than Problem 83, but much deeper. Two transformations related in the way here described are called **similar**, and similarity is the right, the geometric, way to classify linear transformations. To say that two linear transformations are similar is to say, in effect, that they are "essentially the same". Similarity is the single most important possible relation between linear transformations—it lies at the heart of linear algebra.

87. Similarity: matrices   87

A change of basis can be looked at in two ways: geometrically (what does it do to vectors?—Problem 83) and numerically (what does it do to coordinates?—Problem 84). The same two points of view are available in the study of the effect of a change of basis on the higher level: geometric (linear transformations) and numerical (matrices). The first of these was treated by Problem 85; here is the second.

Problem 87. *If one basis of a vector space is used to express a linear transformation as a matrix, and then another basis is used for the same purpose (for the same linear transformation), the result is two matrices—what is the relation between them?*

88   88. Inherited similarity

Similarity is sometimes passed on from one pair of transformations to another. Example: if B and C are similar, then so are B^2 and C^2. Indeed: if $TBT^{-1} = C$ (see Solution 86), then

$$TB^2T^{-1} = TB(T^{-1}T)BT^{-1} = (TBT^{-1})(TBT^{-1}) = CC = C^2.$$

Similarly, of course, if B and C are similar, then so are B^n and C^n for all positive integers n, and, therefore, so are $p(B)$ and $p(C)$ for all polynomials p. (Minor comment: if B is similar to a scalar γ, then T is equal to γ. Reason: $T\gamma T^{-1} = \gamma TT^{-1}$.)

The kind of reasoning used here can go a bit further. It proves, for instance, that if B and C are similar, then so are B' and C'. (Form the adjoints of both sides of the equation $TBT^{-1} = C$, and use the results of Problem 80.) It proves also that if B and C are similar and if both are invertible, then B^{-1} and C^{-1} are similar. (Form the inverses of both sides of the similarity equation.) What is true about products?

Problem 88.   *If B and C are linear transformations (on the same vector space), is it always true that BC and CB are similar?*

Question.   Does it make any difference whether B and C are invertible?

89   89. Similarity: real and complex

Anyone who speaks of a vector space must have selected a coefficient field to begin with; anyone who speaks of a matrix has in mind its entries, which belong to some prescribed field. What happens to vectors, and linear transformations, and matrices when the field changes? If, in particular, two different fields, $\mathbb{E}$ and $\mathbb{F}$ say, appear to be pertinent to some study, with $\mathbb{E} \subset \mathbb{F}$, then every vector space over $\mathbb{F}$ is automatically a vector space over $\mathbb{E}$ also (or, more precisely, naturally induces a vector space over $\mathbb{E}$ by just restricting scalar multiplication to $\mathbb{E}$); the general question is how much information an $\mathbb{F}$ fact gives about an $\mathbb{E}$ space. The following special question is a well known, important, and typical instance.

Problem 89.   *If A and B are two real matrices that are similar over $\mathbb{C}$, do they have to be similar over $\mathbb{R}$?*

90. Rank and nullity

Which linear transformation (matrix) is "bigger",

$$B = \begin{pmatrix} 1 & 1 & 1 \\ 1 & 1 & 1 \\ 1 & 1 & 1 \end{pmatrix} \quad \text{or} \quad C = \begin{pmatrix} 0 & 0 & 1 \\ 0 & 1 & 0 \\ 1 & 0 & 0 \end{pmatrix}?$$

The question doesn't make sense—except for the size of a matrix (in the sense in which these 3×3 examples are of size 3) no way of measuring linear transformations has been encountered yet.

The transformation C is invertible, but the transformation B is not: the range of B is a 1-dimensional subspace. In other words, the transformation B collapses the entire space into a proper subspace—that might be one good reason for calling B smaller than C. Another reason might be that B "shrinks" some vectors (sends them to 0) and C does not. The dimension of the range of a linear transformation—called its **rank**—is a measure of size; in the present example

$$\text{rank } B = 1 \quad \text{and} \quad \text{rank } C = 3.$$

Roughly: transformations are large if their rank is large. The dimension of the kernel—called the **nullity**, and abbreviated as "null"—is another kind of measure of size; in the present example

$$\text{null } B = 2 \quad \text{and} \quad \text{null } C = 0.$$

Roughly: transformations are large if their nullities are small.

Is there a relation between these two measures of size? And what about the sizes of a transformation and its adjoint?

Problem 90. *If A is a linear transformation of rank r on a vector space of dimension n, what are the possible values of the rank of A'? What are the possible values of the nullity of A?*

91. Similarity and rank

Problem 91. *If two linear transformations on a finite-dimensional vector space are similar, must they have the same rank?*

92. Similarity of transposes

If two linear transformations on a finite-dimensional vector space have the same rank, must they be similar? That question is the converse of the one

in Problem 91, and it would be silly to expect it to have an affirmative answer. (Are two invertible matrices always similar?) In some special cases,
however, the affirmative answer might be true anyway. So, for instance, every transformation has the same rank as its adjoint—could that statement
be strengthened sometimes to one about similarity?

Problem 92.   *Is every 2×2 matrix $\begin{pmatrix} \alpha & \beta \\ \gamma & \delta \end{pmatrix}$ similar to its transpose*
$\begin{pmatrix} \alpha & \gamma \\ \beta & \delta \end{pmatrix}$?

93   93. Ranks of sums

Problem 93.   *If A and B are linear transformations on a finite-
dimensional vector space, what is the relation of $\mathrm{rank}(A + B)$ to
the separate ranks,* $\mathrm{rank}\, A$ *and* $\mathrm{rank}\, B$?

94   94. Ranks of products

Problem 94.   *If A and B are linear transformations on a finite-
dimensional vector space, what is the relation of* $\mathrm{rank}\, AB$ *to the separate ranks,* $\mathrm{rank}\, A$ *and* $\mathrm{rank}\, B$?

95   95. Nullities of sums and products

Since rank and nullity always add up to the dimension of the space (Problem 90), every relation between ranks is also a relation between nullities.
That is true, in particular, about the sum and product formulas (Problems
93 and 94), but the nullity relations obtained that way are far from thrilling.
There is, however, a nullity relation that comes nearer to a thrill, and that
is not an immediate consequence of the rank relations already available.

Problem 95.   *If A and B are linear transformations on a finite-
dimensional vector space, is there a simple relation involving*
$\mathrm{null}\, (AB)$, $\mathrm{null}\, A$, *and* $\mathrm{null}\, B$?

96. Some similarities

The best way to get a feeling for what similarity means is to look at special cases—sometimes the answer is not what you would expect.

Problem 96.

(a) Is $B = \begin{pmatrix} 1 & 1 & 1 \\ 0 & 2 & 1 \\ 0 & 0 & 3 \end{pmatrix}$ similar to $C = \begin{pmatrix} 1 & 0 & 0 \\ 0 & 2 & 0 \\ 0 & 0 & 3 \end{pmatrix}$?

(b) Is $B = \begin{pmatrix} 0 & 1 & 1 \\ 0 & 0 & 1 \\ 0 & 0 & 0 \end{pmatrix}$ similar to $C = \begin{pmatrix} 0 & 1 & 0 \\ 0 & 0 & 1 \\ 0 & 0 & 0 \end{pmatrix}$?

(c) Is $B = \begin{pmatrix} 2 & 1 & 1 \\ 0 & 3 & 1 \\ 0 & 0 & 3 \end{pmatrix}$ similar to $C = \begin{pmatrix} 2 & 0 & 0 \\ 0 & 3 & 0 \\ 0 & 0 & 3 \end{pmatrix}$?

(d) Is $B = \begin{pmatrix} 0 & 1 & 0 \\ 0 & 0 & 1 \\ 0 & 0 & 0 \end{pmatrix}$ similar to $C = \begin{pmatrix} 0 & 2 & 0 \\ 0 & 0 & 2 \\ 0 & 0 & 0 \end{pmatrix}$?

(e) Is $B = \begin{pmatrix} 1 & 1 & 1 \\ 0 & 1 & 1 \\ 0 & 0 & 1 \end{pmatrix}$ similar to $C = \begin{pmatrix} 1 & 0 & 0 \\ 1 & 1 & 0 \\ 1 & 1 & 1 \end{pmatrix}$?

Comment. Similar questions make sense, and should be asked, for matrices of size larger than 3×3.

97. Equivalence

The construction of a matrix associated with a linear transformation depends on two bases, not one. Indeed, if

$$X = \{x_1, \ldots, x_n\} \quad \text{and} \quad \widehat{X} = \{\widehat{x}_1, \ldots, \widehat{x}_n\}$$

are bases of V, and if A is a linear transformation on V, then the matrix of A with respect to X and $\widehat{X}$, denote it temporarily by

$$A(X, \widehat{X}),$$

should be defined by

$$Ax_j = \sum_i \alpha_{ij} \widehat{x}_i.$$

The definition originally given (see Problems 66 and 68) corresponds to the special case in which $\mathbb{X} = \widehat{\mathbb{X}}$. That special case leads to the concept of similarity—B and C are similar if there exist bases $\mathbb{X}$ and $\mathbb{Y}$ such that the matrix of B with respect to $\mathbb{X}$ is equal to the matrix of C with respect to $\mathbb{Y}$, or, in the notation introduced above, if

$$B(\mathbb{X}, \mathbb{X}) = C(\mathbb{Y}, \mathbb{Y}).$$

The analogous relation suggested by the general case is called equivalence: B and C are called **equivalent** if there exist basis pairs $(\mathbb{X}, \widehat{\mathbb{X}})$ and $(\mathbb{Y}, \widehat{\mathbb{Y}})$ such that

$$B(\mathbb{X}, \widehat{\mathbb{X}}) = C(\mathbb{Y}, \widehat{\mathbb{Y}}).$$

The principal question about equivalence, written out in complete detail, is in spirit the same as the original question (Problem 72) about similarity.

Problem 97. *If $\{x_1, \ldots, x_n\}$, $\{\widehat{x}_1, \ldots, \widehat{x}_n\}$, $\{y_1, \ldots, y_n\}$, and $\{\widehat{y}_1, \ldots, \widehat{y}_n\}$, are bases of a vector space, and if*

$$Bx_j = \sum_i \alpha_{ij} \widehat{x}_i$$

and

$$Cy_j = \sum_i \alpha_{ij} \widehat{y}_i,$$

what is the relation between the linear transformations B and C?

Reformulation. What happens to linear transformations under two simultaneous changes of bases?

Emphasis. The same matrix (α_{ij}) appears in the two displayed equations.

Comment. The question is somewhat vague, just as it was in Problem 85. The relation is that there exist bases with the stated property—and why isn't that an answer? The unformulated reason, at the time Problem 85 was stated, was the hope that the "geometric" definition could be replaced by an "algebraic" necessary and sufficient condition. That hope persists here too.

98. Rank and equivalence

If E is a projection with range $\mathbb{M}$ and kernel $\mathbb{N}$, then there exists a basis $\{x_1, \ldots, x_r, x_{r+1}, \ldots, x_n\}$ of the space such that $\{x_1, \ldots, x_r\}$ is a basis for

M and $\{x_{r+1}, \ldots, x_n\}$ is a basis for N. The matrix of E with respect to that basis is of the form $\begin{pmatrix} 1 & 0 \\ 0 & 0 \end{pmatrix}$, where the top left "1" represents an identity matrix of size $r \times r$ and the bottom right "0" represents a zero matrix of size $(n-r) \times (n-r)$. Consequence: not only do similar projections have the same rank (Problem 91), but the converse is true: projections of the same rank are similar. Since similarity is a much stronger condition than equivalence, it follows in particular that projections of the same rank are equivalent. Is that statement generalizable?

Problem 98.   *If two linear transformations on a finite-dimensional vector space have the same rank, must they be equivalent?*

CHAPTER 7
CANONICAL FORMS

99. Eigenvalues

A large vector space (one of large dimension, that is) is a complicated object, and linear transformations on it are even more complicated. In the study of a linear transformation on a large space it often helps to concentrate attention on the way the transformation acts on small subspaces. The phrase "a linear transformation acting on a subspace" is usually interpreted to mean that the subspace is invariant under the transformation (in the language of Problem 70), and a "small" subspace is one of dimension 1 (surely the smallest that a non-trivial subspace can be). In view of these comments, a promising approach to the study of linear transformations would seem to be to search for invariant subspaces of dimension 1.

If A is a linear transformation and if x is a vector in ker A then, of course, $Ax = 0$, and it follows that $A(\lambda x) = 0$ for every scalar λ. Consequence: the 1-dimensional subspace consisting of all scalar multiples of x is invariant under A. This is an example—an extreme sort of example—of the possibility described above. What A does to this particular x is simply to multiply it by 0:

$$Ax = 0x.$$

A less extreme example might be a linear transformation A and a vector x such that, say,

$$Ax = 7x.$$

Can that happen? Sure—it happens, for instance, when $A = 7 \cdot I$ (the product of the scalar 7 and the identity transformation I). It happens also in less spectacular (but more typical and more useful) cases such as

$$A_1 = \begin{pmatrix} 7 & 0 \\ 0 & 5 \end{pmatrix}, \qquad x_1 = (1, 0).$$

A small modification yields the less special looking example

$$A_2 = \begin{pmatrix} 7 & 8 \\ 0 & 5 \end{pmatrix}, \qquad x_2 = (1, 0).$$

A different looking and perhaps surprising example is

$$A_3 = \begin{pmatrix} 10 & 3 \\ -5 & 2 \end{pmatrix}, \qquad x_3 = (1, -1).$$

(Verifications?)

All right, so $A_3 x_3 = 7x_3$; is that an accident or is it a bad habit that A_3 has? What other vectors x have the property that $A_3 x = 7x$? Unsatisfactory answer: all scalar multiples of x_3 have that property. (Right? $A_3(4x_3) = 4(A_3 x_3) = 4 \cdot 7x_3 = 7 \cdot (4x_3)$.) Are there any others? The question amounts to asking for solutions (α_1, α_2) of the equations

$$10\alpha_1 + 3\alpha_2 = 7\alpha_1$$

$$-5\alpha_1 + 2\alpha_2 = 7\alpha_2.$$

That's a routine question and the easily calculated answer is that all solutions are of the form $(\tau, -\tau)$, and those are exactly the "unsatisfactory" ones already dismissed.

Is there something special about 7 and x_3? Are there other scalars λ and other vectors x such that

$$Ax = \lambda x?$$

This time the question is about the solutions of

$$10\alpha_1 + 3\alpha_2 = \lambda\alpha_1$$

$$-5\alpha_1 + 2\alpha_2 = \lambda\alpha_2,$$

and that requires a little more thought.

There is one dull solution, namely $\alpha_1 = \alpha_2 = 0$—that works for every A and yields no information. If that's dismissed, if, in other words, only non-zero vectors are to be accepted as solutions, then the question becomes this: for which scalar values of λ does the matrix

$$\begin{pmatrix} 10 - \lambda & 3 \\ -5 & 2 - \lambda \end{pmatrix}$$

have a non-trivial kernel? Equivalently: for which scalar values of λ is it true that

$$\det \begin{pmatrix} 10 - \lambda & 3 \\ -5 & 2 - \lambda \end{pmatrix} = 0?$$

The determinant is easy enough to calculate, and when that's done, the question becomes this: what are the roots of the equation

$$\lambda^2 - 12\lambda + 35 = 0?$$

That can be answered by anyone who knows how to solve quadratic equations. The answer is that there are only two values of λ, namely 7 and 5.

Curiouser and curiouser. The value 7 is an old friend, with the corresponding vector $x_3 = (1, -1)$. What vectors work for 5? That is: what are the (non-zero) solutions of the equations

$$10\alpha_1 + 3\alpha_2 = 5\alpha_1$$

$$-5\alpha_1 + 2\alpha_2 = 5\alpha_2?$$

Easily calculated answer: all vectors of the form $(3\tau, -5\tau)$.

The matrix here studied, and its relation to certain special scalars and vectors, exemplifies quite well the theory on which it all rests. General definition: an **eigenvalue** of a linear transformation A is a scalar λ such that

$$Ax = \lambda x,$$

for some non-zero vector x. (With $x = 0$ the equation is totally useless; it is satisfied no matter what A and λ are.) Every non-zero vector x that can be used here is called an **eigenvector** of A corresponding to the eigenvalue λ.

A scalar λ is an eigenvalue of a linear transformation A on a finite-dimensional vector space if and only if $A - \lambda$ has a non-trivial kernel, and that happens if and only if λ satisfies the **characteristic equation**

$$\det(A - \lambda I) = 0.$$

The expression $\det(A - \lambda I)$ is a polynomial of degree n (the dimension of the space) in λ, called the **characteristic polynomial** of A. What's important about the characteristic equation is what its roots are. In much of linear algebra and its applications the main problem is to find characteristic equations and their roots, that is eigenvalues. Here is a small sample.

Problem 99. *What can the characteristic equation of a projection be?*

100   100. Sums and products of eigenvalues

The eigenvalues of "good" matrices can be quite bad. Thus, for instance, the eigenvalues of a matrix of integers are not necessarily integers—they are not even necessarily rational (Example: $\begin{pmatrix} 0 & 2 \\ 1 & 0 \end{pmatrix}$.) Just how bad can the eigenvalues of a good matrix be?

> **Problem 100.** *Can both the sum and the product of the eigenvalues of a matrix of rational numbers be irrational?*

Comment. The question is about "can", not about "must". For the example $\begin{pmatrix} 0 & 2 \\ 1 & 0 \end{pmatrix}$ the sum and the product are 0 and 2.

101   101. Eigenvalues of products

If A and B are linear transformations on the same finite-dimensional vector space, and if AB is invertible, then each of A and B is invertible (det $A \cdot$ det $B \neq 0$), and therefore BA is invertible. Contrapositively (with the roles of A and B interchanged): if AB is not invertible, then BA is not invertible. Another way of stating the result is that if 0 is an eigenvalue of AB, then 0 is an eigenvalue of BA also. For other eigenvalues the situation is not so clear: when $\lambda \neq 0$, there doesn't seem to be any way to pass from information about $\det(AB - \lambda I)$ to information about $\det(BA - \lambda I)$.

> **Problem 101.** *If A and B are linear transformations on the same finite-dimensional vector space, and if λ is a non-zero eigenvalue of AB, must λ be an eigenvalue of BA also?*

102   102. Polynomials in eigenvalues

> **Problem 102.** *If A is a linear transformation on a finite-dimensional vector space and if p is a polynomial, what information do the eigenvalues of A give about $p(A)$?*

103   103. Diagonalizing permutations

For matrices that are simple enough, the theory of eigenvalues works like a charm. For the trivial 1×1 matrix (2), there is of course nothing to do.

The 2×2 diagonal matrix

$$\begin{pmatrix} 2 & 0 \\ 0 & 3 \end{pmatrix}$$

has two eigenvalues, and its study reduces to that of two trivial matrices of size 1×1. The 3×3 matrix

$$\begin{pmatrix} 2 & 0 & 0 \\ 0 & 3 & 0 \\ 0 & 0 & 4 \end{pmatrix}$$

is larger, but it is still a beautiful one. It has three distinct eigenvalues and corresponding to them three disjoint eigenspaces—the notion of applying eigenvalues to reduce a large study to small pieces (Problem 96) still works just fine.

The matrix

$$\begin{pmatrix} 3 & 0 & 0 \\ 0 & 3 & 0 \\ 0 & 0 & 4 \end{pmatrix}$$

is from the present point of view perhaps a shade less beautiful—it has only two eigenvalues but its eigenspaces still have a total dimension 3, and its study presents no difficulties. Matrices such as

$$\begin{pmatrix} 3 & 1 & 0 \\ 0 & 3 & 0 \\ 0 & 0 & 4 \end{pmatrix},$$

or, for that matter,

$$\begin{pmatrix} 3 & 1 \\ 0 & 3 \end{pmatrix}$$

misbehave a little more—the total dimensions of their eigenspaces are not as large as one could wish. It begins to look as if eigenvalue theory might not stretch to give complete information about matrices.

Things get really tough with a matrix such as

$$A = \begin{pmatrix} 0 & 1 \\ -1 & 0 \end{pmatrix}.$$

Its characteristic equation is

$$\lambda^2 + 1 = 0,$$

and since there is no real number λ that satisfies that equation, it looks as if eigenvalue theory might give no information about A.

The disease just noticed is not fatal; a hint to its cure is contained in the diagnosis. Sure, there is no *real* number that satisfies the characteristic equation, but that phenomenon is exactly what complex numbers are designed to deal with—and, indeed, they solve the problem.

The properties of vector spaces and of linear transformations on them are strongly influenced by the underlying coefficient field. That fact was hardly noticeable till now—much of the theory works equally well for every field. The exposition till now has either tacitly or explicitly assumed that the coefficient field was the field $\mathbb{R}$ of real numbers—the field that is probably the most familiar to most students. In the applications however (of linear algebra, and of mathematics in general) the field $\mathbb{C}$ of complex numbers is often more useful. For that reason, from here on, it will be assumed that the vector spaces to be considered are complex ones, and that, correspondingly, the vectors and linear transformations to be studied admit complex linear combinations. The typical coordinatized example will therefore be $\mathbb{C}^n$ (not $\mathbb{R}^n$).

The problem that follows is a small step toward getting used to the appearance of complex numbers.

Problem 103. *What are the eigenvalues and eigenvectors of the linear transformation A defined on $\mathbb{C}^3$ by*

$$A(x_1, x_2, x_3) = (x_2, x_3, x_1)?$$

104   104. Polynomials in eigenvalues, converse

Does the converse of Solution 102 have a chance of being true? According to Solution 102, if λ is an eigenvalue of A and p is a polynomial, then $p(\lambda)$ is an eigenvalue of $p(A)$. The converse might be something like this: if $p(\lambda)$ is an eigenvalue of $p(A)$, must it be true that λ is an eigenvalue of A? No, that's absurd. Counterexample: if $A = I$ (the identity transformation), $\lambda = 1$ (the number), and $p(\lambda) = \lambda^2$, then $p(-1) (= 1)$ is an eigenvalue of A, but -1 is not an eigenvalue of $p(A) (= 1)$.

The negative solution of one possible version of the converse problem doesn't settle the issue—slightly weaker problems can be posed and can have hopes of affirmative solutions. Here is one: is every eigenvalue of $p(A)$ of the form $p(\lambda)$ for some eigenvalue λ of A? The question is one in which the coefficient field matters: it is conceivable that the answers for the real field and the complex field are different.

The answer for the real field is no. If

$$A = \begin{pmatrix} 0 & 1 \\ -1 & 0 \end{pmatrix},$$

then -1 is eigenvalue of A^2 (check?), but -1 is not of the form λ^2 for some eigenvalue λ of A, simply because A has no eigenvalues—no real ones, that is. If the real field is replaced by the complex field in this example, the answer changes from no to yes. Is that a lucky property of this example, or is it always true?

Problem 104. *If A is a linear transformation on a finite-dimensional (complex) vector space and if p is a polynomial, is every eigenvalue of $p(A)$ of the form $p(\lambda)$ for some eigenvalue λ of A?*

105. Multiplicities

If

$$A = \begin{pmatrix} 3 & 0 & 0 \\ 0 & 3 & 0 \\ 0 & 0 & 4 \end{pmatrix},$$

is 3 an eigenvalue of A? Sure, that's obvious: if $x_1 = (1, 0, 0)$, then

$$Ax_1 = 3x_1.$$

It is also true that every non-zero multiple of x_1 is an eigenvector of A ("non-zero" because that's how eigenvectors are defined), but that true statement is universally true and gives no new information. New information is, however, available: it is also true that if $x_2 = (0, 1, 0)$, then

$$Ax_2 = 3x_2.$$

Both of the radically different vectors x_1 and x_2 are eigenvectors of A. What goes on is that the set of all those vectors x that satisfy the equation

$$Ax = 3x$$

(including the vector 0) is a subspace of dimension 2; it is sometimes called the **eigenspace** of A corresponding to the eigenvalue 3.

(The vector 0 is never regarded as an eigenvector, but the vector 0 is always regarded as belonging to the eigenspace corresponding to an eigenvalue λ. This apparently contradictory use of language might take a few seconds to get used to, but it causes no trouble, and, in fact, it is more convenient than being forced to deal with the awkward "punctured" subspace obtained by considering only the non-zero solutions of $Ax = \lambda x$.)

In some plausible sense the number 3 occurs twice as an eigenvalue of the matrix A above. If the multiplicity—in more detail the **geometric multiplicity**—of an eigenvalue λ of a transformation A is defined as the dimension of the set of solutions of $Ax = \lambda x$, then in the example under consideration the number 3 is an eigenvalue of geometric multiplicity 2.

If

$$B = \begin{pmatrix} 3 & 1 & 0 \\ 0 & 3 & 0 \\ 0 & 0 & 4 \end{pmatrix},$$

how do the facts for B compare with the facts for A? The number 3 is an eigenvalue of B,

$$Bx_1 = 3x_1,$$

just as it was for A. The vector x_2, however, is not an eigenvalue of B.

What is the geometric multiplicity of the eigenvalue 3 for B? The question is one about solutions (u, v, w) of the equations

$$\begin{aligned} 3u + v &= 3u \\ 3v &= 3v \\ 4w &= 3w. \end{aligned}$$

If (u, v, w) is a solution, then the last equation implies that $w = 0$ and the first equation implies that $v = 0$. Consequence: the eigenspace corresponding to the eigenvalue 3 for B is the set of all vectors of the form $(u, 0, 0)$—a space of dimension 1.

Isn't that just a little puzzling? The matrices A and B don't look very different: both are upper triangular, they have the same diagonal, and, consequently, they have the same characteristic polynomial, namely

$$(\lambda - 3)^2(\lambda - 4) \quad (= \lambda^3 - 10\lambda^2 + 33\lambda - 36).$$

The geometric multiplicity of 3 as an eigenvalue of A seems to be caused by the exponent 2 on $(\lambda - 3)$— but that exponent is there for B also.

Well, that's life: the concept of multiplicity has in fact two distinct meanings. In the already defined geometric meaning the number 3 has the multiplicity 2 as an eigenvalue of A and the multiplicity 1 as an eigenvalue of B, but it has the same multiplicities for A and B in the other sense. The **algebraic multiplicity** of a number λ_0 as an eigenvalue of a linear transformation is the number of times λ_0 occurs as a root of the characteristic equation—or, better said, it is the exponent of $(\lambda - \lambda_0)$ in the characteristic polynomial.

It might be helpful to look at a natural concrete example.

Problem 105.   *What are the geometric and algebraic multiplicities of the eigenvalues of the differentiation transformation D on the space $\mathbb{P}_3$ of polynomials of degree less than or equal to 3?*

Reminder.   The differentiation transformation was first mentioned as an example in Problem 54.

106. Distinct eigenvalues   106

What made the diagonalization possible in Problem 103? The easiest transformations to diagonalize are the scalars—the matrix of a scalar transformation is diagonal with respect to every basis. If multiplicities are counted, as they always should be, a scalar transformation on a vector space of dimension n has n eigenvalues—that is, one eigenvalue with multiplicity n. The opposite extreme to scalars are the transformations with n distinct eigenvalues—how difficult are they to diagonalize?

Problem 106.   *If a linear transformation on a vector space of dimension n has n distinct eigenvalues, must it be diagonalizable?*

107. Comparison of multiplicities   107

For the examples

$$
A = \begin{pmatrix} 3 & 0 & 0 \\ 0 & 3 & 0 \\ 0 & 0 & 4 \end{pmatrix} \quad \text{and} \quad B = \begin{pmatrix} 3 & 1 & 0 \\ 0 & 3 & 0 \\ 0 & 0 & 4 \end{pmatrix}
$$

of Problem 103, it turned out that 3 was an eigenvalue of algebraic multiplicity 2 for both A and B, and it had geometric multiplicity 2 for A and 1 for B. How difficult is it to find an example where the algebraic multiplicity is the smaller one?

Problem 107.   *Does there exist a linear transformation on a finite-dimensional vector space with an eigenvalue λ whose algebraic multiplicity is less than its geometric multiplicity?*

108. Triangularization   108

Can every matrix be diagonalized? (For a discussion of diagonalization see Problems 103 and 106.) The answer is no, and the almost universal counterexample $\begin{pmatrix} 0 & 1 \\ 0 & 0 \end{pmatrix}$ proves it. Indeed the eigenspaces of a 2×2 diagonal matrix have total dimension 2, but the "universal counterexample" has only one eigenvalue (namely 0) and only one eigenvector (namely $(0, 1)$ and its scalar multiples). The example can be generalized: the matrix

$$\begin{pmatrix} 0 & 1 & 0 & 0 \\ 0 & 0 & 1 & 0 \\ 0 & 0 & 0 & 1 \\ 0 & 0 & 0 & 0 \end{pmatrix}$$

has only one eigenvalue (namely 0) and only one eigenvector (namely $(1, 0, 0, 0)$ and its scalar multiples), whereas the eigenspaces of a diagonalizable 4×4 matrix have total dimension 4. Similar statements apply of course to the obvious generalizations of this 4×4 matrix to $n \times n$ for every positive integer n.

Granted that not every matrix can be diagonalized, what's the next best thing that can be done? The matrix

$$\begin{pmatrix} 3 & 1 & 0 \\ 0 & 3 & 0 \\ 0 & 0 & 4 \end{pmatrix}$$

is not so easy to work with as

$$\begin{pmatrix} 3 & 0 & 0 \\ 0 & 3 & 0 \\ 0 & 0 & 4 \end{pmatrix},$$

but it's not too bad: its eigenvalues (and their multiplicities) can be read off at a glance, and even its powers are easy to compute. (For instance

$$\begin{pmatrix} 3 & 1 \\ 0 & 3 \end{pmatrix}^n = \begin{pmatrix} 3^n & n3^{n-1} \\ 0 & 3^n \end{pmatrix}.$$

Check?) It is tempting to guess that the next best thing to diagonalize is to **triangularize**. Can *that* always be done?

The characteristic property of a matrix in triangular form, for example a 4×4 matrix such as

$$A = \begin{pmatrix} * & * & * & * \\ 0 & * & * & * \\ 0 & 0 & * & * \\ 0 & 0 & 0 & * \end{pmatrix},$$

is that there exists a basis consisting of vectors u, v, w, etc., (in the example they are $(1, 0, 0, 0)$, $(0, 1, 0, 0)$, $(0, 1, 0, 0)$ etc.) such that
(1) Au is a scalar multiple of u (or, in plain English, such that u is an eigenvalue of A),
(2) Av is a linear combination of u and v,

(3) *Aw* is a linear combination of *u*, *v*, and *w*—etc.

In view of this comment a natural approach to trying to triangularize a matrix *A* is (1) to find an eigenvector *u*, (2) to find a vector *v* such that *Av* is a linear combination of *u* and *v*, etc., etc. The answer to the question is yes: every matrix can be triangularized. The proposed proof is by induction on the size *n* of the matrix. The beginning, $n = 1$, is easy enough: if $n = 1$, there is nothing to do.

For an arbtirary *n*, the first step in any event is always possible: every linear transformation on a complex vector space $\mathbb{V}$ has an eigenvector.

The induction step has to be preceded by the observation that if $\mathbb{M}$ is the 1-dimensional space of all multiples of an eigenvector—an eigenspace —then the quotient space $\mathbb{V}/\mathbb{M}$ has dimension $n - 1$ (Problem 52). Recall now that according to one definition of quotient space the elements of $\mathbb{V}/\mathbb{M}$ are the vectors of $\mathbb{V}$ but with equality defined as congruence modulo $\mathbb{M}$ (Problem 51). Use that definition to define a linear transformation $A_{\mathbb{M}}$ on $\mathbb{V}/\mathbb{M}$ (called the quotient transformation induced by *A*) by writing

$$A_{\mathbb{M}}x = Ax$$

for every *x* in $\mathbb{V}$. This definition needs defense: it must be checked that it is unambiguous. The trouble is (could be) that two "equal" vectors (that is vectors that are congruent modulo $\mathbb{M}$) might have unequal images. The defense, in other words, must prove that if $x \equiv y \bmod \mathbb{M}$, then $Ax \equiv Ay \bmod \mathbb{M}$, or, equivalently, that if $x - y$ is in $\mathbb{M}$, then $A(x - y)$ is in $\mathbb{M}$. In that form the implication is obvious—it asserts no more and no less than that $\mathbb{M}$ is invariant under *A*.

The ground is now prepared for the induction step: since

$$\dim \mathbb{V}/\mathbb{M} = n - 1,$$

the transformation $A_{\mathbb{M}}$ on $\mathbb{V}/\mathbb{M}$ can be triangularized. That means, as a first step, that there exist vectors $v, w, \ldots$ in $\mathbb{V}$ (considered here as a photograph of $\mathbb{V}/\mathbb{M}$) such that $A_{\mathbb{M}}v$ is a scalar multiple of *v*, $A_{\mathbb{M}}w$ is a linear combination of *v* and *w*, etc. In different language: *Av* is equal to a scalar multiple of *v* plus an element of $\mathbb{M}$, *Aw* is equal to a linear combination of *v* and *w* plus an element of $\mathbb{M}$, etc. Conclusion: not only is *Au* a scalar multiple of *u*, but also *Av* is a linear combination of *u* and *v*, and *Aw* is a linear combination of *u*, *v*, and *w*, etc.—and that says exactly that *A* has been triangularized. Conclusion: every transformation can be triangularized.

To fix in one's mind this outline of an argument, it might be a good idea to follow it in a couple of concrete numerical cases, and that's what the following problem suggests.

Problem 108. *Triangularize both the matrices*

$$A = \begin{pmatrix} 1 & 1 & 0 \\ -1 & 2 & 1 \\ 3 & -6 & 6 \end{pmatrix} \quad and \quad B = \begin{pmatrix} 1 & 1 & 0 \\ -4 & 5 & 0 \\ -6 & 3 & 3 \end{pmatrix}.$$

Are they similar?

Comment. To "triangularize" a matrix M it is not enough to exhibit a triangular matrix M_0 and to prove that M and M_0 are similar. What is wanted is either the explicit determination of a new basis with respect to which the new matrix of the same linear transformation is triangular, or, equivalently, the explicit determination of an invertible matrix T, the transformer, such that TMT^{-1} is triangular. In the course of looking for a suitable basis the eigenvalues of M should become visible—which is frequently preceded by the determination of the characteristic polynomial and followed by the determination of the eigenvectors.

Reminder. The theories of upper and lower triangularization are boringly alike; in the discussion above, for no especially good reason, upper was emphasized.

109. Complexification

Does every linear transformation on $\mathbb{R}^n$ have an invariant subspace of dimension equal to 1? To ask that is the same as asking whether every linear transformation on $\mathbb{R}^n$ has an eigenvector, and the answer to that is obviously no (see Problem 103). What happens if the question is liberalized a little?

Problem 109. *If $n > 1$, does every linear transformation on $\mathbb{R}^n$ have an invariant subspace of dimension equal to 2?*

110. Unipotent transformations

Problem 110. *If a linear transformation A on a finite-dimensional (complex) vector space is such that $A^k = 1$ for some positive integer k, must A be diagonalizable?*

Comment. Transformations some positive power of which is equal to the identity are sometimes called **unipotent**.

111. Nilpotence

Transformations with many distinct eigenvalues are diagonalizable (Problem 106); does that imply that if a transformation has only a small number of eigenvalues, then it is difficult to diagonalize? A clue to the answer can be found in the triangularization discussions of transformations with just one eigenvalue (Problem 105).

In the study of linear transformations with just one eigenvalue, the actual numerical value of that eigenvalue can't matter much: if A has the unique eigenvalue α, then $A - \alpha I$ has the unique eigenvalue 0, and $(A - \alpha I) + \beta I$ has the unique eigenvalue β. Since the addition of a scalar cannot produce any major changes, the question might as well be restricted to the easiest eigenvalue to work with, namely 0.

How easy is it to find examples of linear transformations whose only eigenvalue is 0? One example is the ubiquitous

$$\begin{pmatrix} 0 & 1 \\ 0 & 0 \end{pmatrix}.$$

Another is

$$A = \begin{pmatrix} 1 & 1 \\ -1 & -1 \end{pmatrix}.$$

Is the latter obvious? The statement is surely not a deep one—a few seconds' calculation shows that the characteristic polynomial of A is λ^2—but a special point of view on it is usefully generalizable. To wit: since an equation such as

$$Ax = \lambda x$$

implies that

$$A^2 x = \lambda^2 x,$$

it follows from $A^2 = 0$ (a few microseconds' calculation) that $\lambda = 0$ (except in the degenerate case $x = 0$). Generalization: if a linear transformation A is such that $A^q = 0$ for some positive integer q, then its only eigenvalue is 0. The proof is the same as the one just seen: if

$$Ax = \lambda x,$$

then

$$A^q x = \lambda^q x,$$

and therefore $\lambda = 0$ (or $x = 0$).

A linear transformation A such that $A^q = 0$ for some positive integer q is called **nilpotent**; the smallest q that works is called its **index** of nilpotence. The observation of the preceding paragraph was that if A is nilpotent, then spec A consists of 0 alone. The converse is also true, but it is slightly deeper. Indeed: if spec $A = \{0\}$, then a triangularization of A (see Problem 105) has zeroes *on* as well as below the main diagonal, and a triangular matrix like that is nilpotent. The reason is that if

$$A = \begin{pmatrix} 0 & * & * & * & * \\ 0 & 0 & * & * & * \\ 0 & 0 & 0 & * & * \\ 0 & 0 & 0 & 0 & * \\ 0 & 0 & 0 & 0 & 0 \end{pmatrix},$$

is squared, then the result has the form

$$A^2 = \begin{pmatrix} 0 & 0 & * & * & * \\ 0 & 0 & 0 & * & * \\ 0 & 0 & 0 & 0 & * \\ 0 & 0 & 0 & 0 & 0 \\ 0 & 0 & 0 & 0 & 0 \end{pmatrix}.$$

Emphasis: the diagonal just above the main one consists of zeroes only. Multiply by A again: for A^3 the two diagonals just above the main one consist of zeroes only. Continue this way, and infer that $A^n = 0$. Conclusion: A is nilpotent (but a calculation of this sort does not reveal its exact index of nilpotence).

Problem 111. *If a linear transformation A on a finite-dimensional vector space is nilpotent of index q, and if, for each vector x in the space, a subspace $\mathbb{M}(x)$ is defined as the span of the vectors*

$$x, Ax, A^2x, \dots, A^{q-1}x,$$

how large can the dimension of $\mathbb{M}(x)$ be?

112. Nilpotent products

An obvious exercise about nilpotence is to ask whether the product of two nilpotent transformations is necessarily nilpotent. The answer is yes if they commute, but the answer is no in general; a standard easy example is given by the two matrices

$$\begin{pmatrix} 0 & 1 \\ 0 & 0 \end{pmatrix} \quad \text{and} \quad \begin{pmatrix} 0 & 0 \\ 1 & 0 \end{pmatrix}$$

That doesn't settle everything though; there are tricky nilpotence questions that arise in some contexts and to which the answer is not predictable just from the existence of examples such as these.

Problem 112. *If A and B are linear transformations on the same vector space such that $ABAB = 0$, does it follow that $BABA = 0$?*

113. Nilpotent direct sums   113

If A, B, and C are nilpotent matrices of sizes 6, 4, and 4 respectively and indices of nilpotence also 6, 4, and 4 respectively, is the matrix

$$M = \begin{pmatrix} A & 0 & 0 \\ 0 & B & 0 \\ 0 & 0 & C \end{pmatrix}$$

also nilpotent? That's a trivial question; the 14×14 matrix M is obviously nilpotent of index 6.

How else can one obtain a nilpotent matrix of size 14 and index 6? One easy answer is just to juggle the numbers: replace the sizes and indices of B and C, for instance, by 5 and 3, or replace the sizes and indices of B and C by 6 and 2, etc. These examples are direct sums; they are obtained by gluing together examples of the same or smaller size. What other way of manufacturing nilpotent matrices is there?

The result of Problem 111 implies that if M is a nilpotent matrix of index 3, say, then there exists a vector x such that the vectors

$$x, Mx, M^2x$$

are linearly independent. Extend that linearly independent set to a basis and write down the matrix of M with respect to that basis. The result looks like

$$\begin{pmatrix} A & X \\ 0 & B \end{pmatrix},$$

where

$$A = \begin{pmatrix} 0 & 1 & 0 \\ 0 & 0 & 1 \\ 0 & 0 & 0 \end{pmatrix},$$

and B is a matrix that must be nilpotent also, with index less than or equal to 3. Question: can X be thrown away? Precisely: is

$$\begin{pmatrix} A & X \\ 0 & B \end{pmatrix}$$

similar to

$$\begin{pmatrix} A & 0 \\ 0 & B \end{pmatrix}?$$

That question in its general form is one of the most important ones in linear algebra and its answer is correspondingly difficult. It isn't all *that* difficult—the methods used so far serve to prove that the answer is yes—but it tends to be longish and complicated. A slight feeling for the spirit of the answer can be obtained by working out a very easy special case; here is one.

Problem 113.   *What is a basis with respect to which the linear transformation defined by the matrix*

$$M = \begin{pmatrix} 0 & 1 & 0 & 1 & 0 \\ 0 & 0 & 1 & 0 & -1 \\ 0 & 0 & 0 & 0 & 0 \\ 0 & 0 & 0 & 0 & 1 \\ 0 & 0 & 0 & 0 & 0 \end{pmatrix}$$

has the matrix

$$M_0 = \begin{pmatrix} 0 & 1 & 0 & 0 & 0 \\ 0 & 0 & 1 & 0 & 0 \\ 0 & 0 & 0 & 0 & 0 \\ 0 & 0 & 0 & 0 & 1 \\ 0 & 0 & 0 & 0 & 0 \end{pmatrix}?$$

114. Jordan form

What happens when the general theorem that exhibits a nilpotent matrix as a direct sum (Problem 108) is applied repeatedly? The theorem says that with respect to a suitable basis every nilpotent matrix of index q, say, has the form

$$M = \begin{pmatrix} A & 0 \\ 0 & B \end{pmatrix},$$

where A and B have the following special properties.
(1) A is a $q \times q$ matrix of the same form (Jordan form) as the 3×3 matrix

$$\begin{pmatrix} 0 & 1 & 0 \\ 0 & 0 & 1 \\ 0 & 0 & 0 \end{pmatrix}$$

described in Problem 108, meaning that the entries on the diagonal just above the main one are 1 and all others are 0, and

(2) B is nilpotent with index less than or equal to q.

To apply that result the second time means to apply it to B. The result is a representation of B as a direct sum of two nilpotent matrices, of which the first is in Jordan form, with index equal to the index of B (and with size same as its index). Application of the method "repeatedly" as often as possible yields a matrix representation for M of the form

$$\begin{pmatrix} A_1 & 0 & 0 & \\ 0 & A_2 & 0 & \\ 0 & 0 & A_3 & \\ & & & \ddots \end{pmatrix}$$

Here each A_j on the diagonal is a nilpotent matrix in Jordan form, of size and index q_j, with $q_1 \geq q_2 \geq q_3 \geq \cdots$. This is called the Jordan form of M.

Could it be true that every matrix can be obtained by gluing together easy ones? Could it, for instance, be true that the "zero part" of every matrix can be split off and studied separately? What might that mean? Well, a possible hope is that every matrix is (or is similar to?) a direct sum, such as

$$\begin{pmatrix} 0 & 0 \\ 0 & 3 \end{pmatrix},$$

of a zero matrix and a non-zero matrix—but that is not true. Example:

$$\begin{pmatrix} 0 & 3 \\ 0 & 0 \end{pmatrix}.$$

The weakest way of saying "zero" and the strongest way of saying "non-zero" suggest a modified hope: could it be that every matrix is (or is similar to) a direct sum of a nilpotent matrix and an invertible matrix? Yes; that is true, and the result is known as **Fitting's lemma**.

The invertible direct summand that Fitting's lemma yields (call it M) does not have 0 in its spectrum, but since it does have some eigenvalue λ, Fitting's lemma is applicable to the matrix $M - \lambda$. Consequence: M is representable as a direct sum that exactly resembles the one displayed in the nilpotent case, but with direct summands A_j

$$\begin{pmatrix} \lambda & 1 & 0 & \\ 0 & \lambda & 1 & \\ 0 & 0 & \lambda & \\ & & & \ddots \end{pmatrix}$$

on whose main diagonal λ appears instead of 0. In an obvious extension of the terminology introduced before, that form is known as the Jordan form of A.

Arguing the same way separately for each eigenvalue of an arbitrary matrix M leads to the grand conclusion—the assertion that every matrix is similar to a direct sum of matrices $M_1, M_2, M_3, \ldots$ in Jordan form, with distinct eigenvalues. That direct sum is called the Jordan form of M (in the second and final broadening of that expression), and the possibility of representing every M that way is the apex of linear algebra. It is difficult to think of an answerable question about linear transformations whose answer is not a consequence of representability in Jordan form; here is a small but pleasant sample.

Problem 114. *Does every matrix have a square root?*

Comment. If $A = B^2$, then, of course, B is called a **square root** of A.

115. Minimal polynomials

Is every matrix algebraic? The language is borrowed from the theory of algebras: an element a of an algebra is called algebraic if there exists a non-zero polynomial p such that $p(a) = 0$. Example: if E is a projection, and if $p(\lambda) = \lambda^2 - \lambda$, then $p(E) = 0$. Another example: if A is nilpotent of index q and if $p(\lambda) = \lambda^q$, then $p(A) = 0$.

The second example can be generalized: if a linear transformation A_0 on a finite-dimensional vector space has only one eigenvalue, λ_0 say, then $A_0 - \lambda_0 I$ is nilpotent of index q_0 say, and therefore the polynomial

$$m_0(\lambda) = (\lambda - \lambda_0)^{q_0}$$

annihilates A_0. Important note: q_0 is the smallest degree that such a polynomial can have, and $m_0(\lambda)$ is the unique monic polynomial of that degree that does the job.

If a linear transformation has not just one eigenvalue but two, λ_1 and λ_2, then its Jordan form looks like

$$M = \begin{pmatrix} M_1 & 0 \\ 0 & M_2 \end{pmatrix},$$

where $M_1 - \lambda_1$ and $M_1 - \lambda_2$ are nilpotent, with some indexes q_1 and q_2. It follows that there is one and only one monic polynomial of minimal degree that annihilates M, namely the polynomial

$$m(\lambda) = (\lambda - \lambda_1)^{q_1}(\lambda - \lambda_2)^{q_2}.$$

The number 2 has nothing to do with this statement or its proof. The general statement is that for every matrix A there exists a unique monic polynomial of minimal degree that annihilates A; it is called the **minimal polynomial** of A. This minimal polynomial is, in fact, equal to the product of the factors of the form $(\lambda - \lambda_j)^{q_j}$ obtained by letting the λ_j's range through the distinct eigenvalues of A; the q_j's are corresponding indexes in the Jordan (or triangular) form.

Since each factor of the minimal polynomial is a factor of the characteristic polynomial also, it follows that if the characteristic polynomial of A is p, then $p(A) = 0$; this famous statement is known as the **Hamilton-Cayley equation.**

If the minimal polynomial of a linear transformation on a space of dimension n has degree n, does it follow that the transformation is diagonalizable? Answer: no—trivially no—a counterexample is $\begin{pmatrix} 0 & 1 \\ 0 & 0 \end{pmatrix}$.

How do the minimal polynomial and the characteristic polynomial of a diagonal matrix compare? Answer: if

$$A = \begin{pmatrix} \lambda_1 & 0 & 0 & \\ 0 & \lambda_2 & 0 & \\ 0 & 0 & \lambda_3 & \\ & & & \ddots \end{pmatrix},$$

then the characteristic polynomial is the product of all the $(\lambda - \lambda_j)$'s, but the minimal polynomial is the product of just one representative of each possible factor. So, for example, if

$$\begin{pmatrix} 1 & 0 & 0 & 0 & 0 \\ 0 & 1 & 0 & 0 & 0 \\ 0 & 0 & 2 & 0 & 0 \\ 0 & 0 & 0 & 2 & 0 \\ 0 & 0 & 0 & 0 & 2 \end{pmatrix},$$

then the characteristic polynomial is $(\lambda - 1)^2(\lambda - 2)^3$, and the minimal polynomial is $(\lambda - 1)(\lambda - 2)$.

These examples were trying to cultivate friendship toward minimal polynomials; the following problem might test the success of the attempt.

Problem 115. *What are the minimal polynomial and the characteristic polynomial of the differentiation transformation D and the translation transformation T on the space $\mathbb{P}_3$ of polynomials of degree less than or equal to 3? (Reminder: $Dx(t) = \frac{dx}{dt}$ and $Tx(t) = x(t + 1)$.)*

116. Non-commutative Lagrange interpolation

Is there a polynomial p such that $p(n) = 10^n$ for every integer n between 1 and 100 inclusive? Sure, why not: in fact there is a polynomial of degree 99 that does that. All you have to do to prove it is to write down the pertinent system of 100 (linear) equations in the 100 unknown coefficients, and note that its determinant (a special instance of a Vandermonde) is not 0. It is easy to formulate a general theorem of which this result is a special case: if $x_1, \ldots, x_n$ are n distinct numbers (the avoidance of repetitions is essential), and if $y_1, \ldots, y_n$ are any n numbers, then there exists a (unique) polynomial p of degree $n - 1$ such that

$$p(x_j) = y_j$$

for $j = 1, \ldots, n$. The celebrated **Lagrange interpolation formula** is an explicit presentation of that polynomial. The "numbers" in the general theorem can be replaced by elements in an arbitrary field, and the result is a polynomial with coefficients in that field.

Once all that is granted, a shallow generalization is easy to come by: if $X_1, \ldots, X_n$ are pairwise disjoint finite sets of numbers (or elements of an arbitrary field), and if $p_1, \ldots, p_n$ are arbitrary polynomials (with coefficients in the same field), then there exists a polynomial p such that

$$p(x) = p_j(x)$$

whenever $x \in X_j$, $j = 1, \ldots, n$. Proof: apply the Lagrange interpolation theorem to the set of (distinct) numbers x in $X_1 \cup \cdots \cup X_n$ with the corresponding values y chosen to be $p_j(x)$ whenever $x \in X_j$. (The smallest possible degree of p is easy to describe in terms of the number n and the sizes of the sets $X_1, \ldots, X_n$.) A statement of this generalization in terms of matrices goes like this: if $M_1, \ldots, M_n$ are diagonal matrices with pairwise disjoint spectra (that is, pairwise disjoint sets of eigenvalues, or, what comes to exactly the same thing, pairwise disjoint sets of diagonal entries), and if $p_1, \ldots, p_n$ are polynomials, then there exists a polynomial p such that

$$p(M) = p_j(M_j)$$

for $j = 1, \ldots, n$.

The most conspicuous fact about diagonal matrices is that they all commute with one another; the matrix Lagrange interpolation theorem that was just formulated belongs to commutative linear algebra. Does its straightforward non-commutative generalization have a chance of being true, or does the conclusion itself imply some kind of partial commutativity? To avoid extraneous trouble with non-existence of eigenvalues, the straightforward generalization will be formulated over the field $\mathbb{C}$ of complex numbers.

Problem 116. *Is it always true that if $A_1, \ldots, A_n$ are linear transformations with pairwise disjoint spectra (that is, pairwise disjoint sets of eigenvalues) on a finite-dimensional vector space $\mathbb{V}$ over $\mathbb{C}$, and if $p_1, \ldots, p_n$ are polynomials with coefficients in $\mathbb{C}$, then there exists a polynomial p with coefficients in $\mathbb{C}$ such that*

$$p(A_j) = p_j(A_j)$$

for $j = 1, \ldots, n$?

CHAPTER 8

INNER PRODUCT SPACES

117. Inner products   **117**

Which of these vectors in $\mathbb{R}^3$ is larger: $(2, 3, 5)$ or $(3, 4, 4)$? Does the question make sense? The only sizes, the only numbers, that have been considered so far are dimensions. Since $(2, 3, 5)$ belongs to $\mathbb{R}^3$ and $(1, 1, 1, 1)$ belongs to $\mathbb{R}^4$, the latter is in some sense larger, but it is a weak sense and not a useful one. The time has come to look at the classical and useful way of "measuring" vectors.

The central concept is that of an inner product in a real or complex vector space. That is, by definition, a

(1) **Hermitian symmetric**,

(2) **conjugate bilinear**,

and

(3) **positive definite**

form—which means that it is a numerically valued function of ordered pairs of vectors x and y such that

$$(x, y) = \overline{(y, x)}, \tag{1}$$

$$(a_1 x_1 + a_2 x_2, y) = a_1(x_1, y) + a_2(x_2, y), \tag{2}$$

$$(x, x) \geqq 0; \ (x, x) = 0 \text{ if and only if } x = 0. \tag{3}$$

Standard examples: for $x = (\xi_1, \xi_2)$ and $y = (\eta_1, \eta_2)$ in $\mathbb{R}^2$, write

$$(x, y) = \xi_1 \overline{\eta}_1 + \xi_2 \overline{\eta}_2$$

(reminiscent of the formula for the cosine of the angle between two segments), and for x and y in $\mathbb{P}$, write

$$(x, y) = \int_0^1 x(t)\overline{y(t)}\, dt$$

(a formal "continuous" analog of the sum in $\mathbb{R}^2$ and of its natural generalization in $\mathbb{R}^n$).

The upper bars here denote complex conjugation; the reason they are necessary has to do with the associated notion of length. The point is that the length (or norm) of a vector x is defined by

$$\|x\| = \sqrt{(x, x)}.$$

If the formula $\xi_1 \eta_1 + \xi_2 \eta_2$ had been used (instead of $\xi_1 \overline{\eta}_1 + \xi_2 \overline{\eta}_2$), then for a vector x in $\mathbb{C}^2$ the consideration of its scalar multiple ix (where $i = \sqrt{-1}$) would lead to an unpleasant surprise. The relation between inner products and scalars would yield

$$\|ix\|^2 = (ix, ix) = i(x, ix) = i^2(x, x) = -\|x\|^2,$$

and that could be regarded as unpleasant. The square of a length shouldn't really be negative—that would lead to a length whose value is an imaginary number, and that is not the sort of thing one normally thinks of as a suitable measure of size.

An inner product space is a vector space with an inner product. The intuitive interpretation of (x, y) is the cosine of the angle between x and y, and, correspondingly, if $(x, y) = 0$—cosine equal to 0—the vectors x and y are called **orthogonal** (= perpendicular). To what extent is this metric concept in harmony with the linear concepts treated so far?

Problem 117. *How large does a finite orthogonal set of non-zero vectors have to be to be linearly dependent?*

Comment. A set of vectors is called orthogonal if each pair of its elements is orthogonal. Recall that when a set of vectors is enlarged, a linear dependence relation between them becomes more likely than it was before.

118   118. Polarization

The norm in an inner product space is defined in terms of the inner product. Is there any hope of going in the other direction?

Problem 118. *Can two different inner products yield the same norm?*

119. The Pythagorean theorem

The Pythagorean theorem says that the sum of the squares of two sides of a right triangle is equal to the square of the hypotenuse. Is anything like that true for vector spaces in general?

Problem 119. *Under what conditions on two vectors x and y is it true that*

$$\|x + y\|^2 = \|x\|^2 + \|y\|^2?$$

120. The parallelogram law

Problem 120. *Under what conditions on two vectors x and y is it true that*

$$\|x + y\|^2 + \|x - y\|^2 = 2\|x\|^2 + 2\|y\|^2?$$

Comment. The equation is not as strange as at first it might appear. Think about pictures: if x and y are two intersecting sides of a parallelogram, then $x + y$ and $x - y$ can be thought of as its two diagonals. The "parallelogram law" of elementary geometry is exactly the equation under consideration.

121. Complete orthonormal sets

How large can an orthogonal set be? One possible interpretation of that question is this: for which values of n is it possible to find an orthogonal set $\{x_1, \ldots, x_n\}$ of vectors in an inner product space? That's not quite a sensible interpretation: the notation allows many (all) of the x_j's to be 0, and in that sense n can be chosen arbitrarily large. An efficient way to rule out that uninformative interpretation of the question is to "normalize" the vectors that are allowed to enter. In the language that is customary in this circle of ideas, to say that a vector x is normal or normalized means that $\|x\| = 1$, and, correspondingly, an orthogonal set $\{x_1, x_2, \ldots\}$ is called **orthonormal** if $(x_i, x_j) = \delta_{ij}$ for all i and j.

An orthonormal set is called **complete** if it is maximal, that is, if it cannot be enlarged, or, in other words, if it is not a subset of any larger orthonormal set. Since orthonormal sets are linearly independent (Problem 117), an inner product space of dimension n cannot have orthonormal sets with more than n elements. Can it always have that many?

Problem 121. *Does every inner product space of dimension n have an orthonormal set of n elements?*

122   122. Schwarz inequality

In what way are orthonormal bases better than just plain bases? A partial answer is that when a vector x is expanded in terms of an orthonormal basis the coefficients give precise information about the size of the vector. If, in fact $\{x_1, \ldots, x_r\}$ is an orthonormal set (not even necessarily a basis), and x is an arbitrary vector, then

$$
0 \leqq \left\| x - \sum_1^r (x, x_j) x_j \right\|^2
$$

$$
= \left(x - \sum_1^r (x, x_i) x_i, \, x - \sum_1^r (x, x_j) x_j \right)
$$

$$
= (x, x) - \sum_i (x, x_i)(x_i, x) - \sum_j (x_j, x)(x, x_j)
$$

$$
+ \sum_i \sum_j (x, x_i)(x, x_j)(x_i, x_j)
$$

$$
= \|x\|^2 - \sum_i |(x, x_i)|^2 - \sum_i |(x, x_i)|^2 + \sum_i |(x, x_i)|^2.
$$

Consequence:

$$
\sum_i |(x, x_i)|^2 \leqq \|x\|^2.
$$

This result is known as **Bessel's inequality**. It has two important consequences.

(1) If x and y are vectors in an inner product space, then

$$
|(x, y)| \leqq \|x\| \cdot \|y\|.
$$

This result is known as the **Schwarz inequality**. It can be derived from Bessel's inequality as follows: if $y = 0$, both sides are 0; if $y \neq 0$, the set consisting of the vector $\dfrac{y}{\|y\|}$ only is orthonormal, and consequently, by Bessel's inequality,

$$
\left| \left(x, \frac{y}{\|y\|} \right) \right|^2 \leqq \|x\|^2.
$$

(2) If x is a vector and $\{x_1, \ldots, x_n\}$ is an orthonormal basis in an inner product space, then

$$\|x\|^2 = \sum_i |(x, x_i)|^2.$$

To prove that, observe that if x is expanded in terms of the x_j's,

$$x = \sum_j \alpha_j x_j,$$

then, forming the inner product of each side with itself yields

$$\|x\|^2 = \left(\sum_i \alpha_i x_i, \sum_j \alpha_j x_j \right) = \sum_i \sum_j \alpha_i \overline{\alpha}_j (x_i, x_j) = \sum_k |\alpha_k|^2,$$

and forming the inner product of both sides with each x_k yields

$$(x, x_k) = \alpha_k.$$

The equation

$$\|x\|^2 = \sum_i |(x, x_i)|^2$$

is known as **Parseval's identity**. Note: a small modification of the technique proves a more general result: if $x = \sum_j \alpha_j x_j$ and $y = \sum_j \alpha_j x_j$, then

$$(x, y) = \sum_j \alpha_j \overline{\beta}_j.$$

Bessel and Schwarz and Parseval are part of the standard lore of this subject. The answer to the next question is equally well known to the experts, but it is slightly more recondite and, perhaps, a little more fun.

Problem 122. *For which pairs of vectors does the Schwarz inequality become an equation?*

123. Orthogonal complements   123

Just how much is orthogonality in "harmony" with linearity? What is already known is that vectors that differ a lot in the metric sense differ a lot in the linear sense too (orthonormal sets are linearly independent, see Problem 117). What if a bunch of vectors all have a common (orthogonal) enemy—does it follow that they are all (linear) friends? A sharp formulation of that vague question has to do with what are called orthogonal complements.

If $\mathbb{E}$ is a set of vectors in an inner product space $\mathbb{V}$, the **orthogonal complement** $\mathbb{E}^\perp$ (pronounced "E perp") of $\mathbb{E}$ is the set of all those vectors in $\mathbb{V}$ that are orthogonal to every vector in $\mathbb{E}$. It is an easy exercise to verify that $\mathbb{E}^\perp$ is a subspace of $\mathbb{V}$ (it doesn't matter whether $\mathbb{E}$ is a subspace or not), and saying the meanings of the symbols slowly ought to convince anyone that $\mathbb{E} \subset \mathbb{E}^{\perp\perp}$. (Is the intended meaning of $\mathbb{E}^{\perp\perp}$ clear? It is $(\mathbb{E}^\perp)^\perp$.) Consequence:

$$\text{span } \mathbb{E} \subset \mathbb{E}^{\perp\perp}.$$

Problem 123. *If $\mathbb{M}$ is a subspace of a finite-dimensional inner product space $\mathbb{V}$, what are the relations among $\mathbb{M}$, $\mathbb{M}^\perp$, and $\mathbb{M}^{\perp\perp}$?*

124   ## 124. More linear functionals

Is the resemblance between the superscripts such as in $\mathbb{M}^0$ (annihilators of subspaces) and the ones in $\mathbb{M}^\perp$ (orthogonal complements of subspaces) a structural one or merely notational? It turns out that the question is really one about linear functionals.

Linear functionals on an inner product space $\mathbb{V}$ are easy enough to come by: fix an element y in $\mathbb{V}$ and then define a function ξ on $\mathbb{V}$ by writing

$$\xi(x) = (x, y)$$

for all x. That ξ is a linear functional is an immediate consequence of the defining properties of inner products. Are the linear functionals obtained in this way typical?

Problem 124. *If ξ is a linear functional on an inner product space $\mathbb{V}$, does there always exist a vector y in $\mathbb{V}$ such that*

$$\xi(x) = (x, y)$$

for all x?

125   ## 125. Adjoints on inner product spaces

Is a vector in an inner product space the same as a linear functional? That may look like a foolish question, but (a) linear algebraists are used to considering linear functionals as vectors (elements of the dual space), and (b) in an inner product space each vector induces (is?) a linear functional, the one defined by inner products. In fact the correspondence that assigns to each vector y in an inner product space V the linear functional ξ that it induces,

$$\xi(x) = (x, y),$$

is a one-to-one correspondence between all of V and all of the dual space V'. One-to-one? Sure: if y_1 and y_2 correspond to the same u, then

$$(x, y_1) = (x, y_2)$$

for all x, so that

$$(x, y_1 - y_2) = 0$$

for all x, so that $y_1 - y_2$ is orthogonal to every vector x, and therefore $y_1 - y_2 = 0$. All of V'? Sure: that's what Solution 123 proves.

The correspondence $y \mapsto x$ is eager to cooperate with the linear structure of V. If $y_1 \rightarrow \xi_1$ and $y_2 \rightarrow \xi_2$, then $y_1 + y_2 \rightarrow \xi_1 + \xi_2$—that's easy—and if y induces ξ, then a scalar multiple αy induces—no, not $\alpha \xi$, but almost—in fact it induces $\overline{\alpha} \xi$. Clear? If $(x, y) = \xi(x)$ for all x, then $(x, \alpha y) = \overline{\alpha}(x, y) = \overline{\alpha} \xi(x)$ for all x. The correspondence $y \mapsto \xi$ doesn't quite deserve to be called an isomorphism: it is a **conjugate isomorphism**.

Is the dual space of an inner product space an inner product space? Well, that depends: how is the inner product of two linear functionals ξ_1 and ξ_2 defined? If $\xi_1(x) = (x, y_1)$ and $\xi_2(x) = (x, y_2)$, the most natural looking definition is probably this:

$$(\xi_1, \xi_2) = (y_1, y_2).$$

Trouble: it doesn't work. Is (ξ_1, ξ_2) linear in ξ_1? Additivity is all right, but is it true that

$$(\alpha \xi_1, \xi_2) = \alpha(\xi_1, \xi_2)?$$

No: since αy_1 induces $\overline{\alpha} \xi_1$, so that $\alpha \xi_1$ is induced by $\overline{\alpha} y_1$,

$$(\alpha \xi_1, \xi_2) = (\overline{\alpha} y_1, y_2) = \overline{\alpha}(y_1, y_2) = \overline{\alpha}(\xi_1, \xi_2),$$

it follows that, once more, a conjugation appears where it wasn't invited.

There is a brute force remedy, but it's far from clear on first glance that it will work: why not define (ξ_1, ξ_2) to be (y_2, y_1)? Does it work? Yes. Indeed, with that definition,

$$(\alpha \xi_1, \xi_2) = (y_2, \overline{\alpha} y_1) = \xi(y_2, y_1) = \alpha(\xi_1, \xi_2).$$

Conclusion: with the inner product so defined, the space of all linear functionals on an inner product space V is itself an inner product space, and, as such, it is denoted by V^*. The isomorphism statement for the pure vector spaces V and V' now extends to the inner product spaces V and V^*: they too are conjugate isomorphic.

If in accordance with that conjugate isomorphism the spaces V and V^* are identified—regarded as the same—then many earlier statements about the relation between V and V' become more interesting and more usable. So, for instance, the assertion that corresponding to each basis of V there is a dual basis in V' becomes the assertion that to each basis $\{x_1, \ldots, x_n\}$ of V there corresponds another basis $\{\xi_1, \ldots, \xi_n\}$ of V, the dual basis, such that $(x_i, \xi_j) = \delta_{ij}$. The correspondence between subspaces of V and their annihilators in V' becomes the correspondence between subspaces M of V and their orthogonal complements $M^\perp$ also in V. Finally, and most importantly, the correspondence between linear transformations on V and their adjoints on V' becomes a correspondence between linear transformations A on V and their adjoints on V^*, denoted in this context by A^*. The purely linear adjoints and the inner product kind differ in minor ways only, all of which have to do with the conjugation that has to be built into the complex theory. The differences are that

$$(\alpha A)^* = \overline{\alpha} A^* \qquad (\text{not } \alpha A^*), \tag{1}$$

that the matrix of A^* (with respect to an orthonormal basis) is the **conjugate transpose** of the matrix of A,

$$(\alpha_{ij}) \text{ becomes } (\overline{\alpha}_{ji}) \qquad (\text{not } (\alpha_{ji})), \tag{2}$$

and that

$$\det A^* = \overline{\det A} \qquad (\text{not } \det A). \tag{3}$$

That's the bad news—and it sure isn't very bad. The good news is that if A is a linear transformation, not only are A and A^{**} comparable but so are A and A^*. For A and A^{**} "comparable" turns out to mean "equal". For A and A^* that *can* happen, but doesn't have to, but, in any case, new and valuable questions can be asked. When is it true that $A^* = A$? How about $A^* = -A$? What can be said about the sum of A and A^*? What about the product; when do A and A^* commute? These questions are at the basis of the most important part of linear algebra. Before beginning their proper study, a couple of problems should be looked at by way of practice—the properties of the correspondence $A \mapsto A^*$ take a little getting used to.

Problem 125. *The direct sum $V \oplus W$ of two inner product spaces is defined to be the direct sum of the vector spaces V and W endowed* *with the inner product defined by*

$$((\langle x_1, y_1 \rangle, \langle x_2, y_2 \rangle)) = (x_1, x_2) + (y_1, y_2).$$

(Check: is this indeed an inner product?)

(a) *If a linear transformation U is defined on* $\mathbb{V} \oplus \mathbb{V}$ *by*

$$U\langle x, y \rangle = \langle y, -x \rangle,$$

what is U*? What are U*U and UU*?

(b) *The graph of a linear transformation A on a vector space* $\mathbb{V}$ *is the set of all those ordered pairs* $\langle x, y \rangle$ *in* $\mathbb{V} \oplus \mathbb{V}$ *for which* $y = Ax$. *Is the graph always a subspace of* $\mathbb{V} \oplus \mathbb{V}$?

(c) *If* $\mathbb{G}$ *is the graph of a linear transformation A on* $\mathbb{V}$, *what is the graph of A*? How are those graphs related?*

126. Quadratic forms 126

Adjoints enter linear algebra through still another door, the back door of quadratic forms (whose name, to be sure, doesn't sound very linear). A quadratic form is a specialization of a bilinear form, which, in turn is a generalization of a linear functional.

The obvious way to generalize linear functionals is in the direction of the functions of several variables called multilinear forms. The easiest but nevertheless typical multilinear forms are the bilinear ones; they are, by definition, functions of two vector variables that are linear in each variable separately for fixed values of the other. Explicitly: if $\mathbb{V}$ is a vector space, a scalar-valued function φ on $\mathbb{V} \oplus \mathbb{V}$ is a **bilinear form** if

$$\varphi(\alpha_1 x_1 + \alpha_2 x_2, y) = \alpha_1 \varphi(x_1, y) + \alpha_2 \varphi(x_2, y)$$

for each y in $\mathbb{V}$, and at the same time

$$\varphi(x, \beta_1 y_1 + \beta_2 y_2) = \beta_1 \varphi(x, y_1) + \beta_2 \varphi(x, y_2)$$

for each x in $\mathbb{V}$. Example: if $\mathbb{V}$ is $\mathbb{R}^1$ and

$$\varphi(x, y) = xy,$$

then φ is a bilinear form. Less trivially: if $\mathbb{V}$ is $\mathbb{R}^2$ and

$$\varphi(\langle x_1, x_2 \rangle, \langle y_1, y_2 \rangle) = x_1 y_1 + x_2 y_2,$$

then φ is a bilinear form. (Check?) A **quadratic form** is a function obtained from a bilinear one by restriction to equal values of the two variables. That is: the quadratic form induced by the bilinear form φ is the function $\widehat{\varphi}$ defined by

$$\widehat{\varphi}(x) = \varphi(x, x).$$

The most valuable example of a bilinear form is one that is not really an example at all. If $\mathbb{V}$ is an inner product space and if φ is defined by

$$\varphi(x, y) = (x, y),$$

then φ is linear in x, to be sure, and it's trying to be linear in y, but complex conjugation ruins it. A curious usage of words has been adopted in connection with this kind of occurrence of complex conjugation. A complex-valued function ξ that is additive,

$$\xi(x + y) = \xi(x) + \xi(y),$$

and fails to be linear because of complex conjugation,

$$\xi(\alpha x) = \overline{\alpha} \xi(x)$$

has come to be called **semilinear**. (Happy acceptance of this language might be difficult for some—it's not obvious that such a function satisfies exactly *half* the conditions for linearity—but it is well established and it's too late to change it.) In accordance with that usage a function φ of two variables that behaves like an inner product (linear in the first variable and semilinear in the second) could be called one-and-a-half linear—and that is almost what it is called. In fact the Latin for one-and-a-half is used, so that the technical word is **sesquilinear**. (Semilinear functions are sometimes called antilinear and sometimes conjugate linear.)

Linear transformations in conspiracy with inner products can be used to get many examples of sesquilinear forms. Here is how: if A is a linear transformation on an inner product space $\mathbb{V}$, and if x and y are in $\mathbb{V}$, write

$$\varphi(x, y) = (Ax, y).$$

All such examples are sesquilinear and therefore they act, in part, like inner products—but usually only in part. There is no reason why they should be Hermitian symmetric—sometimes they are and sometimes they are not—it all depends on properties of A. There is no reason why they should be positive definite—again whether they are or not depends on A. It is good to know just how these properties depend on A, and that will be discussed soon. For now, however, it's best to get back to the general theory of sesquilinear forms.

If φ is a sesquilinear form, what should $\hat{\varphi}$, be called? (Here, as before, $\hat{\varphi}(x) = \varphi(x, x)$.) No accurate word exists (semiquadratic and sesquiquadratic suggest themselves?—but the world hasn't adopted either one), and, therefore, an innacurate one is commonly used. If φ is a sesquilinear form and $\hat{\varphi}(x) = \varphi(x, x)$, then, just as in the bilinear case, $\hat{\varphi}$ is called the **quadratic form** associated with (induced by) φ.

There is a way of making new sesquilinear forms out of old. If φ is a sesquilinear form, and if P and Q are linear transformations, then the expression

$$\varphi(Px, Qy)$$

defines another sesquilinear form—in this sense linear transformations (or, rather, pairs of linear transformations) *act on* sesquilinear forms. If, in particular,

$$\varphi(x, y) = (Ax, y),$$

then

$$\varphi(Px, Qy) = (APx, Qy) = (Q^*APx, y).$$

That is: the action of P and Q on φ replaces the linear transformation A by an equivalent one (in the strict sense of the word, as discussed in Problem 95). If $\hat{\varphi}$ is the quadratic form associated with φ, then the natural way to mix in a linear transformation is to consider $\hat{\varphi}(Px)$. Since

$$\hat{\varphi}(Px) = (APx, Px) = (P^*APx, x),$$

the action of P on $\hat{\varphi}$ replaces A by the unfamiliar construct P^*AP. That construct is a good thing to know about; the concept it defines is called congruence. That is: two linear transformations A and B are called **congruent** if there exists an invertible linear transformation P such that

$$B = P^*AP.$$

Invertibility is essential here; without it the relation is too loose to be of much interest. Congruence is a special case of equivalence. What are its special properties?

Problem 126.   (a) *Is congruence an equivalence relation?*

(b) *If A and B are congruent, are A^* and B^* congruent?*

(c) *Does there exist a linear transformation A such that A is congruent to a scalar α but $A \neq \alpha$?*

(d) *Do there exist linear transformations A and B such that A and B are congruent but A^2 and B^2 are not?*

(e) *Do there exist invertible linear transformations A and B such that A and B are congruent but A^{-1} and B^{-1} are not?*

127   127. Vanishing quadratic forms

To what extent does a quadratic form (Ax, x) determine the linear transformation A? Can it happen for two different transformations A and B that $(Ax, x) = (Bx, x)$ for all x?

> **Problem 127.** *Does there exist a non-zero linear transformation A on an inner product space such that $(Ax, x) = 0$ for all x?*

Comment. It is clear, isn't it?, that this question about zero is the same as the uniqueness question: is it true that $(Ax, x) = (Bx, x)$ for all x if and only if $((A - B)x, x) = 0$ for all x?

128   128. Hermitian transformations

How closely do linear transformations on a vector space resemble complex numbers? Linear transformations can be added and multiplied—that's old stuff, and it says no more than that they form a ring. Transformations on an inner product space admit another operation, one that resembles complex conjugation (adjoint)—that is another, different aspect of the resemblance.

Complex conjugation can be used to define various important sets of complex numbers. The most obvious one among them is the set of real numbers (some complex numbers are real)—they can be defined as the set of those complex numbers z for which $z = \bar{z}$. The transformation analog is the set of those linear transformations A on an inner product space for which $A = A^*$. They are called **Hermitian**, and they are among the ones that occur most frequently in the applications of linear algebra.

If the matrix of a linear transformation A with respect to an orthonormal basis is (α_{ij}), then the matrix of A^* is the conjugate transpose $(\bar{\alpha}_{ji})$ (see Problem 125). The use of orthonormal bases is crucial here. If, for instance, A has the matrix

$$\begin{pmatrix} 0 & 1 \\ 1 & 0 \end{pmatrix}$$

with respect to the orthonormal basis $\{(1,0), (0,1)\}$, then A is Hermitian. If however the non-orthonormal basis $\{u_1, u_2\}$ is used, where

$$u_1 = (1,0) \qquad \text{and} \qquad u_2 = (1,1),$$

then

$$Au_1 = (0,1) = (1,1) - (1,0) = u_2 - u_1,$$

and

$$Au_2 = u_2,$$

so that the matrix of A with respect to $\{u_1, u_2\}$ is

$$\begin{pmatrix} -1 & 0 \\ 1 & 1 \end{pmatrix}.$$

Since it is easy enough to write down as many conjugate symmetric matrices as anyone could desire, it is easy to produce examples of Hermitian transformations. Very special case: a scalar transformation is Hermitian if and only if the scalar in question is real.

What follows is a sequence of problems (puzzles) intended to give their solver the opportunity of getting used to the properties of Hermitian transformations.

Problem 128. *When is the product of two Hermitian transformations Hermitian?*

129. Skew transformations

129

The most unreal numbers are the so-called pure imaginary ones, the real multiples of i. The complex conjugate of such a number is not equal to itself but is equal to its negative: $\bar{\imath} = -i$. The transformation analogs of those numbers are called **skew Hermitian**, or simply **skew**: they are the transformations A for which $A^* = -A$. Various combinations of Hermitian transformations and skew transformations can sometimes turn out to be Hermitian or skew; here are some sample questions.

Problem 129. (a) *If A and B are congruent and A is skew, does it follow that B is skew?*
    (b) *If A is skew, does it follow that A^2 is skew? How about A^3?*
    (c) *If A is either Hermitian or skew, and if B is either Hermitian or skew, what can be said about $AB + BA$? What about $AB - BA$?*

Comment. "Congruent" refers to the concept discussed in Problem 120.

130   130. Real Hermitian forms

How much do Hermitian transformations resemble real numbers? The motivation for introducing them above was the equation $A = A^*$, but that's only a formal analogy. If A is the matrix

$$\begin{pmatrix} 0 & 1 \\ 0 & 0 \end{pmatrix}$$

— as non-Hermitian as any matrix can get—and if $x = (x_1, x_2)$, then

$$(Ax, x) = \overline{x}_1 x_2$$

—the quadratic form associated with A is as non-real as any can get. If, on the other hand, A is

$$\begin{pmatrix} 0 & 1 \\ 1 & 0 \end{pmatrix},$$

then

$$(Ax, x) = \overline{x}_1 x_2 + x_1 \overline{x}_2 = 2Re\overline{x}_1 x_2$$

—as real as any can get. Are these phenomena typical?

Problem 130.   *What is the relation between Hermitian transformations and transformations with real quadratic form?*

131   131. Positive transformations

Does the set of positive real numbers have as nice a transformation analog as the set of all real numbers? A natural attempt to define positiveness for transformations is to imitate the definition of reality via quadratic forms, and that in fact is what is usually done. A linear transformation A is called **positive** if $(Ax, x) \geq 0$ for every vector x; if $(Ax, x) > 0$ for all non-zero x, the phrase positive definite is used. The symbolic way of saying that A is positive is to write

$$A \geq 0$$

and the statement $A - B \geq 0$ can also be written as

$$A \geq B.$$

The weak sign ($\geq$) can be replaced by the strong one ($>$) when the facts permit it, and

$$B \leq A$$

means, of course, the same as $A \geq B$.

Examples: if A is an arbitrary linear transformation, then $A^*A \geq 0$ (because $(A^*Ax, x) = \|Ax\|^2$), and therefore if B is a Hermitian transformation, then $B^2 \geq 0$ (because $B^2 = B^*B$). These statements are transformation analogs of the numerical statements $\bar{z}z \geq 0$ (for every complex number z) and $u^2 \geq 0$ (for every real number u).

To say that a matrix is positive means that the linear transformation it defines is positive; in matrix notation that is expressed by saying that

$$\sum_i \sum_j \alpha_{ij} \bar{\xi}_i \xi_j \geq 0$$

for every vector $\{\xi_1, \ldots, \xi_n\}$. Concrete examples:

$$\begin{pmatrix} 2 & 1 \\ 1 & 1 \end{pmatrix} \geq 0,$$

because

$$2|\xi_1|^2 + \bar{\xi}_1\xi_2 + \xi_1\bar{\xi}_2 + |\xi_2|^2 = |\xi_1 + \xi_2|^2 + |\xi_1|^2,$$

and (easier)

$$\begin{pmatrix} 1 & 0 \\ 0 & 2 \end{pmatrix} \geq 0;$$

but

$$\begin{pmatrix} 2 & 2 \\ 2 & 1 \end{pmatrix} \qquad \text{and} \qquad \begin{pmatrix} 0 & 1 \\ 1 & 0 \end{pmatrix}$$

are not positive, in both cases because the values of their quadratic forms at the vector $(-1, 1)$ are negative.

Some caution is called for in using the symbolism of ordering when complex numbers enter the picture. Everybody agrees that $3 > 2$, and almost everybody sooner or later agrees that $-2 > -3$. What about

$$3 + 4i > 2 + 4i$$

—is that true? What is true is that subtracting the right side from the left yields a positive number, but, nevertheless, most people feel uncomfortable with the inequality. Common sense suggests that such inequalities are best avoided, and experience shows that nothing is lost by avoiding them and using inequalities for real numbers (and Hermitian transformations) only. It is pertinent to recall that if $A \geq 0$, so that $(Ax, x) \geq 0$ for all x, then in particular (Ax, x) is real for all x, and therefore A must be Hermitian (Problem 130).

Problem 131. (a) *Is there an example of a positive matrix not all of whose entries are positive?*

(b) *Is there an example of a non-positive matrix all of whose entries are positive?*

(c) *Is the matrix* $\begin{pmatrix} 1 & 1 & 1 \\ 1 & 1 & 1 \\ 1 & 1 & 1 \end{pmatrix}$ *positive?*

(d) *Is the matrix* $\begin{pmatrix} 1 & 0 & 1 \\ 0 & 1 & 0 \\ 1 & 0 & 1 \end{pmatrix}$ *positive?*

(e) *For which values of α is the matrix* $\begin{pmatrix} \alpha & 1 & 1 \\ 1 & 0 & 0 \\ 1 & 0 & 0 \end{pmatrix}$ *positive?*

132   132. Positive inverses

Problem 133. *If a positive transformation is invertible, must its inverse also be positive?*

133   133. Perpendicular projections

The addition to a vector space of an inner product structure makes the theory more special (and therefore deeper); how does it affect the questions and answers about projections? The answer is that it affects those questions and answers quite a lot. The main reason for the change is that the inner product structure picks out a special one among the many complements that a subspace has, namely (obviously) the orthogonal complement. Recall that if M and N are complementary subspaces of a finite-dimensional vector space V, that is if

$$M \cap N = \{0\} \qquad \text{and} \qquad M + N = V,$$

so that every z in V is uniquely representable as

$$z = x + y$$

with x in M and y in N, then the projection E to M along N is the linear transformation defined by $Ez = x$ (Problem 72). If V is an inner product space, then the projection onto a subspace M along its orthogonal complement $M^\perp$ is called the **perpendicular projection** onto M. When extraordinary caution is needed, that perpendicular projection can be denoted by

P_M, but most frequently, when the context makes notational and terminological fuss unnecessary, even the word "perpendicular" is dropped and people just speak of the projection onto M.

It would be a pity if perpendicular projections were lost in the crowd of all possible projections. Is there a way of recognizing them?

Problem 133. *Which linear transformations on an inner product space are perpendicular projections?*

Comment. The question is vague—the first problem is to look for a non-vague interpretation of it. In slightly less vague terms the challenge is to look for an algebraic characterization of those linear transformations on an inner product space that are perpendicular projections.

134. Projections on $\mathbb{C} \times \mathbb{C}$ 134

Problem 134. *What can the matrix of a projection on $\mathbb{C} \times \mathbb{C}$ $(= \mathbb{C}^2)$ look like?*

Caution. "Matrix" here refers to a matrix with respect to an orthonormal basis.

135. Projection order   135

How is the geometric ordering of projections related to the algebraically defined ordering via positiveness (Problem 131)? The geometric ordering is one that suggests itself naturally: if E and F are projections with ranges M and N, then it is an almost irresisitible temptation to say that E is smaller than F in case M is smaller than N (meaning that $M \subset N$).

Problem 135. *If E and F are perpendicular projections, is there an implication in either direction between the statements*

$$E \leqq F$$

and

$$\operatorname{ran} E \subset \operatorname{ran} F?$$

136   ## 136. Orthogonal projections

How is the orthogonality of two subspaces reflected by their projections?

> **Problem 136.**   *If E and F are perpendicular projections, what alge-*
> *braic relation between E and F characterizes the geometric property*
> *of ran E being orthogonal to ran F?*

Comment.   Whatever the answer turns out to be, it seems reasonable to
use for that algebraic relation the same word as for subspaces. That is,
E and F shall be called orthogonal projections exactly in case ran E and
ran F are orthogonal subspaces.

137   ## 137. Hermitian eigenvalues

How does the spectrum of a transformation reflect its structure? Partial
answers to this vague question have occurred already, as for instance in the
statement that the nilpotence of A is equivalent to spec $A = \{0\}$ (Problem
111).

Another special sample question is this: can a real matrix (meaning
just that all its entries are real) have non-real eigenvalues? Yes. Example:
the eigenvalues of the matrix

$$\begin{pmatrix} 4 & 1 \\ -17 & 2 \end{pmatrix},$$

that is the roots of the quadratic equation

$$\lambda^2 - 6\lambda + 25 = 0,$$

are $3 + 4i$ and $3 - 4i$. (Is a general construction visible here? Can *every*
complex number be an eigenvalue of a real matrix?)

The concept of a "real matrix" is an artificial one—reality is not a
property of a linear transformation but of a conspiracy between a linear
transformation and a basis. (Such conspiracies are usually called matrices.)
What about the notion of reality that *is* a property of a linear transformation —does that behave differently?

> **Problem 137.**   *What can be said about the eigenvalues of Hermitian*
> *transformations? What about positive transformations?*

Question.   Are the conditions on the eigenvalues necessary, or sufficient,
or both?

138. Distinct eigenvalues   **138**

Here is a good question to ask that may not occur to everyone immediately but that does play a role in the theory: is there any relation between eigenvectors belonging to *different* eigenvalues? Example: if

$$A = \begin{pmatrix} 1 & 2 \\ 0 & 3 \end{pmatrix},$$

then $(1, 0)$ is an eigenvector with eigenvalue 1 and $(1, 1)$ is an eigenvector with eigenvalue 3, and there is no obviously discoverable relation between those two eigenvectors. Why not?

Problem 138. *Is there any relation between eigenvectors belonging to distinct eigenvalues of a Hermitian transformation?*

CHAPTER 9

NORMALITY

139. Unitary transformations

The three most obvious pleasant relations that a linear transformation on an inner product space can have to its adjoint are that they are equal (Hermitian), or that one is the negative of the other (skew), or that one is the inverse of the other (not yet discussed). The word that describes the last of these possibilities is unitary: that's what a linear transformation U is called in case it is invertible and $U^{-1} = U^*$. The definition can be expressed in a "less prejudiced" way as $U^*U = 1$—less prejudiced in the sense that it assumes less—but it is not clear that the less prejudiced way yields just as much. Does it?

> **Problem 139.** *If U is a linear transformation such that $U^*U = 1$, does it follow that $U^*U = 1$?*

140. Unitary matrices

It seems fair to apply the word "unitary" to a matrix in case the linear transformation it defines is a unitary one. (Caution: when language that makes sense in inner product spaces only is applied to matrices, the basis that establishes the correspondence between matrices and linear transformations had better be an orthonormal one.) A quick glance usually suffices to tell whether or not a matrix is Hermitian; is there a way to tell by looking at a matrix whether or not it is unitary? The following special cases are a fair test of any proposed answer to the general question.

Problem 140. (a) *For which values of α is $\begin{pmatrix} \alpha & 0 \\ 1 & 1 \end{pmatrix}$ a unitary matrix?*

(b) *For which values of α is $\begin{pmatrix} \alpha & \frac{1}{2} \\ -\frac{1}{2} & \alpha \end{pmatrix}$ a unitary matrix?*

(c) *Is there a 3×3 unitary matrix whose first row is a multiple of $(1, 1, 1)$?*

141. Unitary involutions

The two simplest properties that a linear transformation on an inner product space can have are being Hermitian or being unitary. A pleasantly and interestingly related property is being involutory. (A linear transformation U is called an **involution** or **involutory** if $U^2 = 1$.) What are the relations among these properties?

Problem 141. *What are the implication relations among the conditions $U^* = U$, $U^*U = 1$, and $U^2 = 1$?*

142. Unitary triangles

Problem 142. *Which unitary matrices are triangular?*

143. Hermitian diagonalization

Diagonal (or diagonalizable) matrices are pleasant to work with; it is always good to discover of a class of matrices under study that they can be diagonalized. (Remember, for instance, the diagonalization of permutations, Problem 103, and the diagonalization of transformations with distinct eigenvalues, Problem 106.)

Is every Hermitian transformation diagonalizable? Here is a phony proof that the anwer is yes. Given a Hermitian A, find a basis

$$\{e_1, e_2, \ldots\}$$

such that the matrix of A with respect to that basis is upper triangular (Problem 108). If, to be specific,

$$Ae_j = \sum_i \alpha_{ij} e_i,$$

with $\alpha_{ij} = 0$ whenever $i > j$, then

$$(Ae_j, e_k) = \sum_i \alpha_{ij}(e_i, e_k) = \sum_i \alpha_{ij}\delta_{ik} = \alpha_{kj}.$$

The Hermitian character of A implies that

$$\alpha_{jk} = (Ae_k, e_j) = (e_k, Ae_j) = \overline{(Ae_j, e_k)} = \overline{\alpha_{kj}},$$

and hence that $\alpha_{jk} = 0$ whenever $j > k$. Consequence:

$$a_{jk} = 0$$

whenever $j \neq k$, and therefore the matrix (α_{kj}) is diagonal.

What's wrong with that proof? Answer: it uses the orthonormality of the basis $\{e_1, e_2, \ldots\}$, and that's completely unjustified. All that the triangularization theorem says is that there exists *some* basis that does the job—it leaves open the question of whether or not there exists an orthonormal basis that does it.

It's easy enough to doctor up a basis so that with respect to it the matrix of some Hermitian transformation comes out triangular but not diagonal. For a concrete example, consider the linear transformation on $\mathbb{C}^2$ whose matrix is

$$A = \begin{pmatrix} 2 & -1 \\ -1 & 2 \end{pmatrix},$$

which is of course seen to be Hermitian by a casual glance. Consider now the (non-orthonormal) basis

$$\{(1, 1), (0, 1)\}$$

of $\mathbb{C}^2$. Since

$$A(1, 1) = (1, 1) \qquad \text{and} \qquad A(0, 1) = (-1, 2),$$

so that

$$A(1, 1) = 1 \cdot (1, 1) + 0 \cdot (0, 1)$$

and

$$A(0, 1) = -1 \cdot (1, 1) + 3 \cdot (0, 1),$$

it follows that the matrix of A with respect to that basis is

$$\begin{pmatrix} 1 & -1 \\ 0 & 3 \end{pmatrix}$$

—triangular but not diagonal.

Which Hermitian matrices are triangular? The answer is "just the diagonal ones"—that's what the phony proof above really proves. The original question, however, still stands.

Problem 143. *If A is a Hermitian transformation on a finite-dimensional complex inner product space, does there always exist an orthonormal basis with respect to which the matrix of A is diagonal?*

Comment. To say that a linear transformation is diagonalizable means that its matrix A (with respect to an arbitrary basis) is similar to a diagonal matrix, and that conclusion can be expressed by saying that there exists an invertible matrix T such that $T^{-1}AT$ is diagonal (Problem 86). In the same way, the assertion that a linear transformation A is diagonalizable with respect to an orthonormal basis can be expressed by saying that there exists a unitary matrix U such that U^*AU is diagonal. This assertion is an immediate consequence of its predecessor—all that has to be recalled is that a linear transformation that changes one orthonormal basis into another is necessarily unitary. The present question could therefore have been formulated this way: if A is a Hermitian matrix, does there always exist a unitary matrix U such that U^*AU is diagonal?

144   144. Square roots

If A is a linear transformation, does e^A make sense? Or $\cos A$? or $\sqrt{A}$? The general question is what sense it makes to form functions of a transformation, and whether it does any good to do so.

Yes, functions of transformations sometimes make sense and are sometimes very useful. The most typical and most important special case is the assertion that every invertible linear transformation (on a finite-dimensional complex vector space) has a square root (Problem 114). That is a matrix generalization of the statement that every complex number has a square root—a true statement that happens, however, not to be especially useful. The useful fact is that every positive number has a positive square root. Is there a good matrix generalization of that?

Problem 144. *How many positive square roots can a positive linear transformation on a finite-dimensional inner product space have?*

145. Polar decomposition

Does it make sense to speak of the absolute value of a linear transformation? An answer is suggested by the so-called polar representation

$$\alpha = \rho e^{i\theta}$$

of a complex number. The angle (= real number) θ is between 0 and 2π, and the number ρ is positive—and the latter is the absolute value of α. (It is worthy of note that except when $\alpha = 0$ the polar representation is unique.) Can such a representation be imitated by linear transformations?

What does imitation mean? A good imitation of a positive number is, presumably, a positive linear transformation. The equation

$$\overline{e^{i\theta}} \cdot e^{i\theta} = 1$$

suggests that a possible imitation of the angle part of the polar representation is a linear transformation whose product with its own adjoint is equal to the identity transformation—that is, a unitary transformation.

> **Problem 145.** *Which linear transformations A on a finite-dimensional inner product space are equal to products UP with U unitary and P positive?*

146. Normal transformations

Can every matrix be diagonalized? Why not? After all every A is equal to $B + iC$, with B and C Hermitian; why isn't it true that the diagonalizations of B and C separately yield a diagonalization of A? The answer is, of course, that diagonalization involved finding a suitable orthonormal basis, and there is no reason to expect that a basis that diagonalizes B will have the same effect on C.

All right then—are Hermitian transformations the only ones that can be diagonalized? Nonsense—of course not—for an example just consider a diagonal matrix such as

$$\begin{pmatrix} i & 0 \\ 0 & 1 \end{pmatrix}$$

that has a non-real entry. Emphasis: diagonalization in these questions means orthonormal diagonalization, or, from a different but equivalent point of view, unitary equivalence to a diagonal matrix; see the comment following Problem 143.

To discover the right middle course to steer between the extravagantly large class of all transformations and the relatively too restricted class of

Hermitian ones, the intelligent thing to do is to examine diagonalizable matrices and try to discover what makes them so. If D is a diagonal matrix, then so is D^*, and, therefore, D and D^* commute. That's a special property of diagonal matrices; does it survive under unitary equivalence? That is: if U is unitary and $A = U^*DU$, is it true that A and A^* commute? Sure:

$$
\begin{aligned}
AA^* &= U^*DUU^*D^*U \\
&= U^*DD^*U \qquad \text{(because U is unitary)} \\
&= U^*D^*DU \\
&= U^*D^*UU^*DU \qquad \text{(because U is unitary)} \\
&= A^*A.
\end{aligned}
$$

Linear transformations with the commutativity property here encountered $(A^*A = AA^*)$ are called **normal**, and while at this stage the connection between normality and diagonalizability is rather tenuous they deserve a look. The best thing to look at might be a property of Hermitian transformations that played an important role in the proof that they are diagonalizable (see Solution 143)—do normal transformations have that property?

> **Problem 146.** *Must eigenvectors belonging to distinct eigenvalues of a normal transformation (on a finite-dimensional inner product space) be orthogonal?*

147   147. Normal diagonalizability

Are normal transformations good imitations of Hermitian ones? The useful Hermitian lemma that helped to prove that Hermitian transformations are diagonalizable extends to the normal case (that's what Problem 146 did); does its consequence extend also?

> **Problem 147.** *If A is a normal transformation on a finite-dimensional complex inner product space, does there always exist an orthonormal basis with respect to which the matrix of A is diagonal?*

Comment. In less stuffy language the question is whether normal transformations are diagonalizable.

148. Normal commutativity   148

The most important kind of questions linear algebra can ask and answer concern the relation between the algebra and the geometry of linear transformations.

Here is an example: if two linear transformations A and B commute, and if λ is an eigenvalue of A with eigenvector x (so that $Ax = \lambda x$), then $ABx = BAx = \lambda Bx$. That is: the algebraic assumption of commutativity yields the geometric conclusion that each eigenspace of either transformation is invariant under the other. Is the converse true: does the geometric statement imply the algebraic one?

The answer is no. If, for instance, A and B are defined on $\mathbb{C}^2$ by

$$A = \begin{pmatrix} 0 & 1 \\ 0 & 0 \end{pmatrix} \quad \text{and} \quad B = \begin{pmatrix} 1 & 1 \\ 0 & 0 \end{pmatrix},$$

then the only eigenspace of A is the set of all vectors of the form $\langle \alpha, 0 \rangle$, and that set is invariant under B, but A and B do not commute.

Does the bad news become good if normality enters the picture?

Problem 148. *If the linear transformations A and B on an inner product space are such that every eigenspace of A is invariant under B, and if B is normal, does it follow that $AB = BA$? What if B is not necessarily normal, but A is?*

149. Adjoint commutativity   149

If A, B, and C are linear transformations such that A commutes with B and B commutes with C, does it follow that A commutes with C? In other words: is the relation of commutativity transitive? The suggestion is seen to be absurd almost sooner than it can be made: if $B = 0$, the assumptions are satisfied, but there is no reason on earth for the conclusion to follow.

The strongly negative nature of the answer adds interest to the study of special cases in an attempt to learn when the answer remains negative and when it just happens to be affirmative. As it turns out, moreover, some of those special cases are useful to know about, especially the ones that have to do with adjoints.

Problem 149. *If A, B, and C are linear transformations on an inner product space $\mathbb{V}$ such that A commutes with B and B commutes with C, and if two of the three are adjoints of one another, does it follow that A commutes with C?*

150   ## 150. Adjoint intertwining

The standard way to describe the fact that $AS = SA$ is, of course, to say that S commutes with A. An important generalization of commutativity also has a word associated with it, a less well-known word: if $AS = SB$, then S **intertwines** A and B. Commutativity is rare; intertwining is more common. When commutativity theorems can be extended to intertwining theorems, good applications usually follow.

A good commutativity theorem is the one about adjoint commutativity (Solution 143): if A is normal and $AS = SA$, then $A^*S = SA^*$. (Caution: the assumption of normality is essential in this implication. The point is that, no matter what A is, S can always be taken to be A itself, and the commutativity assumption is satisfied; if, however, A is not normal, then S, that is A, will not commute with A^*.)

Does the adjoint commutativity theorem have an intertwining version? That is: is it sometimes possible to start with $AS = SB$ and infer that $A^*S = SB^*$? The cautionary example above (A not normal and $S = A^*$) shows that the implication is surely not always true, but there are worse examples than that. Indeed, if

$$A = \begin{pmatrix} 1 & 0 \\ 0 & 0 \end{pmatrix}, \quad B = \begin{pmatrix} 0 & 0 \\ 1 & 0 \end{pmatrix}, \quad \text{and} \quad S = B,$$

then $AS = SB = 0$, whereas $A^*S \, (= AS) = 0$ and

$$SB^* = \begin{pmatrix} 0 & 0 \\ 0 & 1 \end{pmatrix}$$

The reason this example is worse is that one of the constituents, namely A, is normal. Interchanging A and B and replacing S by

$$\begin{pmatrix} 0 & 1 \\ 0 & 0 \end{pmatrix}$$

yields an example in which B is normal. Are there bad examples like this even when both A and B are required to be normal?

Problem 150. *If A and B are normal linear transformations on a finite-dimensional inner product space, and if $AS = SB$, does it always follow that $A^*S = SB^*$?*

151   ## 151. Normal products

The product of two self-adjoint transformations may fail to be self-adjoint, but if they commute then the product must be self-adjoint too. The proof is a trivial piece of algebra: if $A = A^*$ and $B = B^*$, then $(AB)^* = B^*A^* =$

BA. Does the result extend to normal transformations? That is, if A and B are normal and commutative, does it follow that AB is normal? The answer is yes, but it is somewhat more subtle. One way to prove it is to use the adjoint commutativity theorem. In view of the assumption $AB = BA$, the normality of A implies that $A^*B = BA^*$. It follows that all four transformations A, A^*, B, and B^* commute with one another, and hence that

$$(AB)(AB)^* = ABB^*A^* = B^*A^*AB = (AB)^*(AB).$$

For self-adjoint transformations, there is a converse theorem: if the product of two of them turns out to be self-adjoint, then they must commute. The proof is obvious: if

$$A = A^*, \ B = B^*, \quad \text{and} \quad AB = (AB)^*,$$

then

$$AB = B^*A^* = BA.$$

Does the converse extend to normal transformations? That question takes most people a few more seconds to answer than the self-adjoint one; the reason is that the answer is different. No, normal transformations with a normal product do not have to commute. Example: take any two non-commutative unitary transformations their product is unitary, therefore normal.

There is still another twist on questions of this type. If A and B are normal and commutative, so that AB is normal, then, of course, BA is normal too (because it is equal to AB). Is commutativity needed to draw that conclusion?

Problem 151. *If the linear transformations A and B on a finite-dimensional inner product space are such that A, B, and AB are normal, does it follow that BA is normal?*

152. Functions of transformations   152

Polynomials of a linear transformation make sense, and (as Problem 144 indicated) sometimes they can be used to define more complicated functions of linear transformations. If, for instance, A is a normal transformation on finite-dimensional vector space, then everything works smoothly. For an arbitrary function f whose domain is at least as large as spec A (the set of eigenvalues of A) a transformation $f(A)$ can be defined by finding a polynomial p that agrees with f on spec$A$ and writing $f(A) = p(A)$.

The process of forming functions of transformations is a powerful tool, and what makes it so is that the algebraic and analytic properties of such functions mirror faithfully the corresponding properties of numerical functions. Thus, for instance, if A is normal and $f(z) = \bar{z}$ (complex conjugate), then $f(A) = A^*$ (adjoint); if, in addition, A is invertible and $f(z) = \frac{1}{z}$ whenever $z \neq 0$, then $f(A) = A^{-1}$; and if A is positive and $f(z) = \sqrt{z}$ whenever $z \geq 0$, then $f(A) = \sqrt{A}$ (the unique positive square root).

Some functions of linear transformations demand attention even in the absence of normality (and sometimes even in the absence of any inner product structure in the underlying vector space); conspicuous among them are $A \mapsto A^*$ and, for invertible transformations, $A \mapsto A^{-1}$. Can the study of their behavior, good or bad, be reduced to the study of polynomials in A?

> **Problem 152.** (a) *If A is a linear transformation on a finite-dimensional inner product space, does there necessarily exist a polynomial p such that $p(A) = A^*$?*
>
> (b) *If A is an invertible linear transformation on a finite-dimensional vector space, does there necessarily exist a polynomial p such that $p(A) = A^{-1}$?*

153. Gramians

153

How easy is it to recognize that a matrix is positive? ("Positive" is used here in the quadratic form sense, as defined in Problem 131.) So, for example is either of the matrices

$$\begin{pmatrix} 5 & 12 \\ 12 & 25 \end{pmatrix} \quad \text{and} \quad \begin{pmatrix} 5 & 11 \\ 11 & 25 \end{pmatrix}$$

positive? The answer is no for the first one (note that its determinant is negative) and yes for the second, but no single answer like that is important. An effectively computable test for positiveness would be pleasant to have, but, failing that, even an abstract characterization would be welcome. Prediction is sometimes more important in mathematics (if you do so and so, you'll get a positive matrix) than recognition (I don't know what you did, but you ended up with a positive matrix).

The challenge to write down a hundred different matrices is a trivial one, even if "different" is intended to suggest radical differences, not just trivial ones such as are possessed by different scalar multiples of one matrix. It's just as easy to write down a hundred different Hermitian matrices. Is it easy to write down a hundred radically different 4×4 positive matrices?

Yes, it's easy. Consider any four vectors x_1, x_2, x_3, x_4 in $\mathbb{C}^4$ and form the matrix $A = (\alpha_{ij})$ whose entry α_{ij} in row i and column j is the inner product

$$(x_i, x_j).$$

Assertion: A is positive. For the proof what must be shown is that if $v = (v_1, v_2, v_3, v_4)$ is any vector in $\mathbb{C}^4$, then $(Av, v) \geq 0$. The proof is a straightforward computation. If $Au = v = (v_1, v_2, v_3, v_4)$, so that

$$v_j = \sum_i (x_i, x_j) u_i,$$

then

$$(Au, u) = (v, u) = \sum_j \sum_i (x_i, x_j) u_i \overline{u}_j$$

$$= \sum_j \left(\sum_i (x_i u_i, x_j) \right) \overline{u}_j = \left(\sum_i x_i u_i, \sum_j x_j u_j \right)$$

$$= \left\| \sum_i x_i u_i \right\|^2 \geq 0.$$

A matrix such as the A here defined is called a **Gramian**, or, more specifically, the Gramian of the vectors x_1, x_2, x_3, x_4. What was just proved is that every Gramian is positive. To what extent is the converse of that statement true?

Problem 153. *Which positive matrices are Gramians?*

154. Monotone functions   154

Problem 154. *If A and B are positive transformations on a finite-dimensional inner product space such that*

$$A \leq B,$$

does it follow that

$$A^2 \leq B^2?$$

How about

$$\sqrt{A} \leq \sqrt{B}?$$

Comment. The question could have been phrased by asking whether square and square root are monotone functions of transformations.

The restriction to positive transformations is advisable, isn't it? After all the function $x \mapsto x^2$ on the real line is not a monotone function (although it is true that $-3 \leq 2$, it is not true that $9 \leq 4$), but its restriction to the positive part of the line is. In other words, for positive real numbers the answers are yes, which can be read as saying that for linear transformations on a vector space of dimension 1 the answers are yes; the question is whether the answers remain yes for spaces of higher dimensions.

155. Reducing ranges and kernels

Invariant subspaces for a linear transformation A on a finite-dimensional vector space V are easy enough to find; the difficulty, usually, is to prove that they are different from the trivial subspace $\{0\}$ and the improper subspace V. If the vector space is equipped with an inner product, it becomes natural to look for reducing subspaces (subspaces invariant under both A and its adjoint A^*); they are harder to find. Thus, for instance, both ran A and ker A are invariant under A, but they may well fail to be reducing. An easy example is furnished by the 2×2 matrix that is an almost universal counterexample. If

$$A = \begin{pmatrix} 0 & 1 \\ 0 & 0 \end{pmatrix}$$

is regarded as a linear transformation on the space $\mathbb{C}^2$, then both ran A and ker A are equal to the set of all vectors of the form $\langle x, 0 \rangle$, but neither one of them is invariant under A^*. Does something more useful happen if intersections and spans of ranges and kernels are allowed?

> **Problem 155.** *If A is a linear transformation on a finite-dimensional inner product space, which of the subspaces obtained from* ran A, *ker A,* ran A^*, *and* ker A^* *by the formation of intersections and spans necessarily reduce A?*

156. Truncated shifts

The larger the domain of a linear transformation, the more likely it is to have invariant and reducing subspaces. The easiest example of an irreducible linear transformation on a vector space V (that is, a transformation with no reducing subspaces other than $\{0\}$ and V) is the one induced on

$\mathbb{C}^2$ by the matrix

$$\begin{pmatrix} 0 & 0 \\ 1 & 0 \end{pmatrix}$$

(compare Problem 74). That example has natural generalizations to higher dimensions, such as

$$\begin{pmatrix} 0 & 0 & 0 & 0 & 0 \\ 1 & 0 & 0 & 0 & 0 \\ 0 & 1 & 0 & 0 & 0 \\ 0 & 0 & 1 & 0 & 0 \\ 0 & 0 & 0 & 1 & 0 \end{pmatrix}$$

on $\mathbb{C}^5$ for instance.

Another way of describing the same phenomenon is to consider any basis $\{x_1, \ldots, x_n\}$ in any vector space $\mathbb{V}$ of dimension n, say, and let A be the **truncated shift** that sends (shifts) x_j to x_{j+1} ($1 \leq j < n$) and sends x_n to 0. The matrix of A with respect to the basis $\{x_1, \ldots, x_n\}$ is just like the one displayed above (with the role of 5 being played by n). Note that $A^{n-1} \neq 0$ but $A^n = 0$; the transformation A is nilpotent of index exactly n.

When n is large, the domain of the linear transformation A is larger than it is when $n = 2$. How effective is the enlargement as a producer of invariant and reducing subspaces?

Problem 156. *How many invariant subspaces does a truncated shift have? How many reducing subspaces does it have?*

157. Non-positive square roots   157

Positive linear transformations are not the only ones for which the problem of square roots makes sense. Every linear transformation has a square, and that shows that many transformations have square roots even though they have nothing to do with positiveness.

Does the matrix

$$A = \begin{pmatrix} 0 & 1 & 0 \\ 0 & 0 & 1 \\ 0 & 0 & 0 \end{pmatrix}$$

have a square root? No, it does not. If it happened that $B^2 = A$, then (since A is nilpotent of index 3), it would follow that $B^6 = 0$, and that would imply that the minimal polynomial of B is a power of λ. Since the degree cannot be greater than 3 (the degree of the characteristic polynomial), it follows that $B^3 = 0$, whence $A^2 = B^4 = 0$, which is a contradiction. This sort of argument can be used often, and it implies, for instance, that no truncated shift can have a square root. There are, however, many transformations about which the argument gives no information, however much they may resemble truncated shifts.

Problem 157. *Does the matrix*

$$A = \begin{pmatrix} 0 & 1 & 0 \\ 0 & 0 & 0 \\ 0 & 0 & 0 \end{pmatrix}$$

have a square root?

158. Similar normal transformations

158

The "right" relation of equivalence between linear transformations on an abstract vector space is similarity; from the point of view of the structure of vector spaces similar transformations are indistinguishable. Inner product spaces have a richer structure; a relation (such as similarity) that ignores that structure does not give the best information. The "right" relation between linear transformations on inner product spaces is unitary similarity (often, somewhat misleadingly, called unitary equivalence).

What information does unitary similarity give that ordinary similarity does not? Since the rich structure of inner product spaces consists of numerical ways of measuring sizes (angles and lengths), the most natural answer to the question is "size". Consider, for instance, the matrices

$$A = \begin{pmatrix} 1 & 1 \\ 0 & 0 \end{pmatrix} \quad \text{and} \quad B = \begin{pmatrix} 1 & 0 \\ 0 & 0 \end{pmatrix}.$$

Computational verification (or a moment's thought about known elementary sufficient conditions for similarity) will establish that A and B are similar. The unit vectors $\langle 1, 0 \rangle$ and $\langle \frac{1}{\sqrt{2}}, -\frac{1}{\sqrt{2}} \rangle$ are eigenvectors of A; since their inner product is $\frac{1}{\sqrt{2}}$, the angle between them is $\frac{\pi}{4}$. The unit vectors $\langle 1, 0 \rangle$ and $\langle 0, 1 \rangle$ are eigenvectors of B; the angle between them is $\frac{\pi}{2}$. The norm of A is $\frac{1}{\sqrt{2}}$ (compute the larger of the square roots of the eigenvalues of A^*A); the norm of B is 1. Conclusion: A and B, though similar, are certainly not unitarily similar; every measure of size indicates a difference between them.

The linear transformation defined by A is pleasantly related to the inner product structure: A is normal. The transformation defined by B is not normal. These facts establish once again that A and B couldn't possibly be unitarily similar, and they suggest a search for an even more powerful counterexample.

Problem 158. *Do there exist two normal transformations that are similar but not unitarily similar?*

159. Unitary equivalence of transposes 159

The hardest and most important problem about many mathematical structures is to determine when two of them are the "same". One such problem in linear algebra is to find out when two linear transformations are similar, or, in case the underlying vector space comes endowed with an inner product, to find out when two linear transformations are unitarily equivalent. There exists something called elementary divisor theory, which frequently yields satisfactory answers to questions of the first of these types, good enough for explicit calculations. The second type of question is usually much harder.

Suppose, for instance, that A is a linear transformation on a finite-dimensional inner product space. The transformations A and A^* have much in common, especially as far as sizes are concerned. A trivial observation along these lines is that the geometric norms of A and A^* are the same. Since $\det A^*$ is the complex conjugate of $\det A$, it follows that if λ is an eigenvalue of A, then $\bar{\lambda}$ (the complex conjugate of λ) is an eigenvalue of A^*, and hence, in particular, that $|\lambda| = |\bar{\lambda}|$; the sizes of the eigenvalues of the two transformations are the same.

Could it be that A and A^* are always unitarily equivalent? No, that's absurd: complex conjugation is in the way. Unitarily equivalent transformations have the same eigenvalues, but it can perfectly well happen that a linear transformation A has λ for an eigenvalue but not $\bar{\lambda}$; in that case A and A^* cannot possibly be unitarily equivalent.

The adjoint of a matrix is the complex conjugate of its transpose. If the conjugation step is omitted, does the difficulty disappear?

Problem 159. *Is every matrix unitarily equivalent to its transpose?*

To interpret the question, identify each $n \times n$ matrix with the linear transformation that it induces on the inner product space $\mathbb{C}^n$. Note that, by the elementary divisor theory referred to above, if unitary equivalence is replaced by similarity, then the answer is yes: a matrix A and its transpose A' obviously have the same elementary divisors.

160. Unitary and orthogonal equivalence

If two real matrices are complex similar, then they are real similar—a construction to prove that was carried out in Problem 89. What happens (special case) when two real matrices are unitarily equivalent?

Is it possible for two real matrices to be unitarily equivalent in a complex way? That is: do there exist real matrices A and B and a complex unitary matrix U such that $U^*AU = B$? The question is not sharp enough; it admits trivial answers such as $A = B = 0$ and $U = 1$. An answer like that would be described as trivial by everyone, but just exactly what deserves to be called non-trivial?

It's not enough to insist that U be genuinely complex; trivial answers still exist. Example: let U be an arbitrary "genuinely complex" unitary matrix and take $A = B = 1$.

Here is an example of a different kind: take

$$A = \begin{pmatrix} 0 & 0 \\ 0 & 1 \end{pmatrix}, \quad B = \begin{pmatrix} 1 & 0 \\ 0 & 0 \end{pmatrix}, \quad S = \begin{pmatrix} 0 & i \\ 1 & 0 \end{pmatrix},$$

and verify that $S^*AS = (S^{-1}AS) = B$. The technique of Problem 89 shows that if P and Q are the real and imaginary parts of S,

$$S = P + iQ,$$

then there exist real numbers λ such that $T = P + \lambda Q$ is invertible, and for any such T it is true that $T^{-1}AT = B$. The matrices T are of the form $\begin{pmatrix} 0 & \lambda \\ 1 & 0 \end{pmatrix}$, and while it's quite possible for such a matrix to be unitary, that pleasant phenomenon occurs only rarely.

Still another example: take an arbitrary real A and an arbitrary real V that is unitary, define $B = V^*AV$, and write $U = \alpha V$, where α is an arbitrary complex number of modulus 1. This latter example leads to others that cannot be spotted so easily. Form the direct sum of two of them, and then transform every transformation that enters by a sufficiently complicated looking real unitary transformation W. (To transform means to replace X by W^*XW.) The new U (a complicated real transform of the direct sum of a couple of complex scalar multiples of two other real U's) looks as genuinely complex as anything can, but it succeeds in transforming the new real A to the new real B.

Are all these examples "artificial"? The following problem is a precise way of putting this somewhat vague question.

Problem 160.   *If two real matrices are (complex) unitarily equivalent, does it follow that they are also (real) orthogonally equivalent?*

161. Null convergent powers

If a finite-dimensional (real or complex) vector space V does not come equipped with an inner product, then there is no natural notion of distance for vectors in it or linear transformations on it, but there are many equally good "unnatural" ones. To get one of them, choose a basis, and, using it, establish an isomorphism between V and $\mathbb{C}^n$ for the appropriate n. The isomorphism transplants the natural inner product structure (and hence the analytic and topological structures, such as distance and convergence) from $\mathbb{C}^n$ to V. It is good to know (and not especially difficult to prove) that while the distances obtained this way may be very different from one another, the topologies are all the same. If, in particular, $\{A_n\}$ is a sequence of linear transformations on a finite-dimensional (real or complex) vector space, then it might make sense to ask whether the sequence converges to something. It might make sense, but it is usually not worth the trouble to ask such questions; it is simpler and more honest to restrict attention to matrices in the first place.

As an illustration of the kind of analytic question that it is often useful to ask about matrices, consider this one: if a complex matrix A is such that $A^n \to 0$ as $n \to \infty$, what can be said about the eigenvalues of A? Answer: every one of them must be strictly less than 1 in absolute value. Reason: if $Ax = \lambda x$, then $A^n x = \lambda^n x$, and therefore, provided only that $x \neq 0$, it follows that $\lambda^n \to 0$. Is the converse true?

Problem 161. *If every eigenvalue of a complex matrix A is strictly less than 1 in absolute value, does it follow that $A^n \to 0$ as $n \to \infty$?*

162. Power boundedness

The powers of a linear transformation A can exhibit several kinds of good and bad behavior. Solution 161 discussed the possibilities $\|A^n\| \to 0$ (good) and $\|A^n\| \to \infty$ (bad). A good possibility between those two extremes is that $\|A^n\|$, as a function of n, is bounded. The possibility is important enough that it deserves to be given its name even before it is adequately studied and characterized; a linear transformation with that property is called **power bounded**. Which ones are?

If $\|A\| \leq 1$ (that can be expressed by the usual technical term: A is a contraction), then surely $\|A^n\| \leq 1$ for all n. Contractions must be power bounded; can anything else be? Is it possible to have a transformation A

with $\|A\| = 2$ and $\|A^n\|$ bounded? Yes, it is. Example:

$$A = \begin{pmatrix} 0 & 2 \\ 0 & 0 \end{pmatrix}.$$

The point is, of course, that $A^n = 0$ whenever $n \geq 2$; the presence of the entry 2 doesn't affect any of the powers of A after the first.

Examples like the last might tend to shake one's faith in the possibility of a good connection between power boundedness and contractions; the next comment might restore some of that faith. If a linear transformation A is not a contraction but is similar to one,

$$A = S^{-1}CS, \quad \|C\| \leqq 1,$$

then

$$\|A^n\| = \|S^{-1}C^n S\| \leqq \|S^{-1}\| \cdot \|C^n\| \cdot \|S\| \leqq \|S^{-1}\| \cdot \|S\|$$

so that A is power bounded. How likely is that faith to be shaken?

Problem 162. *Is every power bounded linear transformation on a finite-dimensional inner product space necessarily similar to a contraction?*

163 163. Reduction and index 2

The easy examples of irreducible transformations turn out to be nilpotent (see Solution 156); it almost looks as if every nilpotent transformation must be irreducible. That's not true, of course: for a counterexample just form the direct sum of two nilpotent transformations to get one that is still nilpotent but definitely not irreducible. Contemplation of such examples suggests a question: are there relations between the index of nilpotence and the dimension of the space that either force or prevent irreducibility? The answer is not obvious even for the lowest possible index.

Problem 163. *Is there an irreducible nilpotent transformation of index 2 on a space of dimension greater than 2?*

164 164. Nilpotence and reduction

If a linear transformation A on a vector space V of dimension n is nilpotent of index k, what can be said about the existence of reducing subspaces for A? If $k < n$, then A can be reducible (trivial, form direct sums); if $k = n$, then A cannot be reducible (that is in effect what Solution 151 shows); and if $k = 2$ and $n > 2$, then A must be reducible (see Solution 163). What can be guessed from that much evidence? A possible guess is that if $k < n$, then A is necessarily reducible. Is that true?

Problem 164. *Can a nilpotent linear transformation A of index 3 on an inner product space V of dimension 4 be irreducible?*

HINTS

Chapter 1. Scalars

Hint 1. Write down the associative law for $\boxed{+}$.

Hint 2. Same as for Problem 1: substitute and look.

Hint 3. How could it be?

Hint 4. Note the title of the problem.

Hint 5. The affine transformation of the line associated with the real numbers α and β is the one that maps each real number ξ onto $\alpha\xi + \beta$.

Hint 6. Does it help to think about 2×2 matrices? If not, just compute.

Hint 7. Let r_6 (for "reduce modulo 6") be the function that assigns to each non-negative integer the number that's left after all multiples of 6 are thrown out of it. Examples: $r_6(8) = 2$, $r_6(183) = 3$, and $r_6(6) = 0$. Verify that the result of multiplying two numbers and then reducing modulo 6 yields the same answer as reducing them first and then multiplying the results modulo 6. Example: the ordinary product of 10 and 11 is 110, which reduces modulo 6 to 2; the reduced versions of 10 and 11 are 4 and 5, whose product modulo 6 is $20 - 18 = 2$.

169

Hint 8.   The answer may or may not be easy to guess, but once it's correctly guessed it's easy to prove. The answer is yes.

Hint 9.   Not as a consequence but as a coincidence the answer is that the associative ones do and the others don't.

Hint 10.   To find $\dfrac{1}{\alpha + i\beta}$, multiply both numerator and denominator by $\alpha - i\beta$. The old-fashioned name for the procedure is "rationalize the denominator".

Hint 11.   The unit is $\langle 1, 0 \rangle$. Caution: non-commutativity.

Hint 12.   The unit is $\begin{pmatrix} 1 & 0 \\ 0 & 1 \end{pmatrix}$.

Hint 13.   (a) and (b). Are the operations commutative? Are they associative? Do the answers change if $\mathbb{R}_+$ is replaced by $[0, 1]$?
   (c) Add $(-x)$ to both sides of the assumed equation.

Hint 14.   An affine transformation $\xi \mapsto \alpha\xi + \beta$ with $\alpha = 0$ has no inverse; a matrix with

$$\begin{pmatrix} \alpha & \beta \\ \gamma & \delta \end{pmatrix}$$

$\alpha\delta - \beta\gamma = 0$ has no inverse.
   The integers modulo 3 form an additive group, and so do the integers modulo anything else. Multiplication is subtler. Note: the number 6 is not a prime, but 7 is.

Hint 15.   If the underlying set has only two elements, then the answer is no.

Hint 16.   Use both distributive laws.

Hint 17.   In the proofs of the equations the distributive law must enter directly or indirectly; if not there's something wrong. The non-equations are different: one of them is true because that's how language is used, and the other is not always true.

Hint 18.   Think about the integers modulo 5.

Hint 19. The answer is yes, but the proof is not obvious. One way to do it is by brute force; experiment with various possible ways of defining $+$ and $\times$, and don't stop till the result is a field.

A more intelligent and more illuminating way is to think about polynomials instead of integers. That is: study the set $\mathbb{P}$ of all polynomials with coefficients in a field of two elements, and "reduce" that set "modulo" some particular polynomial, the same way as the set $\mathbb{Z}$ of integers is reduced modulo a prime number p to yield the field $\mathbb{Z}_p$. If the coefficient field is taken to be $\mathbb{Q}$ and the modulus is taken to be $x^2 - 2$, the result of the process is (except for notation) the field $\mathbb{Q}(\sqrt{2})$. If the coefficient field is taken to be $\mathbb{Z}_2$ and the modulus is taken to be an appropriately chosen polynomial of degree 2, the result is a field with four elements. Similar techniques work for $8, 16, 32, \ldots$ and $9, 27, 81, \ldots$, etc.

Chapter 2. Vectors

Hint 20. The 0 element of any additive group is characterized by the fact that $0 + 0 = 0$. How can it happen that $\alpha x = 0$? Related question worth asking: how can it happen that $\alpha x = x$?

Hint 21. (1) The scalar distributive law; (2) the scalar identity law; (3) the vector distributive law; (4) none; (5) the associative law; (6) none.

Hint 22. Can you solve two equations in two unknowns?

Hint 23. (a): (1), (2), and (4); (b): (2) and (4).

Hint 24. (a) Always. (b) In trivial cases only. Draw pictures. Don't forget finite fields. If it were known that a vector space over an infinite field cannot be the union of any two of its proper subspaces, would it follow that it cannot be the union of any finite number? In any event: whenever $\mathbb{M}_1$ and $\mathbb{M}_2$ are subspaces, try to find a vector x in $\mathbb{M}_1$ but not in $\mathbb{M}_2$ and a vector y not in $\mathbb{M}_1$, and consider the line through y parallel to x.

Hint 25. (a) Can it be done so that no vector in either set is a scalar multiple of a vector in the other set? (b) Can you solve three equations in three unknowns?

Hint 26. Is it true that if x is a linear combination of y and something in $\mathbb{M}$, then y is a linear combination of x and something in $\mathbb{M}$?

Hint 27.   (a) No; that's easy. (b) Yes; that's very easy. (c) No; and that takes some doing, or else previous acquaintance with the subject. (d) Yes; and all it requires is the definition, and minimum acquaintance with the concept of polynomial.

The reader should be aware that the problem was phrased in incorrect but commonly accepted mathematese. Since "span" is a way to associate a subspace of $\mathbb{V}$ with each subset of $\mathbb{V}$, the correct phrasing of (a) is: "is there a singleton that spans $\mathbb{R}^2$?" Vectors alone, or even together with others (as in (b) and (c)), don't span subspaces; spanning is done by *sets* of vectors. The colloquialism does no harm so long as its precise meaning is not forgotten.

Hint 28.   Note that since $\mathbb{L} \cap (\mathbb{L} \cap \mathbb{N}) = \mathbb{L} \cap \mathbb{N}$, the equation is a special case of the distributive law. The answer to the question is yes. The harder half to prove is that the left side is included in the right. Essential step: subtract.

Hint 29.   Look at pictures in $\mathbb{R}^2$.

Hint 30.   No.

Hint 31.   Just look for the correct term to transpose from one side of the given equation to the other.

Hint 32.   Use Problem 31.

Chapter 3. Bases

Hint 33.   Examine the case in which $\mathbb{E}$ consists of a single vector.

Hint 34.   It is an elementary fact that if $\mathbb{M}$ is an m-dimensional subspace of an n-dimensional vector space $\mathbb{V}$, then every complement of $\mathbb{M}$ has dimension $n - m$. It follows that if several subspaces of $\mathbb{V}$ have a simultaneous complement, then they all have the same dimension. Problem 24 is relevant.

Hint 35.   (a) Irrational? (b) Zero?

Hint 36. $\sqrt{2}$?

Hint 37. (a) $\alpha - \beta$? (b) No room. Keep in mind that the natural coefficient field for $\mathbb{C}^2$ is $\mathbb{C}$.

Hint 38. Why not? One way to answer (a) is to consider two independent vectors each of which is independent of $(1, 1)$. One way to answer (b) is to adjoin $(0, 0, 1, 0)$ and $(0, 0, 1, 1)$ to the first two vectors and $(1, 0, 0, 0)$ and $(1, 1, 0, 0)$ to the second two.

Hint 39. (a) Too much room. (b) Are there any?

Hint 40. How many vectors can there be in a maximal linearly independent set?

Hint 41. What information about $\mathbb{V}^{\text{real}}$ does a basis of $\mathbb{V}$ give?

Hint 42. Can a basis for a proper subspace span the whole space?

Hint 43. Use Problems 32 and 33. Don't forget to worry about independence as well as totality.

Hint 44. Given a subspace, look for an independent set in it that is as large as possible.

Hint 45. Note that finite-dimensionality was not explicitly assumed. Recall that a possibly infinite set is called dependent exactly when it has a *finite* subset that is dependent. Contrariwise, a set is independent if every finite subset of it is independent. As for the answer, all it needs is the definitions of the two concepts that enter.

Hint 46. It is tempting to apply a downward induction argument, possibly infinite. People who know about Zorn's lemma might be tempted to use it, but the temptation is not likely to lead to a good result. A better way to settle the question is to use Problem 45.

Hint 47. Omit one vector, express it as a linear combination of remaining vectors, and then omit a new vector different from all the ones used so far.

Hint 48. A few seconds of geometric contemplation will reveal a relatively independent subset of $\mathbb{R}^3$ consisting of 5 vectors (which is $n + 2$ in this case). If, however, $\mathbb{F}$ is the field $\mathbb{Z}_2$ of integers modulo 2, then a few seconds of computation will show that no relatively independent subset of

$\mathbb{F}^3$ can contain more than 4 vectors. Why is that? What is the big difference between $\mathbb{F}$ and $\mathbb{R}$ that is at work here?

Having thought about that question, proceed to use induction.

Hint 49. A slightly modified question seems to be easier to approach: how many ordered bases are there? For the answer, consider one after another questions such as these: how many ways are there of picking the first vector of a basis?; once the first vector has been picked, how many ways are there of picking the second vector?; etc.

Hint 50. The answer is $n + m$.

Hint 51. There is a sense in which the required constructions are trivial: no matter what $\mathbb{V}$ is, let $\mathbb{M}$ be $\mathbb{O}$ and let $\mathbb{N}$ be $\mathbb{V}$. In that case $\mathbb{V}/\mathbb{M}$ is the same as $\mathbb{V}$ and $\mathbb{V}/\mathbb{N}$ is a vector space with only one element, so that, except for notation, it is the same as the vector space $\mathbb{O}$. If $\mathbb{V}$ was infinite-dimensional to begin with, then this construction provides trivial affirmative answers to both parts of the problem. Many non-trivial examples exist; to find one, consider the vector space $\mathbb{P}$ of polynomials (over, say, the field $\mathbb{R}$ of real numbers).

Hint 52. The answer is $n - m$. Start with a basis of $\mathbb{M}$, extend it to a basis of $\mathbb{V}$, and use the result to construct a basis of $\mathbb{V}/\mathbb{M}$.

Hint 53. If M and N are finite subsets of a set, what relation, if any, is always true among the numbers $\operatorname{card} M$, $\operatorname{card} N$, $\operatorname{card}(M \cup N)$, and $\operatorname{card}(M \cap N)$?

Chapter 4. Transformations

Hint 54. Squaring scalars is harmless; trying to square vectors or their parts is what interferes with linearity.

Hint 55. (1) Every linear functional except one has the same range.

(2) Compare this change of variables with the one in Problem 54 (1 (b)).

(3) How many vectors does the range contain?

(4) Compare this transformation with the squaring in Problem 54 (2 (b)).

(5) How does this weird vector space differ from $\mathbb{R}^1$?

Hint 56. (1) What do you know about a function if you know that its indefinite integral is identically 0?

(2) What do you know about a function if you know that its derivative is identically 0?

(3) Solve two "homogeneous" equations in two unknowns. ("Homogeneous" means that the right sides are 0.)

(4) When is a polynomial 0?

(5) What happens to the coordinate axes?

(6) This is an old friend.

Hint 57. (1) What could possibly go wrong?

(2) Neither transformation goes from $\mathbb{R}^2$ to $\mathbb{R}^3$.

(3) What happens when both S and T are applied to the constant polynomial 1? What about the polynomial x?

(4) Do both products make sense?

(5) What happens when both S and T are applied to the constant polynomial 1? What about the polynomial x? What about x^2?

(6) There is nothing to do but honest labor.

Hint 58. Consider complements: for left divisibility, consider a complement of ker B, and for right divisibility consider a complement of ran B.

Hint 59. (1) If the result of applying a linear transformation to each vector in a total set is known, then the entire linear transformation is known.

(2) How many powers does A have?

(3) What is $A^2 x$?

Hint 60. Make heavy use of the linearity of T.

Hint 61. (1) What is the kernel? (2) What is T^2? (3) What is the range?

Hint 62. Choose the entries (γ) and (δ) closely related to the entries (α) and (β).

Hint 63. Direct sum, equal rows, and similarity.

Hint 64. The "conjecturable" answer is too modest; many of the 0's below the diagonal can be replaced by 1's without losing invertibility.

Hint 65. Start with a basis of non-invertible elements and make them invertible.

Hint 66. To say that a linear transformation sends some independent set onto a dependent set is in effect the same as saying that it sends some non-zero vector onto 0.

Hint 67. If the dimension is 2, then there is only one non-trivial permutation to consider.

Hint 68. There is nothing to do but use the general formula for matrix multiplication. It might help to try the 2×2 case first.

Hint 69. Look at diagonal matrices.

Hint 70. Yes.

Hint 71. Consider differentiation.

Hint 72. If E is a projection, what is E^2?

Hint 73. Multiply them.

Hint 74. No and no. Don't forget to ask and answer some other natural questions in this neighborhood.

Chapter 5. Duality

Hint 75. If there were such a scalar, would it be uniquely determined by the prescribed linear functionals ξ and η?

Hint 76. Use a basis of $\mathbb{V}$ to construct a basis of $\mathbb{V}'$.

Hint 77. This is very easy; just ask what information the hypothesis gives about the kernel of T.

Hint 78. Does it help to assume that $\mathbb{V}$ is finite-dimensional?

Hint 79. If $\mathbb{V}$ is $\mathbb{R}^5$ and $\mathbb{M}$ is the set of all vectors of the form $(\xi_1, \xi_2, \xi_3, 0, 0)$, what is the annihilator of $\mathbb{M}$?

Hint 80. What are their dimensions?

Hint 81. Surely there is only one sanely guessable answer.

Hint 82. Can kernels and ranges be used?

Hint 83. There is no help for it: compute with subscripts.

Chapter 6. Similarity

Hint 84. How does one go from x to y?

Hint 85. How does one go from η to ξ?

Hint 86. Transform from the x's to the y's, as before.

Hint 87. Use, once again, the transformation that takes one basis to the other, but this time in matrix form.

Hint 88. If one of B and C is invertible, the answer is yes.

Hint 89. Think about real and imaginary parts. That can solve the problem, but if elementary divisor theory is an accessible tool, think about it: the insight will be both less computational and more deep.

Hint 90. Extend a basis of ker A to a basis of the whole space.

Hint 91. Yes.

Hint 92. Look first at the case in which $\beta\gamma \neq 0$.

Hint 93. The first question should be about the relation between ranges and sums.

Hint 94. The easy relation is between the rank of a product and the rank of its first factor; how can information about that be used to get information about the second factor?

Hint 95. The best relation involves null $A +$ null B.

Hint 96. For numerical calculations the geometric definition of similarity is easier to use than the algebraic one.

Hint 97.   There are two pairs of bases, and, consequently, it is reasonable to expect that two transformations will appear, one for each pair.

Hint 98.   Even though the focus is on the dimensions of ranges, it might be wise to begin by looking at the dimensions of kernels.

Chapter 7. Canonical Forms

Hint 99.   If A is a linear transformation, is there a connection between the eigenvalues of A and A^2?

Hint 100.   The answer is no. Can you tell by looking at a polynomial equation what the sum and the product of the roots has to be?

Hint 101.   This is not easy. Reduce the problem to the consideration of $\lambda = 1$, and then ask whether the classical infinite series formula for $\dfrac{1}{1-x}$ suggests anything.

Hint 102.   What about monomials?

Hint 103. $\lambda^3 = 1$.

Hint 104.   If μ is an eigenvalue of A, consider the polynomial $p(\lambda) - \mu$.

Hint 105.   What are the eigenvalues?

Hint 106.   What does the assumption imply about eigenvectors?

Hint 107.   No.

Hint 108.   Look for the triangular forms that are nearest to diagonal ones —that is the ones for which as many as possible of the entries above the diagonal are equal to 0.

Hint 109.   Think about complex numbers.

Hint 110.   What can the blocks in a triangularization of A look like?

Hint 111. The answer depends on the dimension of the space and on the index of nilpotence; which plays the bigger role?

Hint 112. The answer depends on size; look at matrices of size 2 and matrices of size 3.

Hint 113. Examine what M does to a general vector $(\alpha, \beta, \gamma, \delta, \varepsilon)$ and then force the issue.

Hint 114. Don't get discouraged by minor setbacks. A possible approach is to focus on the case $\lambda = 1$, and use the power series expansion of $\sqrt{1 + \zeta}$.

Hint 115. Use eigenvalues—they are more interesting. Matrices, however, are quicker here.

Hint 116. What turns out to be relevant is the Chinese Remainder Theorem. The version of that theorem in elementary number theory says that if $x_1, \ldots, x_n$ are integers, pairwise relatively prime, and if $y_1, \ldots, y_n$ are arbitrary integers, then there exists an integer z such that

$$x_j \equiv y_j \bmod z$$

for $j = 1, \ldots, n$. A more sophisticated algebraic version of the theorem has to do with sets of pairwise relatively prime ideals in arbitrary rings, which might not be commutative. The issue at hand is a special case of that algebraic theorem, but it can be proved directly. The ideals that enter are the annihilators (in the ring of all complex polynomials) of the given linear transformations.

Chapter 8. Inner Product Spaces

Hint 117. Form the inner product of a linear dependence relation with any one of its terms.

Hint 118. Is there an expression for (x, y) in terms of x and y and norms —one that involves no inner products such as (u, v) with $u \neq v$?

Hint 119. Examine both real and complex vector spaces.

Hint 120. Always.

Hint 121. Keep enlarging.

Hint 122. Evaluate the norms of linear combinations of x and y.

Hint 123. How close does an arbitrary vector in $\mathbb{V}$ come to a linear combination of an orthonormal basis for $\mathbb{M}$?

Hint 124. Look at $\ker^\perp \xi$.

Hint 125. (a) By definition $(U^*v, w) = (v, Uw)$; there is no help for it but to compute with that. (b) Yes. (c) Look at the image under U of the graph of A.

Hint 126. Some of the answers are yes and some are no, but there is only one (namely (d)) that might cause some head scratching.

Hint 127. Is something like polarization relevant?

Hint 128. Always?

Hint 129. Only (c) requires more than a brief moment's thought; there are several cases to look at.

Hint 130. Problem 127 is relevant.

Hint 131. The easy ones are (a) and (b); the slightly less easy but straightforward ones are (c) and (d). The only one that requires a little thought is (e); don't forget that α must be real for the question to make sense.

Hint 132. The answer is short, but a trick is needed.

Hint 133. What is the adjoint of a perpendicular projection?

Hint 134. A little computation never hurts.

Hint 135. If $E \le F$ and x is in ran E, evaluate $\|x - Fx\|$. If ran $E \subset$ ran F, then $EF = E$.

Hint 136. Is the product of two perpendicular projections always a perpendicular projection?

Hint 137. Quadratic forms are relevant.

Hint 138. If $Ax_1 = \lambda_1 x_1$ and $Ax_2 = \lambda_2 x_2$, examine (x_1, x_2).

Chapter 9. Normality

Hint 139. Is the dimension of the underlying vector space finite or infinite? Is U necessarily either injective or surjective?

Hint 140. Can "unitary" be said in matrix language?

Hint 141. The question is whether any of the three conditions implies any of the others, and whether any two imply the third.

Hint 142. Must they be diagonal?

Hint 143. Look at the eigenspaces corresponding to the distinct eigenvalues.

Hint 144. Diagonalize.

Hint 145. Assume the answer and think backward. The invertible case is easier.

Hint 146. Imitate the Hermitian proof.

Hint 147. Imitate the Hermitian proof.

Hint 148. It's a good idea to use the spectral theorem.

Hint 149. Use Solution 148.

Hint 150. Assuming that $AS = SB$, with both A and B normal, use the linear transformations A, B, and S, as entries in 2×2 matrices, so as to be able to apply the adjoint commutativity theorem.

Hint 151. Put $C = B(A^*A) - (A^*A)B$ and study the trace of C^*C.

Hint 152. (a) Consider triangular matrices. (b) Consider the Hamilton-Cayley equation.

Hint 153.   Use square roots.

Hint 154.   The two answers are different.

Hint 155.   Some do and some don't; the emphasis is on *necessarily*. For some of the ones that don't, counterexamples of size 2 are not large enough.

Hint 156.   How many eigenvectors are there? More generally, how many invariant subspaces of dimension k are there, $0 \leq k \leq n$?

Hint 157.   What's the relation between A and the matrix

$$\begin{pmatrix} 0 & 0 & 1 \\ 0 & 0 & 0 \\ 0 & 0 & 0 \end{pmatrix}?$$

Hint 158.   Consider the polar decomposition of a transformation that affects the similarity; a natural candidate for a unitary transformation that affects the equivalence is its unitary factor. Don't be surprised if the argument wants to lean on the facts about adjoint intertwining (Solution 150).

Hint 159.   Most people find it difficult to make the right guess about this question when they first encounter it. The answer turns out to be no, but even knowing that does not make it easy to find a counterexample, and, having found one, to prove that it works. One counterexample is a 3×3 nilpotent matrix, and one way to prove that it works is to compute.

Hint 160.   Solution 89 describes a way of passing from complex similarity to real similarity, and Solution 158 shows how to go from (real or complex) similarity to (real or complex) unitary equivalence. The trouble is that Solution 158 needs the adjoint intertwining theorem (Solution 150), which must assume that the given transformations are normal. Is the assumed unitary equivalence sufficiently stronger than similarity to imply at least a special case of the intertwining theorem that can be used here?

Hint 161.   Look at the Jordan form of A.

Hint 162.   Is a modification of the argument of Solution 161 usable?

Hint 163.   If A is nilpotent of index 2, examine subspaces of the form $\mathbb{N} + A\mathbb{N}$, where $\mathbb{N}$ is a subspace of ker A^*.

Hint 164.   The most obvious nilpotent transformation on $\mathbb{C}^4$ is the truncated shift (see Problem 156), but that has index 4. It's tempting to look at its square, but that has index 2. What along these lines can be done to produce nilpotence of index 3?

SOLUTIONS

Chapter 1. Scalars

Solution 1.

The associative law for $\boxed{+}$ expressed in terms of $+$ looks like this:

$$2(2\alpha + 2\beta) + 2\gamma = 2\alpha + 2(2\beta + 2\gamma),$$

which comes to

$$4\alpha + 4\beta + 2\gamma = 2\alpha + 4\beta + 4\gamma. \tag{$*$}$$

That can be true, but it doesn't have to be; it is true if and only if $\alpha = \gamma$. If, for instance, $\alpha = \beta = 0$ and $\gamma = 1$, then the desired equation becomes the falsehood

$$0 + 0 + 2 = 0 + 0 + 4. \tag{$**$}$$

Conclusion: the associative law for $\boxed{+}$ is false.

Comment. Does everyone agree that an alphabetical counterexample (such as ($*$)) is neither psychologically nor logically as convincing as a numerical one (such as ($**$))?

2   Solution 2.

The associative law for $\boxed{+}$ is false. The equation

$$\alpha \boxed{+} (\beta \boxed{+} \gamma) = (\alpha \boxed{+} \beta) \boxed{+} \gamma$$

says that

$$2\alpha + (2\beta + \gamma) = 2(2\alpha + \beta) + \gamma,$$

which is true if and only if $\alpha = 0$. If, for instance, $\alpha = 1$ and $\beta = \gamma = 0$, then the desired equation becomes the falsehood

$$2 + (0 + 0) = 2(2 + 0) + 0.$$

3   Solution 3.

For both commutativity and associativity it is harder to find instances where they hold than instances where they don't. Thus, for instance,

$$\left(\alpha^{\beta^{\gamma}}\right) = \alpha^{(\beta^{\gamma})}$$

is true only if $\alpha = 1$, or $\gamma = 1$, or $\beta = \gamma = 2$. If, in particular, $\alpha = \gamma = 2$ and $\beta = 1$, then it is false. Exponentiation is neither commutative nor associative.

4   Solution 4.

Both answers are yes, and one way to prove them is to compute. Since

$$\langle \gamma, \delta \rangle \boxdot \langle \alpha, \beta \rangle = \langle \gamma\alpha - \delta\beta, \gamma\beta + \delta\alpha \rangle,$$

the commutativity of $\boxdot$ is a consequence of the commutativity of the ordinary multiplication of real numbers.

The computation for associativity needs more symbols:

$$\big\langle \langle \alpha, \beta \rangle \boxdot \langle \gamma, \delta \rangle \big\rangle \boxdot \langle \varepsilon, \varphi \rangle$$
$$= \big\langle (\alpha\gamma - \beta\delta)\varepsilon - (\alpha\delta + \beta\gamma)\varphi, (\alpha\gamma - \beta\delta)\varphi + (\alpha\delta + \beta\gamma)\varepsilon \big\rangle$$

and

$$\langle \alpha, \beta \rangle \boxdot \big\langle \langle \gamma, \delta \rangle \boxdot \langle \delta\varepsilon, \varphi \rangle \big\rangle$$
$$= \big\langle \alpha(\gamma\varepsilon - \delta\varphi) - \beta(\gamma\varphi + \delta\varepsilon), \alpha(\gamma\varphi + \delta\varepsilon) + \beta(\gamma\varepsilon - \delta\varphi) \big\rangle.$$

By virtue of the associativity of the ordinary multiplication of real numbers the same eight triple products, with the same signs, occur in the right-hand sides of both these equations.

For people who know about complex numbers and know that for them both addition and multiplication are both commutative and associative, Problem 4 takes just as little work as the paragraph that introduced it. Indeed: if $\langle \alpha, \beta \rangle$ is thought of as $\alpha + \beta i$ then $\boxed{+}$ and $\boxed{\cdot}$ become "ordinary" complex addition and multiplication, and after that insight nothing remains to be done.

Solution 5.

5

Straightforward computation shows that the equation

$$\langle \alpha, \beta \rangle \boxdot \langle \gamma, \delta \rangle = \langle \gamma, \delta \rangle \boxdot \langle \alpha, \beta \rangle$$

is a severe condition that is quite unlikely to be satisfied. An explicit counterexample is given by

$$\langle 1, 1 \rangle \boxdot \langle 2, 1 \rangle = \langle 2, 2 \rangle$$

and

$$\langle 2, 1 \rangle \boxdot \langle 1, 1 \rangle = \langle 2, 3 \rangle.$$

The associativity story is quite different; there straightforward computation shows that it is always true. This way of multiplying pairs of real numbers is not a weird invention; it arises in a natural classical context. An **affine transformation** of the real line is a mapping S defined for each real number ξ by an equation of the form $S(\xi) = \alpha\xi + \beta$, where α and β themselves are fixed preassigned real numbers. If T is another such mapping, $T(\xi) = \gamma\xi + \delta$, then the composition ST (for the purist: $S \circ T$) is given by

$$(ST)(\xi) = S(\gamma\xi + \delta) = \alpha(\gamma\xi + \delta) + \beta = (\alpha\gamma)\xi + (\alpha\delta + \beta).$$

In other words, the product ST of the transformations corresponding to

$$\langle \alpha, \beta \rangle \qquad \text{and} \qquad \langle \gamma, \delta \rangle$$

is exactly the transformation corresponding to $\langle \alpha, \beta \rangle \boxdot \langle \gamma, \delta \rangle$. Since the operation of composing transformations is always associative, the associativity of $\boxdot$ can be inferred with no further computation.

Is that all right? Is the associativity of functional composition accepted? If it is not accepted, it can be proved as follows. Suppose that R, S, and T

are mappings of a set into itself, and write $P = RS$, $Q = ST$. Then, for each x in the domain,

$$((RS)T)(x) = (PT)(x) = P(T(x)) \quad \text{[by the definition of } PT]$$
$$= (RS)(T(x)) = R\big(S(T(x))\big) \quad \text{[by the definition of } RS].$$

whereas

$$\big(R(ST)\big)(x) = (RQ)(x) = R(Q(x)) \quad \text{[by the definition of } RQ]$$
$$= R((ST)(x)) = R\big(S(T(x))\big) \quad \text{[by the definition of } ST].$$

Since the last terms of these two chains of equations are equal, the first ones must be also.

6   Solution 6.

In view of the comment about Problem 5 being a special case, it follows immediately that the present $\boxed{\cdot}$ is not commutative. To get a counterexample, take any two pairs that do not commute for the $\boxed{}$ of Problem 5 and use each of them as the beginning of a quadruple whose last two coordinates are 0 and 1. Concretely:

$$\langle 1, 1, 0, 1 \rangle \cdot \langle 2, 1, 0, 1 \rangle = \langle 2, 2, 0, 1 \rangle$$

and

$$\langle 2, 1, 0, 1 \rangle \cdot \langle 1, 1, 0, 1 \rangle = \langle 2, 3, 0, 1 \rangle.$$

Associativity is harder. It was true for Problem 5 and it might conceivably have become false when the domain was enlarged for Problem 6. There is no help for it but to compute; the result is that the associative law is true here.

For those who know about the associativity of multiplication for 2×2 matrices no computation is necessary; just note that if a quadruple $\langle \alpha, \beta, \gamma, \delta \rangle$ is written as

$$\begin{pmatrix} \alpha & \beta \\ \gamma & \delta \end{pmatrix},$$

then the present product coincides with the ordinary matrix product.

7   Solution 7.

The worst way to solve the problem is to say that there are only 36 (six times six) possible ordered pairs and only 216 (six times six times six) possible ordered triples that can be formed with 0, 1, 2, 3, 4, 5—in principle the commutativity and associativity questions can be decided by examining all of them.

A better way, for commutativity for instance, is to note that if each of α and β is one of the numbers 0, 1, 2, 3, 4, 5, and if the largest multiple of 6 that doesn't exceed their ordinary product is, say, 60, so that $\alpha\beta = \gamma + 60$, where γ is one of the numbers 0, 1, 2, 3, 4, 5, then, because ordinary multiplication is commutative, the same conclusion holds for $\beta\alpha$. Consequence:

$$\alpha \boxdot \beta = \gamma$$

and

$$\beta \boxdot \alpha = \gamma.$$

The reasoning to prove associativity works similarly—the language and the notation have to be chosen with care but there are no traps and no difficulties.

The intellectually most rewarding way is to use the hint. If m and n are non-negative integers, then each of them is one of the numbers 0, 1, 2, 3, 4, 5 plus a multiple of 6 (possibly the zero multiple). Establish some notation: say $r(m) = \alpha$ plus a multiple of 6, and $r(n) = \beta$ plus a multiple of 6. (The reason for "r" is to be reminded of "reduce".) Consequence: when mn and $\alpha\beta$ are reduced modulo 6 they yield the same result. (Think about this step for a minute.) Conclusion:

$$r(mn) = r(m) \boxdot r(n),$$

as the hint promised.

This was work, but it uses a standard technique in algebra (it's called **homomorphism** and it will be studied systematically later), and it pays off. Suppose, for instance, that each of α, β, and γ is one of 0, 1, 2, 3, 4, 5, so that $r(\alpha) = \alpha$ and $r(\beta) = \beta$, $r(\gamma) = \gamma$. The proof of the associative law can be arranged as follows:

$$(\alpha \boxdot \beta) \boxdot \gamma = \big(r(\alpha) \boxdot r(\beta)\big) \boxdot r(\gamma)$$

$$= r(\alpha\beta) \boxdot r(\gamma) \quad \text{[by the preceding paragraph]}$$

$$= r\big((\alpha\beta)\gamma\big) \quad \text{[ditto]}$$

$$= r\big(\alpha(\beta\gamma)\big) \quad \text{[because ordinary multiplication is associative]}$$

$$= r(\alpha) \boxdot r(\beta\gamma) = r(\alpha) \boxdot \big(r(\beta) \boxdot r(\gamma)\big)$$

$$= \alpha \boxdot (\beta \boxdot \gamma)$$

—the last three equalities just unwind what the first three wound up.

An important difference between the modular arithmetic of 6 and 7 will become visible later, but for most of the theory they act the same way, and that is true, in particular, as far as commutativity and associativity are concerned.

8   Solution 8.

The answer may or may not be easy to guess, but once it's correctly guessed it's easy to prove. The answer is yes, and anyone who believes that and sets out to construct an example is bound to succeed.

Call the three elements for which multiplication is to be defined α, β, and γ; the problem is to construct a multiplication table that is commutative but not associative.

Question: what does commutativity say about the table? Answer: symmetry about the principal diagonal (top left to bottom right). That is: if the entry in row α and column β is, say, γ, then the entry in row β and column α must also be γ.

How can associativity be avoided? How, for instance, can it be guaranteed that

$$(\alpha \times \beta) \times \gamma \neq \alpha \times (\beta \times \gamma)?$$

Possible approach: make $\alpha \times \beta = \gamma$ and $\beta \times \gamma = \alpha$; then the associative law will surely fail if $\gamma \times \gamma$ and $\alpha \times \alpha$ are different. That's easy enough to achieve and the following table is one way to do it:

$\times$	α	β	γ
α	α	γ	β
β	γ	β	α
γ	β	α	γ .

Here, for what it's worth, is a verbal description of this multiplication: the product of two distinct factors is the third element of the set, and the product of any element with itself is that element again.

This is not the only possible solution of the problem, but it's one that has an amusing relation to the double addition in Problem 1. Indeed, if the notation is changed so as to replace α by 0, β by 2, and γ by 1, then the present $\times$ satisfies the equation

$$\alpha \times \beta = 2\alpha + 2\beta,$$

where the plus sign on the right-hand side denotes addition modulo 3.

Solution 9.

(1) How could a real number ε be an identity element for double addition? That is, can it be that

$$2\alpha + 2\varepsilon = \alpha$$

for all α? Clearly not: the equation holds only when $\alpha = -2\varepsilon$, so that, in particular, it does not hold when $\alpha = 1$ and $\varepsilon = 0$.

(2) The answer is slightly different for half double addition. It is still true that for no ε does

$$2\alpha + \varepsilon = \alpha$$

hold for all α, but since this operation is not commutative at least a glance at the other order is called for. Could it be, for some ε, that

$$2\varepsilon + \alpha = \alpha$$

for all α? Sure: just put $\varepsilon = 0$. That is: half double addition has no **right identity** element but it does have a **left identity**.

(3) Exponentiation behaves similarly but backward. There is a right identity, namely 1 ($\alpha^1 = \alpha$ for all α), but there is no left identity ($\varepsilon^\alpha = \alpha$ for all α is impossible no matter what ε is).

(4) The ordered pair $\langle 1, 0 \rangle$ (or, if preferred, the complex number $1 + 0 \cdot i$) is an identity for complex multiplication (both left and right, since multiplication is commutative).

(5) The ordered pair $\langle 1, 0 \rangle$ does the job again, but this time, since multiplication is not commutative, the pertinent equations have to be checked both ways:

$$\langle \alpha, \beta \rangle \times \langle 1, 0 \rangle = \langle \alpha, \beta \rangle$$

and

$$\langle 1, 0 \rangle \times \langle \alpha, \beta \rangle = \langle \alpha, \beta \rangle.$$

Equivalently: the identity mapping I, defined by $I(\alpha) = \alpha$, is an affine transformation that is both a right and a left unit for functional composition. That is: if S is an affine transformation, then

$$I \circ S = S \circ I = S.$$

(6) The quadruple $\langle 1, 0, 0, 1 \rangle$ is a unit for matrix multiplication (both left and right), or, if preferred, the identity matrix

$$\begin{pmatrix} 1 & 0 \\ 0 & 1 \end{pmatrix}$$

is an identity element.

Since complex multiplication and and affine multiplication are known to be special cases of matrix multiplication (see Problem 6), it should come as no surprise to learn that the identity elements described in (4) and (5) above are special cases of the one described in (6).

(7) Modular addition and multiplication cause the least trouble: 0 does the job for $+$, and 1 does it for $\times$.

10   Solution 10.

Given α and β, can one find γ and δ so that the product of $\langle \alpha, \beta \rangle$ and $\langle \gamma, \delta \rangle$ is $\langle 1, 0 \rangle$? The problem reduces to the solution of two equations in the unknowns γ and δ:

$$\alpha\gamma - \beta\delta = 1,$$
$$\alpha\delta + \beta\gamma = 0.$$

The standard elementary techniques for doing that yield an answer in every case, provided only that

$$\alpha^2 + \beta^2 \neq 0,$$

or in other words (since α and β are real numbers) provided only that not both α and β are 0.

Alternatively: since in the customary complex notation

$$\frac{1}{\alpha + \beta i} = \frac{\alpha - \beta i}{(\alpha + \beta i)(\alpha - \beta i)} = \frac{\alpha}{\alpha^2 + \beta^2} - \frac{\beta i}{\alpha^2 + \beta^2},$$

it follows that $\langle \alpha, \beta \rangle$ is invertible if and only if $\alpha^2 + \beta^2 \neq 0$, and, if that condition is satisfied, then

$$\langle \alpha, \beta \rangle^{-1} = \left(\frac{\alpha}{\alpha^2 + \beta^2}, \frac{-\beta}{\alpha^2 + \beta^2} \right).$$

11   Solution 11.

The equations to be solved are almost trivial in this case. The problem is, given $\langle \alpha, \beta \rangle$, to find $\langle \gamma, \delta \rangle$ so that

$$\alpha\gamma = 1 \qquad \text{and} \qquad \alpha\delta + \beta = 0.$$

The first equation has a solution if and only if $\alpha \neq 0$, and, if that is so, then the second equation is solvable also. Conclusion: $\langle \alpha, \beta \rangle$ is invertible if and only if $\alpha \neq 0$, and, if so, then

$$\langle \alpha, \beta \rangle^{-1} = \left\langle \frac{1}{\alpha}, -\frac{\beta}{\alpha} \right\rangle.$$

Caution: this multiplication is not commutative, and the preceding computation guarantees a **right inverse** only. Does it work on the left too? Check it:

$$\left\langle \frac{1}{\alpha}, -\frac{\beta}{\alpha} \right\rangle \times \langle \alpha, \beta \rangle = \left\langle \left(\frac{1}{\alpha}\right) \alpha, \frac{1}{\alpha}\beta - \frac{\beta}{\alpha} \right\rangle.$$

Solution 12.   12

It is time to abandon the quadruple notation and the symbol $\times$; from now write

$$\begin{pmatrix} \alpha & \beta \\ \gamma & \delta \end{pmatrix}$$

instead of $\langle \alpha, \beta, \gamma, \delta \rangle$ and indicate multiplication by juxtaposition (placing the two symbols next to one another) instead of by $\times$. The problem is, given a matrix

$$\begin{pmatrix} \alpha & \beta \\ \gamma & \delta \end{pmatrix},$$

to determine whether or not there exists a matrix

$$\begin{pmatrix} \alpha' & \beta' \\ \gamma' & \delta' \end{pmatrix}$$

such that

$$\begin{pmatrix} \alpha & \beta \\ \gamma & \delta \end{pmatrix} \begin{pmatrix} \alpha' & \beta' \\ \gamma' & \delta' \end{pmatrix} = \begin{pmatrix} 1 & 0 \\ 0 & 1 \end{pmatrix}.$$

What is asked for is a solution of four equations in four unknowns. The standard solution techniques are easy enough to apply, but they are, of course, rather boring. There is no help for it, for the present; an elegant general context into which all this fits will become visible only after some of the theory of linear algebra becomes known. The answer is that $\begin{pmatrix} \alpha & \beta \\ \gamma & \delta \end{pmatrix}$ is invertible if and only if $\alpha\delta - \beta\gamma \neq 0$, and, if that is so, then

$$\begin{pmatrix} \alpha' & \beta' \\ \gamma' & \delta' \end{pmatrix} = \begin{pmatrix} \alpha & \beta \\ \gamma & \delta \end{pmatrix}^{-1} = \begin{pmatrix} \frac{\delta}{\alpha\delta-\beta\gamma} & \frac{-\beta}{\alpha\delta-\beta\gamma} \\ \frac{-\gamma}{\alpha\delta-\beta\gamma} & \frac{\alpha}{\alpha\delta-\beta\gamma} \end{pmatrix}.$$

Readers reluctant to derive the result stand to gain something by at least checking it, that is by carrying out the two multiplications

$$\begin{pmatrix} \alpha & \beta \\ \gamma & \delta \end{pmatrix} \begin{pmatrix} \alpha' & \beta' \\ \gamma' & \delta' \end{pmatrix}$$

and

$$\begin{pmatrix} \alpha' & \beta' \\ \gamma' & \delta' \end{pmatrix} \begin{pmatrix} \alpha & \beta \\ \gamma & \delta \end{pmatrix},$$

and noting that they yield the same answer, namely

$$\begin{pmatrix} 1 & 0 \\ 0 & 1 \end{pmatrix}.$$

Comment. The present result applies, in particular, to the special matrices $\begin{pmatrix} \alpha & \beta \\ -\beta & \alpha \end{pmatrix}$, which are, except for notation, the same as the complex numbers discussed in Problem 4. It follows that such a special matrix is invertible if and only if $\alpha\alpha - \beta(-\beta) \neq 0$—which is of course the same condition as $\alpha^2 + \beta^2 \neq 0$. (The awkward form is intended to serve as a reminder of how it arose this time.) If that condition is satisfied, then the inverse is

$$\begin{pmatrix} \frac{\alpha}{\alpha^2+\beta^2} & \frac{-\beta}{\alpha^2+\beta^2} \\ \frac{\beta}{\alpha^2+\beta^2} & \frac{\alpha}{\alpha^2+\beta^2} \end{pmatrix},$$

and that is exactly the matrix that corresponds to the complex number

$$\left(\frac{\alpha}{\alpha^2 + \beta^2}, \frac{-\beta}{\alpha^2 + \beta^2} \right),$$

in perfect harmony with Solution 10.

Similarly the special matrices $\begin{pmatrix} \alpha & \beta \\ 0 & 1 \end{pmatrix}$ are the same as the affine transformations $\langle \alpha, \beta \rangle$ discussed in Problem 5. According to the present result such a special matrix is invertible if and only if $\alpha \cdot 1 - \beta \cdot 0 \neq 0$, and in that case the inverse is $\langle \frac{1}{\alpha}, -\frac{\beta}{\alpha} \rangle$, in perfect harmony with Solution 11.

It is a consequence of these comments that not only is Problem 6 a generalization of Problems 4 and 5, but, correspondingly, Solution 12 has Solutions 10 and 11 as special cases.

13   Solution 13.

(a) The verification that min is both commutative and associative is straight-forward. If anything goes wrong, it must have to do with the existence of a neutral element, an identity element, that plays the role of 0. The question is this: does there exist a positive real number z such that

$$\min(x, z) = x$$

for every positive real number x? The equation demands that z be greater than or equal to every positive real number x—in other words that z be "the largest real number". That's nonsense—there is no such thing; the present candidate fails to be a group.

(b) The verification of commutativity and associativity is easy again. The search for 0 this time amounts to the search for a number z in the set $\{1, 2, 3, 4, 5\}$ with the property that

$$\max(x, z) = x$$

for every number x in the set. The equation demands that z be less than or equal to every positive integer between 1 and 5, and that's easy; the number 1 does the job. It remains to look for inverses. Given x, can we find y so that $\max(x, y) = 1$? No—that's impossible—the equation can never be satisfied unless $x = y = 1$.

(c) Given that $x + y = y$, add $(-y)$ to both sides of the equation. The right side becomes 0, and the left side becomes

$$(x + y) + (-y) = x + \big(y + (-y)\big) = x + 0 = x,$$

and, consequently, $x = 0$.

Comment. What went wrong in (a) was caused by the non-existence of a largest positive real number. What happens if $\mathbb{R}_+$ is replaced by a *bounded* set of positive real numbers, such as the closed unit interval $[0, 1]$? Does the operation min produce a group then? Commutativity, associativity, and the existence of a zero element are satisfied (the role of 0 being played by 1); the question is about inverses. Is it true that to every number x in $[0, 1]$ there corresponds a number y in $[0, 1]$ such that $\min(x, y) = 1$? Certainly not; that can happen only if $x = 1$.

Does the argument for (c) use the commutativity of $+$? Associativity? Both the defining properties of 0?

Solution 14.   14

The set of those affine transformations

$$\xi \mapsto \alpha\xi + \beta$$

(discussed in Problem 5) for which $\alpha \neq 0$ does not have the first of the defining properties of abelian groups (commutativity), but it has all the others (the associative law, the existence of an identity element, and the existence of an inverse for every element)—see Problem 11; it is a group.

The set of invertible 2×2 matrices is not commutative, but has the other properties of abelian groups (see Problem 12); it is a group.

The product 2×3 is equal to 0 modulo 6. That is: multiplication modulo 6 is not defined in the domain in question, or, in other words, the set $\{1, 2, 3, 4, 5\}$ is not closed under the operation. Conclusion: the non-zero integers modulo 6 do not form a multiplicative group.

If α is any one of the numbers 1, 2, 3, 4, 5, 6 what can be said about the numbers

$$\alpha \times 1, \ \alpha \times 2, \ \alpha \times 3, \ \alpha \times 4, \ \alpha \times 5, \ \alpha \times 6$$

(multiplication modulo 7)? First answer: none of them is 0 (modulo 7). (Why? This is important, and it requires a moment's thought.) Second (as a consequence of the first): they are all different. (Why?) Third (as a consequence of the second): except possibly for the order in which they appear, they are the same as the numbers 1, 2, 3, 4, 5, 6, and therefore, in particular, one of them is 1. That is: for each number α there is a number β such that $\alpha \times \beta = 1$: this is exactly the assertion that every α has a multiplicative inverse. Conclusion: the non-zero integers modulo 7 form a multiplicative group.

15   Solution 15.

If there are only two distinct elements, an identity element 1 and another one, say α, then the "multiplication table" for the operation looks like

0	1	α
1	1	α
α	α	?

If the question mark is replaced by 1, the operation is associative; if it is replaced by α, then the element α has no inverse. Conclusion: two elements are not enough to provide a counterexample.

If there are three distinct elements, an identity 1, and two others, α and β, then there is more elbow room, and, for instance, one possibility is

	1	α	β
1	1	α	β
α	α	x	1
β	β	1	y

No matter what x and y are (among 1, α, and β) the operation that the table defines has an identity and every element has an inverse. If $x = \beta$ and $y = \alpha$, the result is associative, so that it does not serve as an example of the sort of thing wanted. If, however, $x = \alpha$, then

$$(\alpha\alpha)\beta = \alpha\beta = 1$$

and

$$\alpha(\alpha\beta) = \alpha 1 = \alpha,$$

so that the operation is not associative (and the same desired negative conclusion follows if $y = \beta$).

Solution 16.

16

Yes, everything is fine, multiplication in a field must be commutative, and, in particular, $0 \cdot x = x \cdot 0 = 0$ for every x, but it's a good idea to look at the sort of thing that can go wrong if not both distributive laws are assumed. Question: if $\mathbb{F}$ is an abelian group with $+$, and if $\mathbb{F}^*$ is an abelian group with $\times$, and if the distributive law

$$\alpha(x + y) = \alpha x + \alpha y$$

is true for all α, x and y, does it follow that multiplication in $\mathbb{F}$ is commutative? Answer: no. Here is an artificial but illuminating example.

Let $\mathbb{F}$ be the set of two integers 0 and 1 with addition defined modulo 2, and with multiplication defined so that $x \cdot 0 = 0$ for all x (that is, for $x = 0$ and for $x = 1$) and $x \cdot 1 = 1$ for all x. (Recall that in addition modulo 12 multiples of 12 are discarded; in addition modulo 2 multiples of 2 are discarded. The only thing peculiar about addition modulo 2 is that $1 + 1 = 0$.) It is clear that $\mathbb{F}$ with $+$ is an abelian group, and it is even clearer that $\mathbb{F}^*$ (which consists of the single element 1) with $\times$ is an abelian group. The distributive law

$$\alpha(x + y) = \alpha x + \alpha y$$

is true; to prove it, just examine the small finite number of possible cases. On the other hand the distributive law

$$(\alpha + \beta)x = \alpha x + \beta x$$

is not true; indeed

$$(0+1) \cdot 1 = 1$$

and

$$0 \cdot 1 + 1 \cdot 1 = 1 + 1 = 0.$$

Irrelevant side remark: the associative law $\alpha(\beta\gamma) = (\alpha\beta)\gamma$ is true—straightforward verification. The commutative law is false, by definition: $0 \cdot 1 = 1$ and $1 \cdot 0 = 0$.

If, however, both distributive laws are assumed, in other words, if the system under consideration is a bona fide field, then all is well. Indeed, since

$$(0+1)x = 0 \cdot x + 1 \cdot x$$

for all x, and since the left side of this equation is x whereas the right side is

$$0 \cdot x + x,$$

it follows (from Problem 1) that

$$0 \cdot x = 0$$

for all x. A similar use of the other distributive law,

$$x(0+1) = x \cdot 0 + x \cdot 1,$$

implies that

$$x \cdot 0 = 0$$

for all x. In other words, every product that contains 0 as a factor is equal to 0, and that implies everything that's wanted, and it implies, in particular, that multiplication is both associative and commutative.

17   Solution 17.

(a) It is to be proved that $0 \times \alpha$ acts the way 0 does, so that what must be shown is that $0 \times \alpha$ added to any β yields β. It must in particular be true that $(0 \times \alpha) + \alpha = \alpha \, (= 0 + \alpha)$, and, in fact, that's enough: if that is true then the additive cancellation law implies that $0 \times \alpha = 0$. The proof therefore can be settled by the following steps:

$$(0 \times \alpha) + \alpha = (0 \times \alpha) + (1 \times \alpha) \quad \text{(because 1 is the multiplicative unit)}$$

$$= (0 + 1) \times \alpha \quad \text{(by the distributive law)}$$

$$= 1 \times \alpha \quad \text{(because 0 is the additive unit)}$$

$$= \alpha.$$

(b) It is to be proved that $(-1)\alpha$ acts the way $-\alpha$ does, so that what must be shown is that $\alpha + (-1)\alpha = 0$. Proof:

$$\alpha + (-1)\alpha = (1 \times \alpha) + ((-1) \times \alpha) = (1 + (-1)) \times \alpha = 0 \times \alpha = 0.$$

(c) It helps to know "half" of the asserted equation, namely

$$(-\alpha)\beta = -(\alpha\beta),$$

and the other, similar, half

$$\alpha(-\beta) = -(\alpha\beta).$$

The first half is true because

$$\alpha\beta + (-\alpha)\beta = (\alpha + (-\alpha))\beta \quad \text{(distributive law)}$$

$$= 0 \times \beta = 0,$$

which shows that $(-\alpha)\beta$ indeed acts just the way $-(\alpha\beta)$ is supposed to. The other half is proved similarly. The proof of the main assertion is now an easy two step deduction:

$$(-\alpha)(-\beta) = -(\alpha(-\beta)) = -(-(\alpha\beta)) = \alpha\beta.$$

(d) This is not always true. Counterexample: integers modulo 2. (See Problem 18.)

(e) By definition the non-zero elements of $\mathbb{F}$ constitute a multiplicative group, which says, in particular, that the product of two of them is again one of them.

Solution 18.   18

The answer is yes. The example illustrates the possible failure of the distributive law and hence emphasizes the essential role of that law.

Let $\mathbb{F}$ be $\{0, 1, 2, 3, 4\}$, with $+$ being addition modulo 5 and $\times_1$ being multiplication modulo 5. In this case all is well; $(\mathbb{F}, +, \times_1)$ is a field.

An efficient way of defining a suitable $\times_2$ is by a multiplication table, as follows:

$\times_2$	0	1	2	3	4
0	0	0	0	0	0
1	0	1	2	3	4
2	0	2	1	4	3
3	0	3	4	1	2
4	0	4	3	2	1

A verbal description of the multiplication of the elements 2, 3, and 4 is this: the product of two distinct ones among them is the third. Compare Problem 8. The distributive law does indeed break down:

$$2 \times_2 (3 + 4) = 2 \times_2 2 = 1,$$

but

$$(2 \times_2 3) + (2 \times_2 4) = 4 + 3 = 2.$$

Comment.   This is far from the only solution. To get another one, let $\mathbb{F}$ be $\{0, 1\}$ with $+$ being addition and $\times_1$ being multiplication modulo 2; in this case $(\mathbb{F}, +, \times_1)$ is a field. If, on the other hand, $\times_2$ is defined by the ridiculous equation

$$\alpha \times_2 \beta = 1$$

for all α and β, then

$$1 \times_2 (1 + 1) = 1$$

but

$$(1 \times_2 1) + (1 \times_2 1) = 1 + 1 = 0.$$

19   Solution 19.

The answer is yes, there does exist a field with four elements, but the proof is not obvious. An intelligent and illuminating approach is to study the set $\mathbb{P}$ of all polynomials with coefficients in a field and "reduce" that set "modulo" some particular polynomial, the same way as the set $\mathbb{Z}$ of integers is reduced modulo a prime number p to yield the field $\mathbb{Z}_p$.

Logically, the right coefficient field to start with for the purpose at hand is $\mathbb{Z}_2$, but to get used to the procedure it is wise to begin with a more familiar situation, which is not directly relevant.

Let $\mathbb{P}$ be the set of all polynomials with coefficients in the field $\mathbb{Q}$ of rational numbers, and let p be the particular polynomial defined by

$$p(x) = x^2 - 2.$$

Important observation: the polynomial p is **irreducible**. That means non-factorable, or, more precisely, it means that if p is the product of two polynomials with coefficients in $\mathbb{Q}$, then one of them must be a constant.

Let $\mathbb{F}$ be the result of "reducing $\mathbb{P}$ modulo p". A quick way of explaining what that means is to say that the elements of $\mathbb{F}$ are the same as the elements of $\mathbb{P}$ (polynomials with rational coefficients), but the concept of equality is redefined: for present purposes two polynomials f and g are to be regarded as the same if they differ from one another only by a multiple of p. The customary symbol for "equality except possibly for a multiple of p" is $\equiv$, and the relation it denotes is called **congruence**. In more detail: to say that f is congruent to g modulo p, in symbols

$$f \equiv g \text{ modulo } p,$$

means that there exists a polynomial q (with rational coefficients) such that

$$f - g = pq.$$

What happens to the "arithmetic" of polynomials when equality is interpreted modulo p? That is: what can be said about sums and products modulo p?

As far as the addition of polynomials of degree 0 and degree 1 is concerned, nothing much happens:

$$(\alpha x + \beta) + (\gamma x + \delta) = (\alpha + \gamma)x + (\beta + \delta),$$

just as it should be. When polynomials of degree 2 or more enter the picture, however, something new happens. Example: if

$$f(x) = x^2 \quad \text{and} \quad g(x) = -2,$$

then

$$f(x) + g(x) \equiv 0 \text{ (modulo } p).$$

Reason: $f + g$ is a multiple of p (namely $p \cdot 1$) and therefore

$$(f + g) - 0 \equiv 0 \text{ modulo } p.$$

Once that is accepted, then even multiplication offers no new surprises. If, for instance,

$$f(x) = g(x) = x,$$

then

$$f \cdot g \equiv 2 \ (\text{modulo } p);$$

indeed, $f \cdot g - 2 = p$.

What does a polynomial look like, modulo p? Since x^2 can always be replaced by 2 (is "equal" to 2), and, consequently, $x^3 \ (= 2x^2)$ can be replaced by $2x$, and $x^4 \ (= 2 \cdot x^3)$ can be replaced by 4, etc., it follows that every polynomial is "equal" to a polynomial of degree 0 or 1. Once that is agreed to, it follows with almost no pain that $\mathbb{F}$ is a field. Indeed, the verification that $\mathbb{F}$ with addition (modulo p) is an abelian group takes nothing but a modicum of careful thinking about the definitions. The same statement about the set of non-zero elements of $\mathbb{F}$ with multiplication (modulo p) takes a little more thought: where do inverses come from? The clue to the answer is in the following computation:

$$\frac{1}{\alpha + \beta x} = \frac{\alpha - \beta x}{\alpha^2 - 2\beta^2}.$$

Familiar? Of course it is: it is the same computation as the rationalization of the denominator that was needed to prove that $\mathbb{Q}(\sqrt{2})$ is a field. All the hard work is done; the distributive laws give no trouble, and the happy conclusion is that $\mathbb{F}$ is a field, and, in fact, except for notation it is the same as the field $\mathbb{Q}(\sqrt{2})$.

The same technique can be applied to many other coefficient fields and many other moduli. Consider, to be specific, the field $\mathbb{Z}_2$, and let $\mathbb{P}$ this time be the set of all polynomials

$$\alpha_0 + \alpha_1 + \alpha_2 x^2 + \cdots + \alpha_n x^n$$

of all possible degrees, with coefficients in $\mathbb{Z}_2$. (Caution: $5x + 3$ means

$$(x + x + x + x + x) + (1 + 1 + 1);$$

it is a polynomial, and it is equal to $x + 1$ modulo 2. It is dangerous to jump to the conclusion that the polynomial $x^5 + x^3$, which means $xxxxx + xxx$, can be reduced similarly.) The set $\mathbb{P}$ of all such polynomials is an abelian group with respect to addition (modulo 2, of course); thus, for example, the sum of

$$x^5 + x^3 + x + 1$$

and

$$x^3 + x^2 + x$$

is

$$x^5 + x^2 + 1.$$

Polynomials admit a natural commutative multiplication also (example:

$$(x^2 + 1)x^3 + x = x^5 + x),$$

with a unit (the constant polynomial 1), and addition and multiplication together satisfy the distributive laws. Not all is well, however; multiplicative inverses cause trouble. Example: there is no polynomial f such that $xf(x) = 1$; the polynomial x (different from 0) has no reciprocal. In this respect polynomials behave as the integers do: the reciprocal of an integer n is not an integer (unless $n = 1$ or $n = -1$). Just as for integers, reduction by a suitable modulus can cure the disease. A pertinent modulus for the present problem is $x^2 + x + 1$.

Why is it pertinent? Because reduction modulo a polynomial of degree k, say, converts every polynomial into one of degree less than k, and modulo 2 there are, for each k, exactly 2^k polynomials of degree less than k. That's clear, isn't it?—to determine a polynomial of degree $k - 1$ or less, the number of coefficients that has to be specified is k, and there are two choices, namely 0 and 1, for each coefficient. If we want to end up with exactly four polynomials that constitute a field with four elements, the value of k must therefore be 2. Modulo 2 the four polynomials of degree less than 2 are 0, 1, x, and $x + 1$. Just as the modulus by which the integers must be reduced to get a field must be a prime—an unfactorable, irreducible number—the modulus by which the polynomials must be reduced here should be an unfactorable, irreducible polynomial. Modulo 2 there are exactly four polynomials of degree exactly 2, namely the result of adding one of 0, 1, x, or $x + 1$ to x^2. Three of those, namely

$$x^2 = x \cdot x,$$

$$x^2 + 1 = (x + 1)(x + 1),$$

and

$$x^2 + x = x(x + 1)$$

are factorable; the only irreducible polynomial of degree 2 is $x^2 + x + 1$.

The reduced objects, the four polynomials

$$0, \ 1, \ x, \ x + 1$$

are added (modulo 2) the obvious way; the modulus does not enter. It does enter into multiplication. Thus, for instance, to multiply modulo $x^2 + x + 1$, first multiply the usual obvious way and then throw away multiples of $x^2 + x + 1$. Example: $x^3 = 1$ (modulo $x^2 + x + 1$). Reason:

$$x^3 = x(x^2) = x\big((x^2 + x + 1) + (x + 1)\big) = x(x + 1)$$
$$= x^2 + x = (x^2 + x + 1) + 1 = 1.$$

The multiplication table looks like this:

$\times$	0	1	x	$x + 1$
0	0	0	0	0
1	0	1	x	$x + 1$
x	0	x	$x + 1$	1
$x + 1$	0	$x + 1$	1	x

The inspiration is now over; what remains is routine verification. The result is that with addition and multiplication as described the four polynomials $0, 1, x, x + 1$ do indeed form a field.

To construct a field with nine elements, proceed similarly: use polynomials with coefficients in the field of integers modulo 3 and reduce modulo the polynomial $x^3 + 2x + 2$.

Is there a field with six elements? The answer is no. The proof depends on a part of vector space theory that will be treated later, and the fact itself has no contact with the subject of this book. The general theorem is that the number of elements in a finite field is always a power of a prime, and that for every prime power there is one (and except for change of notation only one) finite field with that many elements.

Chapter 2. Vectors

20   **Solution 20.**

The scalar zero law is a consequence of the other conditions; here is how the simple proof goes. If x is in $\mathbb{V}$, then

$$0x + 0x = (0 + 0)x \quad \text{(by the vector distributive law)}$$
$$= 0x,$$

and therefore, simply by cancellation in the additive group $\mathbb{V}$, the forced conclusion is that $0x = 0$.

As for the vector zero law, the scalar distributive law implies that $\alpha 0$ is always zero. Indeed:

$$\alpha 0 + \alpha 0 = \alpha(0 + 0) = \alpha 0,$$

and therefore, simply by cancellation in the additive group $\mathbb{V}$, the forced conclusion is that $\alpha 0 = 0$.

It is good to know that these two results about 0 are in a sense best possible. That is: if $\alpha x = 0$, then either $\alpha = 0$ or $x = 0$. Reason: if $\alpha x = 0$ and $\alpha \neq 0$, then

$$x = 1x = \left(\frac{1}{\alpha}\alpha\right)x = \left(\frac{1}{\alpha}\right)(\alpha x) \quad \text{(by the associative law),}$$

which implies that

$$x = \left(\frac{1}{\alpha}\right)0 = 0.$$

Comment. If a scalar multiplication satisfies all the conditions in the definition of a vector space, how likely is it that $\alpha x = x$? That happens when $x = 0$ and it happens when $\alpha = 1$; can it happen any other way? The answer is no, and, by now, the proof is easy: if $\alpha x = x$, then $(\alpha - 1)x = 0$, and therefore either $\alpha - 1 = 0$ or $x = 0$.

A pertinent comment is that every field is a vector space over itself. Isn't that obvious? All it says is that if, given $\mathbb{F}$, and if the space $\mathbb{V}$ is defined to be $\mathbb{F}$ itself, with addition in $\mathbb{V}$ being what it was in $\mathbb{F}$ and scalar multiplication being ordinary multiplication in $\mathbb{F}$, then the conditions in the definition of a vector space are automatically satisfied. Consequence: if $\mathbb{F}$ is a field, then the equation $0\alpha = 0$ in $\mathbb{F}$ is an instance of the scalar zero law. In other words, the solution of Problem 17 (a) is a special case of the present one.

Solution 21.   21

(1) The scalar distributive law fails: indeed

$$2 * 1 = 2^2 \cdot 1 = 4,$$

but

$$1 * 1 + 1 * 1 = 1 \cdot 1 + 1 \cdot 1 = 2.$$

The verifications that all the other axioms of a vector space are satisfied are painless routine.

(2) The scalar identity law fails; all other conditions are satisfied.

(3) Since the mapping $\alpha \mapsto \alpha^2$ is multiplicative $((\alpha\beta)^2 = \alpha^2\beta^2)$, the associative law for the new scalar product is true (this should be checked, and it is fun to check). The new scalar identity law follows from the fact that $1^2 = 1$. The verification of the new scalar distributive law depends on the fact that if α and β are scalars (in the present sense, a very special case), then

$$(\alpha + \beta)^2 = \alpha^2 + \beta^2.$$

(That identity holds, in fact, if and only if the field has "characteristic 2", which means that $\alpha + \alpha = 0$ for every α in $\mathbb{F}$. An equivalent way of expressing that condition is just to say that $2 = 0$, where "2" means $1 + 1$, of course.) The scalar distributive law, however, is false. Indeed:

$$1\big((1,0) + (0,1)\big) = 1(1,1) = (1,1),$$

whereas

$$1(1,0) + 1(0,1) = (1+1,0) + (0,1) = (1+1,1).$$

(4) Nothing is missing; the definitions of $\boxed{+}$ and $\boxed{\cdot}$ do indeed make $\mathbb{R}_+$ into a real vector space.

(5) In this example the associative law fails. Indeed, if $\alpha = \beta = i$, then

$$(\alpha\beta) \cdot 1 = (-1)1 = -1,$$

whereas

$$\alpha \cdot (\beta \cdot 1) = 0 \cdot (0) = 0.$$

The verifications of the distributive laws (vector or scalar), and of the scalar identity law, are completely straightforward; all that they depend on (in addition to the elementary properties of the addition of complex numbers) is that Re does the right thing with 0, 1, and $+$. (The right thing is $Re\,0 = 0$, $Re\,1 = 1$, and $Re(\alpha + \beta) = Re\,\alpha + Re\,\beta$.)

(6) Here, once more, nothing is missing. The result is a special case of the general observation that if $\mathbb{F}$ is a field and $\mathbb{G}$ is a subfield, then $\mathbb{F}$ is a vector space over $\mathbb{G}$.

Question.   What is the status of the zero laws (scalar and vector) in these examples? The proof that they held (Problem 20) depended on the truth of the other conditions; does the failure of some of those conditions make the zero laws fail also?

Comment. Examples (1), (2), (3), and (5) show that the definition of vector spaces by four axioms contains no redundant information. A priori it is conceivable that some cleverly selected subset of those conditions (consisting of three, or two, or even only one) might be strong enough to imply the others. There are 15 non-empty subsets, and a detailed study of all possibilities threatens to be more than a little dull. An examination of some of those possibilities can, however, be helpful in coming to understand some of the subtleties of the algebra of scalars and vectors, and that's what the examples (1), (2), (3), and (5) have provided. Each of them shows that some particular one of the four conditions is independent of the other three: they provide concrete counterexamples (of $\mathbb{F}$, $\mathbb{V}$, and a scalar multiplication defined between them) in which three conditions hold and the fourth fails.

Despite example (5), the associative law is almost a consequence of the others. If, to be specific, the underlying field is $\mathbb{Q}$, and if $\mathbb{V}$ is a candidate for a vector space over $\mathbb{Q}$, equipped with a scalar multiplication that satisfies the two distributive laws and the scalar identity law, then it satisfies all the other conditions, and, in particular, it satisfies the associative law also, so that $\mathbb{V}$ is an honest vector space over $\mathbb{Q}$. The proof is not especially difficult, but it is of not much use in linear algebra; what follows is just a series of hints.

The first step might be to prove that $2x$ is necessarily equal to $x+x$, and that, more generally, for each positive integer m, the scalar product mx is the sum of m summands all equal to x. This much already guarantees that $(\alpha\beta)x = \alpha(\beta x)$ whenever α and β are positive integers. To get the general associative law two more steps are necessary. One: recall that $0 \cdot x = 0$ and $(-1)x = -x$ (compare the corresponding discussions of the status of the other vector space axioms)—this yields the associative law for all integers. Two: $\frac{1}{2}x + \frac{1}{2}x = x$, and, more generally, the sum of n summands all equal to $\frac{1}{n}x$ is equal to x—this yields the associative law for all reciprocals of integers. Since every rational number has the form $m \cdot \frac{1}{n}$, where m and n are integers, the associative law follows for all elements of $\mathbb{Q}$. Caution: the reader who wishes to flesh out this skeletal outline should be quite sure that the lemmas needed (for example $(-1)x = -x$) can be proved without the use of the associative law.

A similar argument can be used to show that if the underlying field is the field of integers modulo a prime p, then, again, the associative law is a consequence of the others. These facts indicate that for a proof of the independence of the associative law the field has to be more complicated than $\mathbb{Q}$ or $\mathbb{Z}_p$. (Reminder: fields such as $\mathbb{Z}_p$ occurred in the discussion preceding Problem 19.) A field that is complicated enough is the field $\mathbb{C}$ of complex numbers—that's what the counterexample (5) shows.

22   Solution 22.

It's easy enough to verify that

$$3(1, 1) - 1(1, 2) = (2, 1)$$

and

$$-1(1, 1) + 1(1, 2) = (0, 1),$$

so that $(2, 1)$ and $(0, 1)$ are indeed linear combinations of $(1, 1)$ and $(1, 2)$, but these equations don't reveal any secrets; the problem is where do they come from—how can they be discovered?

The general question is this: for which vectors (α, β) can real numbers ξ and η be found so that

$$\xi(1, 1) + \eta(1, 2) = (\alpha, \beta)?$$

In terms of coordinates this vector equation amounts to two numerical equations:

$$\xi + \eta = \alpha$$

$$\xi + 2\eta = \beta.$$

To find the unknowns ξ and η, subtract the top equation from the bottom one to get

$$\eta = \beta - \alpha,$$

and then substitute the result back in the top equation to get

$$\xi + \beta - \alpha = \alpha,$$

or, in other words,

$$\xi = 2\beta - \alpha.$$

That's where the unknown coefficients come from, and, once derived, the consequence is easy enough to check:

$$(2\alpha - \beta)(1, 1) + (\beta - \alpha)(1, 2) = (2\alpha - \beta + \beta - \alpha, 2\alpha - \beta + 2\beta - 2\alpha)$$

$$= (\alpha, \beta).$$

Conclusion: every vector in $\mathbb{R}^2$ is a linear combination of $(1, 1)$ and $(1, 2)$.

The process of solving two linear equations in two unknowns (eliminate one of the unknowns and then substitute) is itself a part of linear algebra. It is used here without any preliminary explanation because it is almost self-explanatory and most students learn it early. (Incidentally: in this context the phrase **linear equations** means **equations of first degree**, that is, typically, equations of the form

$$\alpha\xi + \beta\eta + \gamma = 0$$

in the two unknowns ξ and η.)

Solution 23.   23

For (a) the sets described by (1), (2), and (4) are subspaces and the sets described by (3), (5), and (6) are not. The proofs of the positive answers are straightforward applications of the definition; the negative answers deserve at least a brief second look.

(3) The vector 0 $(= (0,0,0))$ does not satisfy the condition.

(5) The vector $(1,1,1)$ satisfies the condition, but its product by i $(= \sqrt{-1})$ does not.

(6) The vector $(1,1,1)$ satisfies the condition, but its product by i does not.

For (b) the sets described by (2) and (4) are subspaces and the sets described by (1) and (3) are not. The proofs of the positive answers are straightforward. For the negative answers:

(1) The polynomials $x^3 + x$ and $-x^3 + 2$ satisfy the condition, but their sum does not.

(3) The polynomial x^2 satisfies the condition, but its product by i $(= \sqrt{-1})$ does not.

Comment. The answers (5) for (a) and (3) for (b) show that the sets $\mathbb{M}$ involved are not subspaces of the *complex* vector spaces involved—but what would happen if $\mathbb{C}^3$ in (a) were replaced by $\mathbb{R}^3$, and, similarly, the complex vector space $\mathbb{P}$ in (b) were replaced by the corresponding real vector space? Answer: the results would stay the same (negative): just replace "i" by "-1".

Solution 24.   24

(a) The intersection of any collection of subspaces is always a subspace. The proof is just a matter of language: it is contained in the meaning of the word "intersection". Suppose, indeed, that the subspaces forming a collection are distinguished by the use of an index γ; the problem is to prove that if each M_γ is a subspace, then the same is true of $M = \bigcap_\gamma M_\gamma$. Since every M_γ contains 0, so does M, and therefore M is not empty. If x and y belong to M (that is to every M_γ), then $\alpha x + \beta y$ belongs to every M_γ (no matter what α and β are), and therefore $\alpha x + \beta y$ belongs to M. Conclusion: M is a subspace.

(b) If one of two given subspaces is the entire vector space V, then their union is V; the question is worth considering for proper subspaces only. If M_1 and M_2 are proper subspaces, can $M_1 \cup M_2$ be equal to V? No, never. If one of the subspaces includes the other, then their union is equal to the larger one, which is not equal to V. If neither includes the other, the reasoning is slightly more subtle; here is how it goes.

Consider a vector x in M_1 that is not in M_2, and consider a vector y that is not in M_1 (it doesn't matter whether it is in M_2 or not). The set of all scalar multiples of x, that is the set of all vectors of the form αx, is a line through the origin. (The geometric language doesn't have to be used, but it helps.) Translate that line by the vector y, that is, form the set of all vectors of the form $\alpha x + y$; the result is a parallel line (not through the origin). Being parallel, the translated line has no vectors in common with M_1. (To see the geometry, draw a picture; to understand the algebra, write down a precise proof that $\alpha x + y$ can never be in M_1.) How many vectors can the translated line have in common with M_2? Answer: at most one. Reason: if both $\alpha x + y$ and $\beta x + y$ are in M_2, with $\alpha \neq \beta$, then their difference $(\alpha - \beta)x$ would be in M_2, and division by $\alpha - \beta$ would yield a contradiction. It is a consequence of these facts that the set $\mathbb{L}$ of all vectors of the form $\alpha x + y$ (a line) has at most one element in common with $M_1 \cup M_2$. Since there are as many vectors in $\mathbb{L}$ as there are scalars (and that means at least two), it follows that $M_1 \cup M_2$ cannot contain every vector in V.

Granted that V cannot be the union of two proper subspaces, how about three? As an example of the sort of thing that can happen, consider the field $\mathbb{F}$ of integers modulo 2; the set $\mathbb{F}^2$ of all ordered pairs of elements of $\mathbb{F}$ is a vector space in the usual way. The subset

$$\{(0,0),(0,1)\}$$

is a subspace of $\mathbb{F}^2$, and so are the subsets

$$\{(0,0),(1,0)\}$$

and

$$\{(0,0),(1,1)\}.$$

The set-theoretic union of these three subspaces is all of $\mathbb{F}^2$; this is an example of a vector space that is the union of three proper subspaces of itself. The example looks degenerate, in a sense: the vector space has only a finite number of vectors in it, and it should come as no surprise that it can be the union of a finite number of proper subspaces. Every vector space is the union of its "lines", and in the cases under consideration there are only a finite number of them.

Under these circumstances, the intelligent thing to do is to ask about infinite fields, and, sure enough, it turns out that a vector space over an infinite field is never the union of a finite number of proper subspaces; the proof is just a slight modification of the one that worked for $n = 2$ and all fields (infinite or not).

Suppose, indeed, that $M_1, \ldots, M_n$ are proper subspaces such that none of them is included in the union of the others. From the present point of view that assumption involves no loss of generality; if one of them is included in the union of the others, just omit it, and note that the only effect of the omission is to reduce the number n to $n - 1$. It follows that there exists a vector x_1 in M_1 that does not belong to M_j for $j \neq 1$, and (since M_1 is not the whole space) there exists a vector x_0 that does not belong to M_1.

Consider the line through x_0 parallel to x_1. Precisely: let $\mathbb{L}$ be the set of all vectors of the form $x_0 + \alpha x_1$ (α a scalar). How large can the intersections $L \cap M_j$ be (where $j = 1, \ldots, n$)? Since x_1 belongs to M_1 it follows that $x_0 + \alpha x_1$ cannot belong to M_1 (for otherwise x_0 would also); this proves that $\mathbb{L} \cap M_1 = \varnothing$. As for the sets $\mathbb{L} \cap M_j$ with $j \neq 1$, they can contain no more than one vector each. Reason: if both $x_0 + \alpha x_1$ and $x_0 + \beta x_1$ belong to M_j, then so does their difference, $(\alpha - \beta) x_1$, and, since x_1 is not in M_j, that can happen only when $\alpha = \beta$.

Since (by hypothesis) there are infinitely many scalars, the line $\mathbb{L}$ contains infinitely many vectors. Since, however, by the preceding paragraph, the number of elements in $\mathbb{L} \cap (M_1 \cup \cdots \cup M_n)$ is less than n, it follows that $M_1 \cup \cdots \cup M_n$ cannot cover the whole space; the proof is complete.

What the argument depends on is a comparison between the cardinal number of the ground field and a prescribed cardinal number n. Related theorems are true for certain related structures. One example: a group is never the union of *two* proper subgroups. Another example: a Banach space is never the union of a finite or countable collection of **closed** proper subspaces.

Caution. Even if the ground field is uncountable (has cardinal number greater than $\aleph_0$, as does $\mathbb{R}$ for instance), it is possible for a vector space to be the union of a countably infinite collection of proper subspaces. Example: the vector space $\mathbb{P}$ of all real polynomials is the union of the subspaces $\mathbb{P}_n$ consisting of all polynomials of degree less than or equal to n, $n = 1, 2, 3, \ldots$.

25   Solution 25.

(a) Sure, that's easy; just consider, for instance, the sets $\{(1,0), (0,1)\}$ and $\{(2,0), (0,2)\}$. That answers the question, but it seems dishonest—could a positive answer have been obtained so that no vector in either set is a scalar multiple of a vector in the other set? Yes, and that's easy too, but it requires a few more seconds of thought. One example is $\{(1,0), (0,1)\}$ and $\{(1,1), (1,-1)\}$.

(b) The span of $\{(1,1,1), (0,1,1), (0,0,1)\}$ is $\mathbb{R}^3$, or, in other words, every vector in $\mathbb{R}^3$ is a linear combination of the three vectors in the set.

Why? Because no matter what vector (α, β, γ) is prescribed, coefficients ξ, η, and ζ can be found so that

$$\xi(1,1,1) + \eta(0,1,1) + \zeta(0,0,1) = (\alpha, \beta, \gamma).$$

In fact this one vector equation says the same thing as the three scalar equations

$$\xi = \alpha,$$
$$\xi + \eta = \beta,$$
$$\xi + \eta + \zeta = \gamma,$$

and those are easy equations to solve. The solution is

$$\xi = \alpha,$$
$$\eta = \beta - \xi = \beta - \alpha,$$
$$\zeta = \gamma - \xi - \eta = \gamma - \alpha - (\beta - \alpha) = \gamma - \beta.$$

Check:

$$\alpha(1,1,1) + (\beta - \alpha)(0,1,1) + (\gamma - \beta)(0,0,1) = (\alpha, \beta, \gamma).$$

Comment.   The span of the two vectors $(0,1,1)$ and $(0,0,1)$ is the set of all $(0, \xi, \xi + \eta)$, which is in fact the (η, ζ)-plane. The span of the two vectors $(1,1,1)$ and $(0,1,1)$ is the plane consisting of the set of all $(\xi, \xi + \eta, \xi + \eta)$, and the span of $(1,1,1)$ and $(0,0,1)$ is still another plane.

Solution 26.   26

Yes, it follows. To say that $x \in \bigvee\{M, y\}$ means that there exists a vector z in M and there exist scalars α and β such that

$$x = \alpha y + \beta z.$$

It follows, of course, that

$$\alpha y = x - \beta z,$$

and, moreover, that $\alpha \neq 0$—the latter because otherwise x would belong to M, contradicting the assumption. Conclusion:

$$y \in \bigvee\{M, x\},$$

and that implies the equality of the spans of $\{M, x\}$ and $\bigvee\{M, y\}$.

Solution 27.   27

(a) No, there is no vector that spans $\mathbb{R}^2$. Indeed, for each vector (x, y) in $\mathbb{R}^2$, its span is the set of all scalar multiples of it, and that can never contain every vector. Reason: if $x = 0$, then $(1, 0)$ is not a multiple of (x, y), and if $x \neq 0$, then $(x, y + 1)$ is not a multiple of (x, y).

(b) Yes, there are two vectors that span $\mathbb{R}^2$, many ways. One obvious example is $(1, 0)$ and $(0, 1)$; another is $(1, 1)$ and $(1, -1)$—see Problem 25.

(c) No, no two vectors can span $\mathbb{R}^3$. Suppose, indeed, that

$$x = (x_1, x_2, x_3) \qquad \text{and} \qquad y = (y_1, y_2, y_3)$$

are any two vectors in $\mathbb{R}^3$; the question is whether for an arbitrary $z = (z_1, z_2, z_3)$ coefficients α and β can be found so that $\alpha x + \beta y = z$. In other words, for given (x_1, x_2, x_3) and (y_1, y_2, y_3) can the equations

$$\alpha x_1 + \beta y_1 = z_1,$$

$$\alpha x_2 + \beta y_2 = z_2,$$

$$\alpha x_3 + \beta y_3 = z_3,$$

be solved for the unknowns α and β, no matter what z_1, z_2, and z_3 are? The negative answer can be proved either by patiently waiting till the present discussion of linear algebra reaches the pertinent discussion of dimension theory, or by making use of known facts about the solution of three equations in two unknowns (which belongs to the more general context of systems with more equations than unknowns). In geometric language the facts can be expressed by saying that all linear combinations of x and y are contained in a single plane.

(d) No, no finite set of vectors spans the vector space $\mathbb{P}$ of all polynomials (no matter what the underlying coefficient field is). The reason is that polynomials have degrees. In a finite set of polynomials there is one with maximum degree; no linear combination of the set will produce a polynomial with greater degree than that. Since $\mathbb{P}$ contains polynomials of all degrees, the span of the finite set cannot exhaust $\mathbb{P}$. Compare the cautionary comment at the end of Solution 24.

28   Solution 28.

The modular identity does hold for subspaces.

The easy direction is $\supset$: the right side is included in the left. Reason: $\mathbb{L} \cap \mathbb{M} \subset \mathbb{L}$ (obviously) and $\mathbb{L} \cap \mathbb{N} \subset \mathbb{M} + (\mathbb{L} \cap \mathbb{N})$. In other words, both summands on the right are included in the left, and, therefore, so is their sum.

The reverse direction takes a little more insight. If x is a vector in the left side, then $x \in \mathbb{L}$ and $x = y + z$ with $y \in \mathbb{M}$ and $z \in \mathbb{L} \cap \mathbb{N}$. Since $y = x - z$, and since $-z$ belongs to $\mathbb{L} \cap \mathbb{N}$ along with z, so that, in particular, $-z \in \mathbb{L}$, it follows that $y \in \mathbb{L}$. Since by the choice of notation, $y \in \mathbb{M}$, it follows that $y \in \mathbb{L} \cap \mathbb{M}$, and hence that

$$x \in (\mathbb{L} \cap \mathbb{M}) + (\mathbb{L} \cap \mathbb{N}),$$

as promised.

29   Solution 29.

The question is when do addition and intersection satisfy the distributive law. Half the answer is obvious: the right side is included in the left. Reason: both $\mathbb{L} \cap \mathbb{M}$ and $\mathbb{L} \cap \mathbb{N}$ are included in both $\mathbb{L}$ and $\mathbb{M} + \mathbb{N}$.

As for the other half, if every vector in $\mathbb{V}$ is a scalar multiple of a particular vector x, then $\mathbb{V}$ has very few subspaces—in fact, only two, $\mathbb{O}$ and $\mathbb{V}$. In that case the distributive law for subspaces is obviously true; in all other cases it's false.

Suppose, indeed, that $\mathbb{V}$ contains two vectors x and y such that neither one is a scalar multiple of the other. (Look at a picture in $\mathbb{R}^2$.) If $\mathbb{L}$, $\mathbb{M}$, and $\mathbb{N}$ are the sets of all scalar multiples of $x + y$, x, and y, respectively, then $\mathbb{L} \cap \mathbb{M}$ and $\mathbb{L} \cap \mathbb{N}$ are $\mathbb{O}$, so that the right side is $\mathbb{O}$, whereas $\mathbb{M} + \mathbb{N}$ includes $\mathbb{L}$, so that the left side is $\mathbb{L}$.

Solution 30.   **30**

For most total sets $\mathbb{E}$ in a vector space $\mathbb{V}$ it is easy to find a subspace $\mathbb{M}$ that has nothing in common with $\mathbb{E}$. For a specific example, let $\mathbb{V}$ be $\mathbb{R}^2$ and let $\mathbb{E}$ be $\{(1,0),(0,1)\}$; the subspace $\mathbb{M}$ spanned by $(1,1)$ is disjoint from $\mathbb{E}$.

Solution 31.   **31**

The answers are yes, and the proofs are easy.

If $x_0 = \sum_{j=1}^{n} \alpha_j x_j$, then put $\alpha_0 = -1$ and note that $\sum_{j=0}^{n} \alpha_j x_j = 0$. Since not all the scalars $\alpha_0, \alpha_1, \ldots, \alpha_n$ are 0 (because at least α_0 is not), it follows that the enlarged set $\{x_0, x_1, \ldots, x_n\}$ is dependent.

In the converse direction, if $\sum_{j=0}^{n} \alpha_j x_j = 0$, with not every α_j equal to 0, then there is at least one index i such that $\alpha_i \neq 0$. Solve for x_i to get $x_i = \sum_{j \neq i} \frac{\alpha_j}{\alpha_i} x_j$. (The symbol $\sum_{j \neq i}$ indicates the sum extended over the indices j different from i.) That's it: the last equation says that x_i is a linear combination of the other x's.

It is sometimes convenient to regard a finite set $\{x_0, x_1, \ldots, x_n\}$ of vectors as presented in order, the order of indices, and then to ask about the dependence of the initial segments $\{x_0\}$, $\{x_0, x_1\}$, $\{x_0, x_1, x_2\}$, etc. The proof given above yields the appropriate result. A more explicit statement is this corollary: a set $\{x_0, x_1, \ldots, x_n\}$ of non-zero vectors is dependent if and only if at least one of the vectors $x_1, \ldots, x_n$ is a linear combination of the preceding ones. The important word is "preceding". The proof of "if" is trivial. The proof of "only if" is obtained from the second half of the proof given above by choosing x_i to be the first vector after x_0 for which the set $\{x_1, \ldots, x_i\}$ is linearly dependent. (Caution: is it certain that there is such an x_i?) The desired result is obtained by solving such a linear dependence relation for x_i.

Solution 32.   **32**

Yes, every finite-dimensional vector space has a finite basis; in fact, if $\mathbb{E}$ is a finite total set for $\mathbb{V}$, then there exists an independent subset $\mathbb{F}$ of $\mathbb{E}$ that is a basis for $\mathbb{V}$. The trick is to use Problem 31.

If $\mathbb{V} = \mathbb{O}$, the result is trivial; there is no loss of generality in assuming that $\mathbb{V} \neq \mathbb{O}$. In that case suppose that $\mathbb{E}$ is a finite total set for $\mathbb{V}$ and begin by asking whether 0 belongs to $\mathbb{E}$. If it does, discard it; the resulting set (which might as well be denoted by $\mathbb{E}$ again) is still total for $\mathbb{V}$. If $\mathbb{E}$ is independent, there is nothing to do; in that case $\mathbb{F} = \mathbb{E}$. If $\mathbb{E}$ is dependent, then, by Problem 31, there exists an element of $\mathbb{E}$ that is a linear combination of the others. Discard that element, and note that the resulting set (which might as well be denoted by $\mathbb{E}$ again) is still total for $\mathbb{V}$. Keep repeating the argument of the preceding two sentences as long as necessary; since $\mathbb{E}$ is finite, the repetitions have to stop in a finite number of steps. The only thing that can stop them is arrival at an independent set, and that completes the proof.

Chapter 3. Bases

33   Solution 33.

If $\mathbb{T}$ is a total set for a vector space $\mathbb{V}$, and $\mathbb{E}$ is a finite *independent* set in $\mathbb{V}$, then there exists a subset $\mathbb{F}$ of $\mathbb{T}$, with the same number of elements as $\mathbb{E}$ such that $(\mathbb{T} - \mathbb{F}) \cup \mathbb{E}$ is total.

The proof is simplest in case $\mathbb{E}$ consists of a single non-zero vector x. All that has to be done then is to express x as a linear combination $\sum_i \alpha_i x_i$ of vectors in $\mathbb{T}$ and find a coefficient α_i different from 0. From $x = \sum_j \alpha_j y_j$ it follows that $y_i = \frac{1}{\alpha_i}\left(x - \sum_{j \neq i} \alpha_j y_j\right)$. If y_i is discarded from $\mathbb{T}$ and replaced by x, the result is just as total as it was before, because each linear combination of vectors in $\mathbb{T}$ is equal to a linear combination of x and of vectors in $\mathbb{T}$ different from y_i.

In the general case, $\mathbb{E} = \{x_1, \ldots, x_n\}$, apply the result of the preceding paragraph inductively to one x at a time. Begin, that is, by finding y_1 in $\mathbb{T}_1$ $(= \mathbb{T})$ so that $\mathbb{T}_2 = (\mathbb{T}_1 - \{y_1\}) \cup \{x_1\}$ is total. For the second step, find y_2 in $\mathbb{T}_2$ so that $\mathbb{T}_3 = (\mathbb{T}_2 - \{y_2\}) \cup \{x_2\}$ is total, and take an additional minute to become convinced that $\mathbb{T}_3$ contains x_1, that is that y_2 couldn't have been x_1. The reason for the latter is the assumed independence of the x's; if x_1 had been discarded from $\mathbb{T}_2$, no linear combination of x_2, together with the vectors that have not been discarded, could recapture it. Keep going the same way, forming

$$\mathbb{T}_k + 1 = (\mathbb{T}_k - \{y_k\}) \cup \{x_k\},$$

till $\mathbb{T}_n$ is reached. The result is a new total set obtained from $\mathbb{T}$ by changing a subset $\mathbb{F} = \{y_1, \ldots, y_n\}$ of $\mathbb{T}$ into the prescribed set

$$\mathbb{E} = \{x_1, \ldots, x_n\}.$$

The name of the result is the **Steinitz exchange theorem**.

The result has three useful corollaries.

Corollary 1.   *If* $\mathbb{E}$ *is an independent set and* $\mathbb{T}$ *is a total set in a finite-dimensional vector space, then the number of elements in* $\mathbb{E}$ *is less than or equal to the number of elements in* $\mathbb{T}$.

Corollary 2.   *Any two bases for a finite-dimensional vector space have the same number of elements.*

The **dimension** of a finite-dimensional vector space $\mathbb{V}$, abbreviated $\dim \mathbb{V}$, is the number of elements in a basis of $\mathbb{V}$.

Corollary 3.   *Every set of more than* n *vectors in a vector space* $\mathbb{V}$ *of dimension* n *is dependent. A set of* n *vectors in* $\mathbb{V}$ *is a basis if and only if it is independent, or, alternatively, if and only if it is total.*

Note that these considerations answer, in particular, a question asked long before (Problem 27), namely whether two vectors can span $\mathbb{R}^3$. Since $\dim \mathbb{R}^3 = 3$, the answer is no.

Solution 34.   34

If several subspaces of a space $\mathbb{V}$ of dimension n have a simultaneous complement, then they all have the same dimension, say m, so that that is at least a necessary condition. Assertion: if the coefficient field is infinite, then that condition is sufficient also: finite collections of subspaces of the same dimension m necessarily have simultaneous complements.

If the common dimension m is equal to n, then each of the given subspaces is equal to $\mathbb{V}$ (is it fair in that case to speak of "several" subspaces?), and the subspace $\{0\}$ is a simultaneous complement—a thoroughly uninteresting degenerate case. If $m < n$, then the given subspaces $M_1, \ldots, M_k$ are proper, and it follows from Problem 24 that there exists a vector x in $\mathbb{V}$ that doesn't belong to any of them. If $\mathbb{L}$ is the 1-dimensional space spanned by x, then $M_j \cap \mathbb{L} = \{0\}$ for each j, and, moreover, all the subspaces

$$M_1 + \mathbb{L}, \ldots, M_k + \mathbb{L}$$

have dimension $m + 1$. Either $m + 1 = n$ (in which case $M_j + \mathbb{L} = \mathbb{V}$ for each j, and, in fact, $\mathbb{L}$ is a simultaneous complement of all the M_j's), or $m + 1 < n$, in which case the reasoning can be applied again. Applying it inductively a total of $n - m$ times produces the promised simultaneous complement.

The generalization of Problem 24 to uncountable ground fields and countable collections of proper subspaces is just as easy to apply as the ungeneralized version. Conclusion: if the ground field is uncountable, then countable collections of subspaces of the same dimension m necessarily have simultaneous complements.

35　　Solution 35.

(a) If x and 1 are linearly *dependent*, then there exist rational numbers α and β, not both 0, such that $\alpha \cdot 1 + \beta \cdot \xi = 0$. The coefficient β cannot be 0 (for if it were, than α too would have to be), and, consequently, this dependence relation implies that $x = -\dfrac{\alpha}{\beta}$, and hence that x is rational. The reverse implication is equally easy: x and 1 are linearly dependent if and only if x is rational.

(b) The solution of two equations in two unknowns is involved, namely the equations

$$\alpha(1+\xi) + \beta(1-\xi) = 0$$
$$\alpha(1-\xi) + \beta(1+\xi) = 0$$

in the unknowns α and β. If $\xi \neq 0$, then α and β must be 0; the only case of linear dependence is the trivial one, $(1,1)$ and $(1,1)$.

36　　Solution 36.

How about $(x,1,0)$, $(1,x,1)$, and $(0,1,x)$? The assumption of linear dependence leads to three equations in three unknowns that form a conspiracy: they imply that $x(x^2 - 2) = 0$. Consequence: x must be 0 or else $\pm\sqrt{2}$, and, indeed, in each of those cases, linear dependence does take place. That makes sense for $\mathbb{R}$, but not for $\mathbb{Q}$; in that case linear dependence can take place only when $x = 0$.

37　　Solution 37.

(a) If $(1,\alpha)$ and $(1,\beta)$ are to be linearly independent, then clearly α cannot be equal to β, and, conversely, if $\alpha \neq \beta$, then linear independence does take place.

(b) No, there is not enough room in $\mathbb{C}^2$ for three linearly independent vectors; the trouble is that three equations in two unknowns are quite likely to have a non-trivial solution. Better: $\mathbb{C}^2$ has dimension 2, and the existence of three linearly independent vectors would imply that the dimension is at least 3.

Solution 38.   38

Why not?

For (a) consider, for instance, two independent vectors in $\mathbb{C}^2$, such as $(1, 0)$ and $(1, -1)$, each of which is independent of $(1, 1)$, and use them to doctor up the two given vectors. One possibility is to adjoin

$$(0, 0, 1, 0) \quad \text{and} \quad (1, 0, 0, 0)$$

to the first given pair and adjoin

$$(0, 0, 1, -1) \quad \text{and} \quad (1, -1, 0, 0)$$

to the second given pair.

For (b), adjoin

$$(0, 0, 1, 0) \quad \text{and} \quad (0, 0, 1, 1)$$

to the first two vectors and adjoin

$$(-1, 1, 0, 0,) \quad \text{and} \quad (0, 1, 0, 0)$$

to the second two.

Solution 39.   39

(a) Never—there is too much room in $\mathbb{C}^3$. Better: since the dimension of $\mathbb{C}^3$ is 3, two vectors can never constitute a basis in it.

(b) Never—the sum of the first two is the third—they are linearly dependent.

Solution 40.   40

How many vectors can there be in a maximal linearly independent set? Clearly not more than 4, and it doesn't take much work to realize that any four of the six prescribed vectors are linearly independent. Conclusion: the answer is the number of 4-element subsets of a 6-element set, that is $\binom{6}{4}$.

41   Solution 41.

If x is an arbitrary non-zero vector in $\mathbb{V}$, then x and ix $(= \sqrt{-1}x)$ are linearly independent over $\mathbb{R}$. (Reason: if α and β are real numbers and if

$$\alpha x + \beta(ix) = 0,$$

then

$$(\alpha + \beta i)x = 0,$$

and since $x \neq 0$, it follows that $\alpha + \beta i = 0$.) Consequence: if the vectors $x_1, x_2, x_3, \ldots$ constitute a basis in $\mathbb{V}$, then the same vectors, together with their multiples by i, constitute a basis in $\mathbb{V}^{\text{real}}$. Conclusion: the "real dimension" of $\mathbb{V}$ is $2n$. Unsurprising corollary: the real dimension of $\mathbb{C}$ is 2.

42   Solution 42.

Suppose, more generally, that $\mathbb{M}$ and $\mathbb{N}$ are finite-dimensional subspaces of a vector space, with $\mathbb{M} \subset \mathbb{N}$. If $\mathbb{M} \neq \mathbb{N}$, then a basis for $\mathbb{M}$ cannot span $\mathbb{N}$. Take a basis for $\mathbb{M}$ and adjoin to it a vector in $\mathbb{N}$ that is not in $\mathbb{M}$. The result is a linearly independent set in $\mathbb{N}$ containing more elements than the dimension of $\mathbb{M}$—which implies that $\mathbb{M}$ and $\mathbb{N}$ do not have the same dimension. Conclusion: if a subspace of $\mathbb{N}$ has the same dimension as $\mathbb{N}$, then it must be equal to $\mathbb{N}$.

43   Solution 43.

The answer is yes; every finite independent set in a finite-dimensional vector space can be extended to a basis. The assertion (Problem 32) that in a finite-dimensional vector space there always exists a finite basis is a special case: it just says that the empty set (which is independent) can be extended to a basis.

The proof of the general answer has only one small trap. Given a finite independent set $\mathbb{E}$, consider an arbitrary finite basis $\mathbb{B}$, and apply the Steinitz exchange theorem (see Solution 33). The result is that there exists a total set that includes $\mathbb{E}$ and has the same number of elements as $\mathbb{B}$; but is it obvious that that set must be independent? Yes, it is obvious. If it were dependent, then (see Problem 32) a proper subset of it would be a basis, contradicting the fact (Corollary 2 in Solution 33) that any two bases have the same number of elements.

Note that the result answers the sample question about the set $\{u, v\}$ described before the statement of the problem: there does indeed exist a basis of $\mathbb{C}^4$ containing u and v. One such basis is $\{u, v, x_1, x_2\}$.

Solution 44.   44

If $\mathbb{V}$ is a vector space of dimension n, say, and if $\mathbb{M}$ is a subspace of $\mathbb{V}$, then $\mathbb{M}$ is indeed finite-dimensional, and, in fact, the dimension of $\mathbb{M}$ must be less than or equal to n. If $\mathbb{M} = \mathbb{O}$, then the dimension of $\mathbb{M}$ is 0, and the proof is complete. If $\mathbb{M}$ contains a non-zero vector x_1, let $\mathbb{M}_1$ ($\subset \mathbb{M}$) be the subspace spanned by x_1. If $\mathbb{M} = \mathbb{M}_1$, then $\mathbb{M}$ has dimension 1, and the proof is complete. If $\mathbb{M} \neq \mathbb{M}_1$, let x_2 be an element of $\mathbb{M}$ not contained in $\mathbb{M}_1$, and let $\mathbb{M}_2$ be the subspace spanned by x_1 and x_2; and so on. After no more than n steps the process reaches an end. Reason: the process yields an independent set, and no such set can have more than n elements (since every independent set can be extended to a basis, and no basis can have more than n elements). The only way the process can reach an end is by having the x's form a set that spans $\mathbb{M}$—and the proof is complete.

Solution 45.   45

A total set is minimal if and only if it is independent. The most natural way to approach the proofs of the two implications involved seems to be by contrapositives. That is: $\mathbb{E}$ is *not* minimal if and only if it is *dependent*.

Suppose, indeed, that $\mathbb{E}$ is not minimal, which means that $\mathbb{E}$ has a non-empty subset $\mathbb{F}$ such that the relative complement $\mathbb{E} - \mathbb{F}$ is total. If x is any vector in $\mathbb{F}$, then there exist vectors $x_1, \ldots, x_n$ in $\mathbb{E} - \mathbb{F}$ and there exist scalars $\alpha_1, \ldots, \alpha_n$ such that

$$x = \sum_{j=1}^{n} \alpha_j x_j,$$

which implies, of course, that the subset $\{x, x_1, \ldots, x_n\}$ of $\mathbb{E}$ is dependent.

If, in reverse, $\mathbb{E}$ is dependent, then there exist vectors $x_1, \ldots, x_n$ in $\mathbb{E}$ and there exist scalars $\alpha_1, \ldots, \alpha_n$ not all zero such that

$$\sum_{j=1}^{n} \alpha_j x_j = 0.$$

Find i so that $\alpha_i \neq 0$, and note that

$$x = -\sum_{j \neq i} \frac{\alpha_j}{\alpha_i} x_j.$$

This implies that the set $\mathbb{F} = \mathbb{E} - \{x_i\}$ is just as total as $\mathbb{E}$, and hence that $\mathbb{E}$ is not minimal.

46   Solution 46.

If $\mathbb{E}$ is a total subset of a finite-dimensional vector space $\mathbb{V}$, express each vector in a basis of $\mathbb{V}$ as a linear combination of vectors in $\mathbb{E}$. The vectors actually used in all these linear combinations form a finite total subset of $\mathbb{E}$. That subset has an independent subsubset with the same span (see Problem 33), and, therefore, that subsubset is total. Since an independent total set is minimal, the reasoning proves the existence of a minimal total subset of $\mathbb{E}$.

The conclusion remains true for spaces that are not finite-dimensional, but at least a part of the technique has to be different. What's needed, given $\mathbb{E}$, is an independent subset of $\mathbb{E}$ with the same span. A quick way to get one is to consider the set of all independent subsets of $\mathbb{E}$ and to find among them a *maximal* one. (That's the same technique as is used to prove the existence of bases.) The span of such a maximal independent subset of $\mathbb{E}$ has to be the same as the span of $\mathbb{E}$ (for any smaller span would contradict maximality). Since the span of $\mathbb{E}$ is $\mathbb{V}$, that maximal independent subset is itself total. Since an independent total set is a minimal total set (Problem 45), the proof is complete: every total set has minimal total subset.

47   Solution 47.

An infinitely total set $\mathbb{E}$ always has an infinite subset $\mathbb{F}$ such that $\mathbb{E} - \mathbb{F}$ is total. Here is one way to construct an $\mathbb{F}$.

Consider an arbitrary vector x_1 in $\mathbb{E}$. Since, by assumption, $\mathbb{E} - \{x_1\}$ is total, there exists a finite subset $\mathbb{E}_1$ of $\mathbb{E} - \{x_1\}$ whose span contains x_1. Let x_2 be a vector in the relative complement $\mathbb{E} - (\{x_1\} \cup \mathbb{E}_1)$. Since, by assumption, $\mathbb{E} - (\{x_1, x_2\} \cup \mathbb{E}_1)$ is total, it has a finite subset $\mathbb{E}_2$ whose span contains x_2. Keep iterating the procedure. That is, at the next step, let x_3 be a vector in

$$\mathbb{E} - (\{x_1, x_2\} \cup \mathbb{E}_1 \cup \mathbb{E}_2),$$

note that that relative complement is total, and that, therefore, it has a finite subset $\mathbb{E}_3$ whose span contains x_3. The result of this iterative procedure is an infinite set $\mathbb{F} = \{x_1, x_2, x_3, \ldots\}$ with the property that $\mathbb{E} - \mathbb{F}$ is total. Reason: $\mathbb{E}_j$ is a subset of $\mathbb{E} - \mathbb{F}$ for each j, and therefore x_j belongs to the span of $\mathbb{E} - \mathbb{F}$ for each j.

Solution 48.

48

Assertion: if $\{x_1, \ldots, x_k\}$ is a relatively independent subset of $\mathbb{R}^n$, where $k \geq n$, then there exists a vector x_{k+1} such that $\{x_1, \ldots, x_k, x_{k+1}\}$ is relatively independent.

For the proof, form all subsets of $n-1$ vectors of $\{x_1, \ldots, x_k\}$, and, for each such subset, form the subspace they span. (Note that the dimension of each of those subspaces is exactly $n-1$, not less. The reason is the assumed relative independence. This fact is not needed in the proof, but it's good to know anyway.) The construction results in a finite number of subspaces that, between them, certainly do not exhaust $\mathbb{R}^n$; choose x_{k+1} to be any vector that does not belong to any of them. (The property of the field $\mathbb{R}$ that this argument depends on is that $\mathbb{R}$ is infinite.)

Why is the enlarged set relatively independent? To see that, suppose that $y_1, \ldots, y_{n-1}$ are any $n-1$ distinct vectors of the set $\{x_1, \ldots, x_k\}$. In a non-trivial dependence relation connecting the y's and x_{k+1}, that is, in a relation of the form

$$\sum_i \beta_i y_i + \alpha x_{k+1} = 0,$$

the coefficient α cannot be 0 (for otherwise the y's would be dependent). Any such non-trivial dependence would, therefore, imply that x_{k+1} belongs to the span of the y's, which contradicts the way that x_{k+1} was chosen. This completes the proof of the assertion.

Inductive iteration of the assertion (starting with an independent set of n vectors) yields a relatively independent set $\{x_1, x_2, x_3, \ldots\}$ with infinitely many elements.

A student familiar with cardinal numbers might still be unsatisfied. The argument proves, to be sure, that there is no finite upper bound to the possible sizes of relatively independent sets, but it doesn't completely answer the original question. Could it be, one can go on to ask, that there exist relatively independent sets with uncountably many elements? The answer is yes, but its proof seems to demand transfinite techniques (such as Zorn's lemma).

Solution 49.

49

Let q be the number of elements in the coefficient field $\mathbb{F}$ and let n be the dimension of the given vector space over $\mathbb{F}$. Since a basis of $\mathbb{F}^n$ is a set of exactly n independent n-tuples of elements of $\mathbb{F}$, the question is (or might as well be): how many independent sets of exactly n vectors in $\mathbb{F}^n$ are there?

Any non-zero n-tuple can be the first element of a basis; pick one, and call it x_1. Since the number of vectors in $\mathbb{F}^n$ is q^n, and since only the zero vector is to be avoided, the number of possible choices at this stage is $q^n - 1$. Any n-tuple that is not a scalar multiple of x_1 can follow x_1 as the second element of a basis; pick one and call it x_2. Since the number of vectors in $\mathbb{F}^n$ is q^n, and since only the scalar multiples of x_1 are to be avoided, the number of possible choices at this stage is $q^n - q$. (Note that the number of scalar multiples of x_1 is the same as the number of scalars, and that is q.) The next step in this inductive process is typical of the most general step. Any n-tuple that is not a linear combination of x_1 and x_2 can follow x_1 and x_2 as the third element of a basis; pick one and call it x_3. Since the number of vectors in $\mathbb{F}^n$ is q^n, and since only the linear combinations of x_1 and x_2 are to be avoided, the number of possible choices at this stage is $q^n - q^2$. (The number of linear combinations of two independent vectors is the number of the set of all pairs of scalars, and that is q^2.) Keep going the same way a total of n times altogether; the final answer is the product

$$(q^n - 1)(q^n - q)(q^n - q^2) \cdots (q^n - q^{n-1})$$

of the partial answers obtained along the way.

Caution: this product is not the number of bases, but the number of *ordered* bases, the ones in which a basis obtained by permuting the vectors of one already at hand is considered different from the original one. (Emphasis: the permutations here referred to are *not* permutations of co-ordinates in an n-tuple, but permutations of the vectors in a basis.) To get the number of honest (unordered) bases, divide the answer by $n!$.

A curious subtlety arises in this kind of counting. If $\mathbb{F} = \mathbb{Z}_2$, and the formula just derived is applied to $\mathbb{F}^3$ (that is, $q = 2$ and $n = 3$), it yields

$$(8 - 1)(8 - 2)(8 - 4)$$

ordered bases, and, therefore, 28 unordered ones. Related question: how many bases for $\mathbb{R}^3$ are there in which each vector (ordered triple of real numbers) is permitted to have the coordinates 0 and 1 only? A not too laborious count yields the answer 29. What accounts for the difference? Answer: the set $\{(0, 1, 1), (1, 0, 1), (1, 1, 0)\}$ is a basis for $\mathbb{R}^3$, but the same symbols interpreted modulo 2 describe a subset of $\mathbb{F}^3$ that is not a basis. (Why not?)

50   Solution 50.

The wording of the question suggests that the direct sum of two finite-dimensional vector spaces is finite-dimensional. That is true, and the best way to prove it is to use bases of the given vector spaces to construct a basis of their direct sum.

If $\{x_1, \ldots, x_n\}$ and $\{y_1, \ldots, y_m\}$ are bases for $\mathbb{U}$ and $\mathbb{V}$ respectively, then it seems natural to look at the set $\mathbb{B}$ of vectors

$$(x_1, 0), \ldots, (x_n, 0), (0, y_1), \ldots, (0, y_m),$$

and try to prove that it is a basis for $\mathbb{U} \oplus \mathbb{V}$.

The easiest thing to see is that $\mathbb{B}$ spans $\mathbb{U} \oplus \mathbb{V}$. Indeed, since every vector x in $\mathbb{U}$ is a linear combination of the x_i's, it follows that every vector of the form $(x, 0)$ in $\mathbb{U} \oplus \mathbb{V}$ is a linear combination of the $(x_i, 0)$'s. Similarly, every vector of the form $(0, y)$ is a linear combination of the $(0, y_j)$'s, and those two conclusions together imply that every vector (x, y) in $\mathbb{U} \oplus \mathbb{V}$ is a linear combination of the vectors in $\mathbb{B}$.

Is it possible that the set $\mathbb{B}$ is dependent? If

$$\alpha_1(x_1, 0) + \cdots + \alpha_n(x_n, 0) + \beta_1(0, y_1) + \cdots + \beta_m(0, y_m) = (0, 0),$$

then

$$\left(\sum_i \alpha_i x_i, \sum_j \beta_j x_j \right) = (0, 0),$$

and it follows from the independence of the x_i's and of the y_j's that $\alpha_1 = \cdots = \alpha_n = \beta_1 = \cdots = \beta_m = 0$, and the proof is complete.

Solution 51.   51

(a) Let the role of $\mathbb{V}$ be played by the vector space $\mathbb{P}$ of all real polynomials, and let $\mathbb{M}$ be the subspace of all even polynomials (see Problem 25). When are two polynomials equal (congruent) modulo $\mathbb{M}$? Answer: when their difference is even. When, in particular, is a polynomial equal to 0 modulo $\mathbb{M}$? Answer: when it is even. Consequence: if $p_n(x) = x^{2n+1}$, for $n = 0, 1, 2, \ldots$, then a non-trivial linear combination of a finite set of these p_n's can never be 0 modulo $\mathbb{M}$. Reason: in any linear combination of them, let k be the largest index for which the coefficient of p_k is not 0, and note that in that case the degree of the linear combination will be $2k + 1$ (which is not even). Conclusion: the quotient space $\mathbb{V}/\mathbb{M}$ has an infinite independent subset, which implies, of course, that it is not finite-dimensional.

(b) If, on the other hand, $\mathbb{N}$ is the subspace of all polynomials p for which $p(0) = 0$ (the constant term is 0), then the equality of two polynomials modulo $\mathbb{N}$ simply means that they have the same constant term. Consequence: every polynomial is congruent modulo $\mathbb{N}$ to a scalar multiple of the constant polynomial 1, which implies that the dimension of $\mathbb{V}/\mathbb{M}$

is 1. If bigger examples are wanted, just make N smaller. To be specific, let N be the set of all those polynomials in which not only the constant term is required to be 0, but the coefficients of the powers x, x^2, and x^3 are required to be 0 also. Consequence: every polynomial is congruent modulo N to a polynomial of degree 3 at most, which implies that the dimension of V/M is 4.

52   Solution 52.

If M is an m-dimensional subspace of an n-dimensional vector space V, then V/M has dimension $n - m$. Only one small idea is needed to begin the proof—after that everything becomes mechanical. The assumption that $\dim V = n$ means that a basis of V has n elements; the small idea is to use a special kind of basis, the kind that begins as a basis of M. To say that more precisely, let $\{x_1, \ldots, x_m\}$ be a basis for M, and extend it, by adjoining suitable vectors $x_{m+1}, \ldots, x_n$, so as to make it a basis of V. From now on no more thinking is necessary; the natural thing to try to do is to prove that the cosets

$$x_{m+1} + M, \ldots, x_n + M$$

form a basis for V/M.

Do they span V/M? That is: if $x \in V$, is the coset $x + M$ necessarily a linear combination of them? The answer is yes, and the reason is that x is a linear combination of $x_1, \ldots, x_n$, so that

$$x = \sum_{i=1}^{b} \alpha_i x_i$$

for suitable coefficients. Since

$$\sum_{j=1}^{m} \alpha_j x_j$$

is congruent to 0 modulo M, it follows that

$$x + M = \sum_{i>m} \alpha_i (x_i + M)$$

and that's exactly what's wanted.

Are the cosets $x_{m+1} + M, \ldots, x_n + M$ independent? Yes, and the reason is that the vectors $x_{m+1}, \ldots, x_n$ are independent modulo M. Indeed, if a linear combination of these vectors turned out to be equal to a vector, say z, in M, then z would be a linear combination of $x_1, \ldots, x_m$, and the only possible linear combination it could be is the trivial one (because the totality of the x's is independent).

The proof is complete, and it proved more than was promised: it concretely exhibited a basis of $n - m$ elements for V/M.

Solution 53.

The answer is easy to guess, easy to understand, and easy to prove, but it is such a frequently occurring part of mathematics that it's well worth a few extra minutes of attention. The reason it is easy to guess is that span and dimension behave (sometimes, partially) the same way as union and counting. The number of elements in the union of two finite sets is not the sum of their separate numbers—not unless the sets are disjoint. If they are *not* disjoint, then adding the numbers counts twice each element that belongs to both sets—the sum of the numbers of the separate sets is the number of elements in the union *plus* the number of elements in the intersection. The same sort of thing is true for spans and dimensions; the correct version of the formula in that case is

$$\dim(M + N) + \dim(M \cap N) = \dim M + \dim N.$$

The result is sometimes known as the **modular equation**.

To prove it, write $\dim(M \cap N) = k$, and choose a basis

$$\{z_1, \ldots, z_k\}$$

for $M \cap N$. Since a basis for a subspace can always be extended to a basis for any larger space, there exist vectors $x_1, \ldots, x_m$ such that the set

$$\{x_1, \ldots, x_m, z_1, \ldots, z_k\}$$

is a basis for M; in this notation

$$\dim M = m + k.$$

Similarly, there exist vectors $y_1, \ldots, y_n$ such that the set

$$\{y_1, \ldots, y_n, z_1, \ldots, z_k\}$$

is a basis for N; in this notation

$$\dim N = n + k.$$

The span of the x's is disjoint from N (for otherwise the x's and z's together couldn't be independent), and, similarly, the span of the y's is disjoint from

M. It follows that the set

$$\{x_1, \ldots, x_m, y_1, \ldots, y_n, z_1, \ldots, z_k\}$$

is a basis for $M + N$. The desired equation, therefore, takes the form

$$(m + n + k) + k = (m + k) + (n + k),$$

which is obviously true. (Note how the intersection $M \cap N$ is "counted twice" on both sides of the equation.)

That's all there is to it, but the proof has a notational blemish that is frequent in mathematical exposition. It is quite possible that some of the dimensions under consideration in the proof are 0; the case

$$\dim(M \cap N) = 0,$$

for instance, is of special interest. In that special case the notation is inappropriate: the suffix on z_1 suggests that $M \cap N$ has a non-empty basis, which is false. It is not difficult to cook up a defensible notational system in such situations, but usually it's not worth the trouble; it is easier (and no less rigorous) just to remember that in case something is 0 a part of the argument goes away.

Chapter 4. Transformations

54   **Solution 54.**

(a) The definitions (1) and (3) yield linear transformations; the definition (2) does not. The verification of linearity in (1) is boring but easy; just replace (x, y) by an arbitrary linear combination

$$\alpha_1(\xi_1, \eta_1) + \alpha_2(\xi_2, \eta_2),$$

apply T, and compare the result with the result of doing things in the other order. Here it is, for the record. Do NOT read it till after trying to write it down independently, and, preferably, do not ever read it.

First:

$$T\big(\alpha_1(\xi_1, \eta_1) + \alpha_2(\xi_2, \eta_2)\big)$$

$$= T(\alpha_1\xi_1 + \alpha_2\xi_2, \alpha_1\eta_1 + \alpha_2\eta_2)$$

$$= \big(\alpha(\alpha_1\xi_1 + \alpha_2\xi_2) + \beta(\alpha_1\eta_1 + \alpha_2\eta_2), \gamma(\alpha_1\xi_1 + \alpha_2\xi_2) + \delta(\alpha_1\eta_1 + \alpha_2\eta_2)\big).$$

Second:

$$\alpha_1 T(\xi_1, \eta_1) + \alpha_2 T(\xi_2, \eta_2)$$

$$= \alpha_1(\alpha\xi_1 + \beta\eta_1, \gamma\xi_1 + \delta\eta_1) + \alpha_2(\alpha\xi_2 + \beta\eta_2, \gamma\xi_2 + \delta\eta_2).$$

Third and last: compare the second lines of these equations.

As for (2): its linearity was already destroyed by the squaring counterexample in the discussion before the statement of the problem. Check it.

The example (3) is the same as (1); the only difference is in the names of the fixed scalars.

(b) As before, the definitions (1) and (3) yield linear transformations and the definition (2) does not. To discuss (1), look at any typical polynomial, such as, say

$$9x^3 - 3x^2 + 2x - 5,$$

and do what (1) says to do, namely, replace x by x^2. The result is

$$9x^6 - 3x^4 + 2x^2 - 5.$$

Then think of doing this to two polynomials, that is, to two elements of $\mathbb{P}$, and forming the sum of the results. Is the outcome the same as if the addition had been performed first and only then was x replaced by x^2? Do this quite generally: think of two arbitrary polynomials, think of adding them and then replacing x by x^2, and compare the result with what would have happened if you had replaced x by x^2 first and added afterward. It's not difficult to design suitable notation to write this down in complete generality, but thinking about it without notation is more enlightening—and the answer is yes. Yes, the results are the same. That's a statement about addition, which is a rather special linear combination, but the scalars that enter into linear combinations have no effect on the good outcome.

The definition (2) is the bad kind of squaring once more. Counterexample: consider the polynomial (vector) $p(x) = x$ and the scalar 2, and compare $T(2p(x))$ with $2Tp(x)$. The first is $(2p(x))^2$, which is $4x^2$, and the second is $2x^2$. Question: what happens if $p(x)$ is replaced by the even simpler polynomial $p(x) = 1$—is that a counterexample also?

The discussion of (3) can be carried out pretty much the same way as the discussion of (1): instead of talking about linear combinations and replacing x by x^2, talk about linear combinations and multiply them by x^2. It doesn't make any difference which is done first—the formula (6) does indeed define a linear transformation.

55   Solution 55.

(1) If F is a linear functional defined on a vector space V, then either $F(v)$ is 0 for every vector v in V, or it is not. (The possibility is a realistic one: the equation $F(v) = 0$ does indeed define a linear functional on every vector space.) If $F(v) = 0$ for all v, then ran F just consists of the vector 0 (in $\mathbb{R}^1$), and nothing else has to be said. If that is not the case, then the range of F contains some vector x_0 in $\mathbb{R}^1$ (a real number) different from 0. To say that x_0 is in the range means that V contains some vector v_0 such that $F(v_0) = x_0$. Since F is a linear functional (linear transformation), it follows in particular that

$$F(xv_0) = xx_0$$

for every real number x. As x ranges over all real numbers, so does the product xx_0. Conclusion: the range of F is all of $\mathbb{R}^1$.

(2) The replacement of x by $x + 2$ is a change of variables similar to (but simpler than) the replacement of x by x^2 considered in Problem 54 (1 (b)), and the proof that it is a linear transformation is similar to (but simpler than) what it was there. Squaring the variable can cause trouble because it usually raises the degree of the polynomial to which it is done (usually?—does it ever not do so?); the present simple change of variables does not encounter even that difficulty.

(3) The range of this transformation contains only one vector, namely $(0, 0)$; it is indeed a linear transformation.

(4) The equation does not define a linear transformation. Counterexamples are not only easy to find—they are hard to miss. For a special one, consider the vector $(0, 0, 0)$ and the scalar 2. Is it true that

$$T\bigl(2 \cdot (0, 0, 0)\bigr) = 2 \cdot T(0, 0, 0)?$$

The left side of the equation is equal to $T(0, 0, 0)$, which is $(2, 2)$; the right side, on the other hand, is equal to $2 \cdot (2, 2)$, which is $(4, 4)$.

(5) The "weird" vector space, call it W for the time being, is really the easy vector space $\mathbb{R}^1$ in disguise; they differ in notation only. That statement is worth examining in detail.

Suppose that two people, call them P and Q, play a notation game. Player P is thinking of the vector space $\mathbb{R}^1$, but as he plays the game he never *says* anything about the vectors that are in his thoughts—he *writes* everything. His first notational whimsy is to enclose every vectorial symbol in a box; instead of writing a vector x (in the present case a real number), he writes $\boxed{x}$, and instead of writing something like $2+3 = 5$ or something like $2 \cdot 3 = 6$, he writes

$$\boxed{2} \; \boxed{+} \; \boxed{3} \; = \boxed{5} \quad \text{or} \quad 2 \; \boxed{\cdot} \; \boxed{3} \; = \; \boxed{6}.$$

(Note: "2" in the last equation is a scalar, not a vector; that's why its symbol is not, should not be, in a box.) Player Q wouldn't be seriously mystified by such a thin disguise.

Suppose next that the notational change is a stranger one—the operational symbols $+$ and $\cdot$ continue to appear in boxes, but the symbols for vectors appear as exponents with the base 2. (Caution: vectors, not scalars.) In that case every time P thinks of a vector x what he writes is the number s obtained by using x as an exponent on the base 2. Example: P thinks 1 and writes 2; P thinks 0 and writes 1; P thinks 2 and writes 4; P thinks $\frac{1}{2}$ and writes $\sqrt{2}$; P thinks -3 and writes $\frac{1}{8}$. What will Q ever see? Since s is positive no matter what x is (that is, 2^x is positive no matter what real number x is), all the numbers that Q will ever see are positive. As x ranges over all possible real numbers, the exponential s (that is, 2^x) ranges over all possible positive real numbers. When P adds two real numbers (vectors), x and y say, what he reports to Q is $s \boxed{+} t$, where $s = 2^x$ and $t = 2^y$. Example: when P adds 1 and 2 and gets 3, the report that Q sees is $2 \boxed{+} 4 = 8$. As far as Q is concerned the numbers he is looking at were multiplied.

Scalar multiplication causes a slight additional notational headache. Both P and Q are thinking about a real vector space, which means that both are thinking about vectors, but P's vectors are numbers in $\mathbb{R}^1$ and Q's vectors are numbers in $\mathbb{R}_+$. Scalars, however, are the same for both, just plain real numbers. When P thinks of multiplying a real number x (a vector) by a real number y (a scalar), the traditional symbol for what he gets is yx, but what he writes is

$$y \boxed{\cdot} s = t,$$

where $s = 2^x$ and $t = 2^{yx}$. Notice that $2^{yx} = (2^x)^y$, or, in other words, $t = s^y$. Example: when P is thinking (in traditional notation) about $3 \cdot 2 = 6$, what Q sees is $3 \boxed{\cdot} 4 = 64$, which he interprets to mean that the scalar multiple of 4 by 3 has to be obtained by raising 4 to the power 3.

That's it—the argument shows (doesn't it?) that $\mathbb{R}^1$ and $\mathbb{R}_+$ differ in notation only. Yes, $\mathbb{R}_+$ is indeed a vector space. If T is defined on $\mathbb{R}_+$ by $T(s) = \log_2 s$ (note: log to the base 2), then T in effect decodes the notation that P encoded. When Q applies T to a vector s in $\mathbb{R}_+$, and gets $\log_2 s$, he recaptures the notation that P disguised. Thus, in particular, when $s = 2^x$ and $t = 2^y$ and T is applied to $s \boxed{+} t$, the result is $\log_2(2^x \cdot 2^y)$, which is $x + y$. In other words,

$$T(s \boxed{+} t) = Ts + Tt,$$

which is a part of the definition of a linear transformation. The other part, the one about scalar multiples goes the same way: if $s = 2^x$ and $t = 2^{yx}$, then

$$T(y \boxed{\cdot} s) = T(t) = yx = yT(s).$$

There is nothing especially magical about $\log_2$; logarithms to other bases could have been used just as well. Just remember that $\log_{10} s$, for instance, is just a constant multiple of $\log_2 s$—in fact

$$\log_{10} s = (\log_{10} 2) \cdot \log_2 s$$

for every positive real number s. If T had been defined by

$$T(s) = \log_{10} s,$$

the result would have been the same; the constant factor $\log_{10} s$ just goes along for the ride.

56   Solution 56.

(1) What do you know about a function if you know that its indefinite integral is identically 0? Answer: the function must have been 0 to start with. Conclusion: the kernel of the integration transformation is $\{0\}$.

(2) What do you know about a function if you know that its derivative is identically 0? Answer: the function must be a constant. Conclusion: ker D is the set of all constant polynomials.

(3) How can it happen that

$$2x + 3y = 0$$

and

$$7x - 57 = 0?$$

To find out, eliminate x. Since

$$7 \cdot 2x + 7 \cdot 3y = 0$$

and

$$2 \cdot 7x - 2 \cdot 5y = 0,$$

therefore

$$21y - 10y = 0,$$

or $y = 0$, and from that, in turn, it follows that

$$2x + 3y = 2x + 3 \cdot 0 = 2x = 0,$$

and hence that $x = 0$. Conclusion: $\ker T = \{(0,0)\}$.

(4) How can it happen for a polynomial p that $p(x^2) = 0$? Recall, for instance, that if

$$p(x) = 9x^3 - 3x^2 + 2x - 5,$$

then

$$p(x^2) = 9x^6 - 3x^4 + 2x^2 - 5;$$

the only way that can be 0 is by having all its coefficients equal to 0, which happens only when all the coefficients of p were 0 to begin with. (See Problem 54 (2 (a)).) Conclusion: the kernel of this change of variables is $\{0\}$.

(5) To say that $T(x, y) = (0, 0)$ is the same as saying that $(x, 0) = (0, 0)$, and that is the same as saying that $x = 0$. In other words, if (x, y) is in the kernel of T, then $(x, y) = (0, y)$. Conclusion: $\ker T$ is the y-axis.

(6) This is an old friend. The question is this: for which vectors (x, y) in $\mathbb{R}^2$ is it true that $x + 2y = 0$? Answer: the ones for which it is true, and nothing much more intelligent can be said about them, except that the set was encountered before and given the name $\mathbb{R}_0^2$. (See Problem 22.)

Solution 57.   57

(1) The answer is yes: the stretching transformation, which is just scalar multiplication by 7, commutes with *every* linear transformation. The computation is simple: if v is an arbitrary vector, then

$$(ST)v = S(Tv) \quad \text{by the definition of composition}$$

$$= 7(Tv) \quad \text{by the definition of } S$$

and

$$(TS)v = T(Sv) \quad \text{by the definition of composition}$$

$$= T(7v) \quad \text{by the definition of } S$$

$$= 7(Tv) \quad \text{by the linearity of } T.$$

The number 7 has, of course, nothing to do with all this: the same conclusion is true for every scalar transformation. (For every scalar γ the linear transformation S defined for every vector v by $Sv = \gamma v$ is itself called a scalar. Words are stretched by this usage, but in a harmless way, and breath is saved.) The proof is often compressed into one line (slightly artificially) as follows:

$$(ST)v = S(Tv) = c(Tv) = T(cv) = T(Sv) = (TS)v.$$

(2) The question doesn't make sense; $S \colon \mathbb{R}^3 \to \mathbb{R}^3$ (that is, S is a transformation from $\mathbb{R}^3$ to $\mathbb{R}^3$) and $T \colon \mathbb{R}^3 \to \mathbb{R}^2$, so that TS can be formed, but ST cannot.

(3) If $p(x) = x$, then $STp(x)$ (a logical fussbudget would write $((ST)p)(x)$, but the fuss doesn't really accomplish anything) $= STx = x^2 \cdot x = x^3$ and $TSx = Tx^2 = x^2 \cdot x^2 = x^4$—and that's enough to prove that S and T do not commute.

A student inexperienced with thinking about the minimal, barebones, extreme cases that are usually considered mathematically the most elegant might prefer to examine a more complicated polynomial (not just x, but, say, $1 + 2x + 3x^2$). For the brave student, however, there is an even more extreme case to look at (more extreme than x): the polynomial $p(x) = 1$. The action of T on 1 is obvious: $T1 = x^2$. What is the action of S on 1? Answer: the result of replacing the variable x by x^2 throughout—and since x does not explicitly appear in 1, the consequence is that $S1 = 1$. Consequence: $ST1 = Sx^2 = x^4$ and $TS1 = T1 = x^2$. Conclusion: (as before) S and T do not commute.

To say that S and T do not commute means, of course, that the compositions ST and TS are not the same linear transformation, and that, in turn, means that they disagree at at least one vector. It might happen that they agree at many vectors, but just one disagreement ruins commutativity. Do the present ST and TS agree anywhere? Sure: they agree at the vector 0. Anywhere else? That's a nice question, and it's worth a moment's thought here. Do ST and TS agree at any polynomial other than 0? Since

$$STp(x) = S\big(x^2 p(x)\big) = x^4 p(x^2)$$

and

$$TSp(x) = Tp(x^2) = x^2 p(x^2),$$

the question reduces to this: if $p \neq 0$, can $x^4 p(x^2)$ and $x^2 p(x^2)$ ever be the same polynomial? The answer is obviously no: if that equation held for $p \neq 0$, it would follow that $x^2 = x^4$, which is ridiculous. (Careful: $x^2 = x^4$ is not an equation to be solved for an unknown x. It offers itself as an equation, an identity, between two polynomials, and *that's* what's ridiculous.)

(4) Since $S: \mathbb{R}^2 \to \mathbb{R}^1$ and $T: \mathbb{R}^1 \to \mathbb{R}^2$, both products ST and TS make sense, and, in fact

$$ST: \mathbb{R}^1 \to \mathbb{R}^1 \qquad \text{and} \qquad TS: \mathbb{R}^2 \to \mathbb{R}^2.$$

It may be fun to calculate what ST and TS are, but for present purposes it is totally unnecessary. The point is that ST is a linear transformation on $\mathbb{R}^1$ and TS is a linear transformation on $\mathbb{R}^2$; the two have different domains and it doesn't make sense to ask whether they are equal. No, that's not correctly said: it makes sense to ask, but the answer is simply no.

(5) To decide whether $STp(x) = TSp(x)$ for all p, look at a special case, and, in particular, look at an extreme one such as $p(x) = 1$, and hope that it solves the problem. Since $S1 = 1$ and $T1 = 1$, it follows that $ST1 = TS1 = 1$. Too bad—that doesn't settle anything. What about $p(x) = x$? Since $STx = S0 = 0$ and $TSx = T(x+2) = 2$, that does settle something: S and T do not commute.

(6)-(1) The scalar 7 doesn't affect domain, range, or kernel: the question is simply about $\operatorname{dom} T$, $\operatorname{ran} T$, and $\ker T$. Answer:

$$\operatorname{dom} T = \operatorname{ran} T = \mathbb{R}^2, \qquad \text{and} \qquad \ker T = \{0\}.$$

(6)-(2) Since $TS(x, y, z) = T(7x, 7y, 7z) = (7x, 7y)$, it follows easily that $\operatorname{dom} TS = \mathbb{R}^3$, $\operatorname{ran} TS = \mathbb{R}^2$, and $\ker TS$ is the set of all those vectors (x, y, z) in $\mathbb{R}^3$, for which $x = y = 0$, that is, the z-axis. (Look at the whole question geometrically.) Since there is no such thing as ST, the part of the question referring to it doesn't make sense.

(6)-(3) The domains are easy: $\operatorname{dom} ST = \operatorname{dom} TS = \mathbb{P}$. The kernels are easy too: since

$$STp(x) = Sx^2 p(x) = x^4 p(x^2)$$

and

$$TSp(x) = Tp(x^2) = x^2 p(x^2),$$

it follows that $\ker ST = \ker Ts = \{0\}$. The question about ranges takes a minute of thought. It amounts to this: which polynomials are of the form $x^2 p(x^2)$, and which are of the form $x^4 p(x^2)$? Answer: $\operatorname{ran} TS$ is the set of all *even* polynomials with 0 constant term, and $\operatorname{ran} ST$ is the set of all those even polynomials in which, in addition, the coefficient of x^2 is 0 also.

(6)-(4) Now is the time to calculate the products:

$$STx = S(x, x) = x + 2x + 3x$$

and

$$TS(x, y) = T(x + 2y) = (x + 2y, x + 2y).$$

Answers: $\text{dom } ST = \mathbb{R}^1$, $\text{dom } TS = \mathbb{R}^2$, $\text{ran } ST = \mathbb{R}^1$, $\text{ran } TS$ is the "diagonal" consisting of all vectors (x, y) in $\mathbb{R}^2$ with $x = y$, $\ker ST = \{0\}$, and $\ker TS$ is the line with the equation $x + 2y = 0$.

(6)-(5) To find the answer to (5) the only calculations needed were for STx and TSx. To get more detailed information, more has to be calculated, as follows:

$$ST(\alpha + \beta x + \gamma x^2 + \delta x^3) = S(\alpha + \gamma x^2)$$
$$= \alpha + \gamma(x + 2)^2 = (\alpha + 2\gamma) + 4\gamma x + 4\gamma x^2$$

and

$$TS(\alpha + \beta x + \gamma x^2 + \delta x^3) = T\left(\alpha + \beta(x + 2) + \gamma(x + 2)^2 + \delta(x + 2)^3\right)$$
$$= (\alpha + 2\beta + 4\gamma + 8\delta) + (\gamma + 6\delta)x^2$$

There is no trouble with domains (both are $\mathbb{P}_3$). The range of ST is the set of all those quadratic polynomials for which the coefficients of x and x^2 are equal, and the range of TS is the set of all those quadratic polynomials for which the coefficient of x is 0. The kernel of ST is the set of all those cubic polynomials, that is polynomials of the form

$$\alpha + \beta x + \gamma x^2 + \delta x^3,$$

for which $\alpha = \gamma = 0$, and the kernel of TS is the set of all those whose coefficients satisfy the more complicated equations

$$\alpha + 2\beta + 4\gamma + 8\delta = \gamma + 6\delta = 0.$$

58   Solution 58.

Yes, $\text{ran } A \subset \text{ran } B$ implies the existence of a linear transformation T such that $A = BT$. The corresponding necessary condition for right divisibility, $A = SB$, is

$$\ker B \subset \ker A,$$

and it too is sufficient.

The problem is, given a vector x in the vector space $\mathbb{V}$, to define Tx, and, moreover, to do it so that Ax turns out to be equal to BTx. Put $y = Ax$, so that $y \in \text{ran } A$; the assumed condition then implies that $y \in \text{ran } B$. That means that $y = Bz$ for some z, and the temptation is to define Tx

to be z. That might not work. The difficulty is one of ambiguity: z is not uniquely determined by y. It could well happen that y is equal to both Bz_1 and Bz_2); should Tx be z_1 or z_2?

If $Bz_1 = Bz_2$, then $B(z_1 - z_2) = 0$, which says that

$$z_1 - z_2 \in \ker B.$$

The way to avoid the difficulty is to stay far away from $\ker B$, and the way to do that is to concentrate, at least temporarily, on a complement of $\ker B$. Very well: let $\mathbb{M}$ be such a complement, so that

$$\mathbb{M} \cap \ker B = \{0\} \qquad \text{and} \qquad \mathbb{M} + \ker B = \mathbb{V}.$$

Since B maps $\ker B$ to $\{0\}$, the image of $\mathbb{M}$ under B is equal to the entire range of B, and since $\mathbb{M}$ has only 0 in common with $\ker B$, the mapping B restricted to $\mathbb{M}$ is one-to-one. It follows that for each vector x there exists a vector z in $\mathbb{M}$ such that $Ax = Bz$, and, moreover, there is only one such z; it is now safe to yield to temptation and define Tx to be z. The conceptual difficulties are over; the rest consists of a routine verification that the transformation T so defined is indeed linear (and, even more trivially, that $A = BT$).

As for right divisibility, $A = SB$, the implication from there to $\ker B \subset \ker A$ is obvious; all that remains is to prove the converse. A little experimentation with the ideas of the preceding proof will reveal that the right thing to consider this time is a complement $\mathbb{N}$ of $\operatorname{ran} B$. For any vector x in $\operatorname{ran} B$, that is, for any vector of the form By, define Sx to be Ay. Does that make sense? Couldn't it happen that one and the same x is equal to both By_1 and By_2, so that Sx is defined ambiguously to be either Ay_1 or Ay_2? Yes, it could, but no ambiguity would result. The reason is that if $By_1 = By_2$, so that $y_1 - y_2 \in \ker B$, then the assumed condition implies that $y_1 - y_1 \in \ker A$, and hence that $Ay_1 = Ay_1$. Once S is defined on $\operatorname{ran} B$, it is easy to extend it to all of $\mathbb{V}$ just by setting it equal to 0 on $\mathbb{N}$. The rest consists of a routine verification that the transformation S so defined is indeed linear (and, even more trivially, that $A = SB$).

Solution 59.   59

The questions have interesting and useful answers in the finite-dimensional case; it is, therefore, safe and wise to assume that the underlying vector space is finite-dimensional.

(1) If the result of applying a linear transformation A to each vector in a total set is known, then the entire linear transformation is known. It

is instructive to examine that statement in a simple special case; suppose that the underlying vector space is $\mathbb{R}^2$. If

$$A(1,0) = (\alpha, \gamma)$$

and

$$A(0,1) = (\beta, \delta)$$

(there is a reason for writing the letters in a slightly non-alphabetic order here), then

$$A(x,y) = x(\alpha, \gamma) + y(\beta, \delta) = (\alpha x + \beta y, \gamma x + \delta y)$$

(and the alphabet has straightened itself out).

The reasoning works backwards too. Given A, corresponding scalars $\alpha, \beta, \gamma, \delta$ can be found (uniquely); given scalars $\alpha, \beta, \gamma, \delta$, a corresponding linear transformation A can be found (uniquely).

The space $\mathbb{R}^2$ plays no special role in this examination; every 2-dimensional space behaves the same way. And the number 2 plays no special role here; any finite-dimensional space behaves the same way. The only difference between the low and the high dimensions is that in the latter more indices (and therefore more summations) have to be juggled. Here is how the juggling looks.

Given: a linear transformation A on a vector space $\mathbb{V}$ with a prescribed total set, and an arbitrary vector x in $\mathbb{V}$. Procedure: express x as a linear combination of the vectors in the total set, and deduce that the result of applying A to x is the same linear combination of the results of applying A to the vectors of the total set. If, in particular, $\mathbb{V}$ is finite-dimensional, with basis $\{e_1, e_2, \ldots, e_n\}$, then a linear transformation A is uniquely determined by specifying Ae_j for each j. The image Ae_j is, of course, a linear combination of the e_i's, and, of course, the coefficient of e_i in its expansion depends on both i and j. Consequence: Ae_j has the form $\sum_{i=1}^{n} \alpha_{ij} e_i$. In reverse: given an array of scalars α_{ij} $(1 = 1, \ldots, n; j = 1, \ldots, n)$, a unique linear transformation A is defined by specifying that

$$Ae_j = \sum_{i=1}^{n} \alpha_{ij} e_i$$

for each j. Indeed, if

$$x = \sum_{j=1}^{n} \gamma_j e_j,$$

then

$$Ax = \sum_{j=1}^{n} \gamma_j A e_j = \sum_{j=1}^{n} \gamma_j \sum_{i=1}^{n} \alpha_{ij}\gamma_j = \sum_{i=1}^{n} \left(\sum_{j=1}^{n} \alpha_{ij}\gamma_j \right) e_j.$$

The conclusion is that there is a natural one-to-one correspondence between linear transformations A on a vector space of dimension n and square arrays (matrices) $\{\alpha_{ij}\}$ $(1 = 1,\ldots,n; \; j = 1,\ldots,n)$. Important comment: linear combinations of linear transformations correspond to the same linear combinations of arrays. If, that is,

$$A e_j = \sum_{i=1}^{n} \alpha_{ij} e_i \quad \text{and} \quad B e_j = \sum_{i=1}^{n} \beta_{ij} e_i,$$

then

$$(\alpha A + \beta B)e_j = \sum_{i=1}^{n} (\alpha \alpha_{ij} + \beta \beta_{ij})e_i.$$

Each $\{\alpha_{ij}\}$ has n^2 entries; except for the double subscripts (which are hardly more than a matter of handwriting) the α_{ij}'s are the coordinates of a vector in $\mathbb{R}^{n^2}$. Conclusion: the vector space $\mathbb{L}(\mathbb{V})$ is finite-dimensional; its dimension is n^2.

(2) Consider the linear transformations $1, A, A^2, \ldots, A^{n^2}$. They constitute $n^2 + 1$ elements of the vector space $\mathbb{L}(\mathbb{V})$ of dimension n^2, and, consequently, they must be linearly dependent. The assertion of linear dependence is the assertion of the existence of scalars $\alpha_0, \alpha_1, \ldots, \alpha_{n^2}$ such that

$$\alpha_0 + \alpha_1 A + \cdots + \alpha_{n^2} A^{n^2} = 0,$$

and that, in turn, is the assertion of the existence of a polynomial

$$\alpha_0 + \alpha_1 x + \cdots + \alpha_{n^2} x^{n^2}$$

such that $p(A) = 0$. Conclusion: yes, there always exists a non-zero polynomial p such that $p(A) = 0$.

(3) If A is defined by $Ax = y_0(x)x_0$, then

$$A^2 x = A[Ax] = y_0(x)Ax_0 = y_0(x)[y_0(x_0)x_0] = y_0(x_0)Ax.$$

In other words: $A^2 x$ is a scalar multiple (by the scalar $y_0(x_0)$) of Ax, or, simpler said, A^2 is a scalar multiple (by the scalar $y_0(x_0)$) of A. Differently expressed, the conclusion is that if p is the polynomial (of degree 2) defined by

$$p(t) = t^2 - y_0(x_0)t,$$

then $p(A) = 0$; the answer to the question is 2.

60   Solution 60.

Suppose that T is a linear transformation with inverse T^{-1} on a vector space $\mathbb{V}$. If v_1 and v_2 are in $\mathbb{V}$ with $Tu_1 = v_1$ and $Tu_2 = v_2$, then

$$T^{-1}(v_1 + v_2) = T^{-1}(Tu_1 + Tu_2)$$
$$= T^{-1}\big(T(u_1 + u_2)\big) \quad \text{(because } T \text{ is linear)}$$
$$= u_1 + u_2 \quad \text{(by the definition of } T^{-1})$$
$$= T^{-1}v_1 + T^{-1}v_2 \quad \text{(by the definition of } T^{-1}),$$

and, similarly, if v is an arbitrary vector in $\mathbb{V}$, α is an arbitrary scalar, and $v = Tu$, then

$$T^{-1}(\alpha v) = T^{-1}\big(\alpha(Tu)\big) = T^{-1}\big(T(\alpha u)\big) = \alpha u = \alpha(T^{-1}v)$$

—q.e.d.

61   Solution 61.

(1) What is the kernel of T? That is: for which $\begin{pmatrix} \xi \\ \eta \end{pmatrix}$ does it happen that

$$\begin{pmatrix} 2\xi + \eta \\ 2\xi + \eta \end{pmatrix} = \begin{pmatrix} 0 \\ 0 \end{pmatrix}?$$

Exactly those for which $2\xi + \eta = 0$, or, in other words, $\eta = -2\xi$, and that's a lot of them. The transformation T has a non-trivial kernel, and, therefore, it is not invertible.

(2) The kernel question can be raised again, and yields the answer that both ξ and η must be 0; in other words the only $\begin{pmatrix} \xi \\ \eta \end{pmatrix}$ in the kernel is $\begin{pmatrix} 0 \\ 0 \end{pmatrix}$. That suggests very strongly that T is invertible, but a really satisfying answer to the question is obtained by forming T^2. Since all that T does is interchange the two coordinates of whatever vector it is working on, T^2 interchanges them twice—which means that T^2 leaves them alone. Consequence: $T^2 = 1$, or, in other words, $T^{-1} = T$.

(3) The differentiation transformation D on $\mathbb{P}_5$ is not invertible. Reason (as twice before in this problem): D has a non-trivial kernel. That is: there exist polynomials p different from 0 for which $Dp = 0$—namely, all constant polynomials (except 0).

Solution 62.

Both assertions are false.

For (1), take (α) and (β) to be invertible, and put $(\gamma) = (\beta)^{-1}$, $(\delta) = (\alpha)^{-1}$. In that case

$$M = \begin{pmatrix} (\alpha) & (\beta) \\ (\beta)^{-1} & (\alpha)^{-1} \end{pmatrix},$$

which makes it obvious that all four formal determinants are equal to the matrix 0. If, in particular,

$$(\alpha) = \begin{pmatrix} 1 & 0 \\ 1 & 1 \end{pmatrix} \qquad \text{and} \qquad (\beta) = \begin{pmatrix} 1 & 1 \\ 0 & 1 \end{pmatrix},$$

then

$$(\alpha)^{-1} = \begin{pmatrix} 1 & 0 \\ -1 & 1 \end{pmatrix} \qquad \text{and} \qquad (\beta)^{-1} = \begin{pmatrix} 1 & -1 \\ 0 & 1 \end{pmatrix},$$

so that

$$M = \begin{pmatrix} 1 & 0 & 1 & 1 \\ 1 & 1 & 0 & 1 \\ 1 & -1 & 1 & 0 \\ 0 & 1 & -1 & 1 \end{pmatrix}.$$

The point is that M is invertible. Such a statement is never obvious—something must be proved. The simplest proof is concretely to exhibit the inverse, but the calculation of matrix inverses is seldom pure joy. Be that as it may, here it is; in the present case

$$M^{-1} = \begin{pmatrix} -1 & 1 & 1 & 0 \\ 0 & 1 & -1 & -1 \\ 1 & 0 & -1 & -1 \\ 1 & -1 & 0 & 1 \end{pmatrix}.$$

For (2), take (α) involutory $((\alpha)^2 = 1)$ and (β) nilpotent of index 2 $((\beta)^2 = 0)$, and put $(\gamma) = (\beta)$, $(\delta) = (\alpha)$. In that case

$$M = \begin{pmatrix} (\alpha) & (\beta) \\ (\beta) & (\alpha) \end{pmatrix},$$

which makes it obvious that all four formal determinants are equal to the identity matrix 1. If, in particular,

$$(\alpha) = \begin{pmatrix} 0 & 1 \\ 1 & 0 \end{pmatrix} \qquad \text{and} \qquad (\beta) = \begin{pmatrix} 0 & 0 \\ 1 & 0 \end{pmatrix},$$

then

$$M = \begin{pmatrix} 0 & 1 & 0 & 0 \\ 1 & 0 & 1 & 0 \\ 0 & 0 & 0 & 1 \\ 1 & 0 & 1 & 0 \end{pmatrix}.$$

Since the first and third columns of M are equal, so that M sends the first and third natural basis vectors to the same vector, the matrix M is not invertible.

63   Solution 63.

The problem of evaluating $\det M_1$ calls attention to a frequently usable observation, namely that the determinant of a direct sum of matrices is the product of their determinants. (The concept of direct sums of matrices has not been defined—is its definition guessable from the present context?) Since

$$\det \begin{pmatrix} 1 & 2 \\ 2 & 1 \end{pmatrix} = -3 \quad \text{and} \quad \det \begin{pmatrix} 3 & 4 \\ 4 & 3 \end{pmatrix} = -7,$$

it follows that $\det M_1 = 21$.

If a matrix has two equal columns (or two equal rows?), then it is not invertible, and, therefore, its determinant must be 0. The matrix M_2 has two equal rows (for instance, the first and the fifth, and also the second and the fourth) and therefore $\det M_2 = 0$.

The simplest trick for evaluating $\det M_3$ is to observe that M_3 is similar to the direct sum of three copies of the matrix $\begin{pmatrix} 3 & 2 \\ 2 & 3 \end{pmatrix}$. (The concept of similarity of matrices has not been defined yet—is its definition guessable from the present context?) The similarity is achieved by a permutation matrix. What that means, in simple language, is that if the rows and columns of M_3 are permuted suitably, M_3 becomes such a direct sum. Since

$$\det \begin{pmatrix} 3 & 2 \\ 2 & 3 \end{pmatrix} = 5,$$

it follows that $\det M_3 = 5^3 = 125$.

64   Solution 64.

If $n = 1$, then (1) is the only invertible 01-matrix and the number of its entries equal to 1 is 1; that's an uninteresting extreme case. When $n = 2$, the optimal example is

$$\begin{pmatrix} 1 & 1 \\ 0 & 1 \end{pmatrix}$$

with three 1's. What happens when $n = 3$?

The invertible matrix

$$\begin{pmatrix} 1 & 1 & 1 \\ 0 & 1 & 1 \\ 0 & 0 & 1 \end{pmatrix}$$

has six 1's; can that be improved? There is one and only one chance. An extra 1 in either the second column or in the second row would ruin invertibility; what about an extra 1 in position $\langle 3, 1 \rangle$? It works: the matrix

$$\begin{pmatrix} 1 & 1 & 1 \\ 0 & 1 & 1 \\ 1 & 0 & 1 \end{pmatrix}$$

is invertible. An efficient way to prove that is to note that its determinant is equal to 1.

Is the general answer becoming conjecturable? The procedure is inductive, and the general step is perfectly illustrated by the passage from 3 to 4. Consider the 4×4 matrix

$$\begin{pmatrix} 1 & 1 & 1 & 1 \\ 0 & 1 & 1 & 1 \\ 1 & 0 & 1 & 1 \\ 1 & 1 & 0 & 1 \end{pmatrix},$$

and expand its determinant in terms of the first column. The cofactor of the $\langle 1, 1 \rangle$ entry is invertible by the induction assumption. The $\langle 2, 1 \rangle$ entry is 0, and, therefore, contributes 0 to the expansion. The cofactor of the $\langle 3, k \rangle$ entry, for $k > 2$, contains two identical rows, namely the first two rows that consist entirely of 1's—it follows that that cofactor contributes 0 also. Consequence (by induction): the matrix is invertible.

The number of 1's in the matrix here exhibited is obtained from n^2 by subtracting the number of entries in the diagonal just below the main one, and that number is $n - 1$. This proves that the number of 1's can always be as great as $n^2 - n + 1$.

Could it be greater? If a matrix has as many as $n^2 - n + 2 \, (= n^2 - (n-2))$ entries equal to 1, then it has at most $n - 2$ entries equal to 0. Consequence: it must have at least two rows that have no 0's in them at all, that is at least two rows with nothing but 1's in them. A matrix with two rows of 1's cannot be invertible.

Comment. Can the desired invertibilities be proved without determinants? Yes, but the proof with determinants seems to be quite a bit simpler, and even, in some sense, less computational.

65   Solution 65.

Yes, $\mathbb{L}(V)$ has a basis consisting of invertible linear transformations. One way to construct such a basis is to start with an easy one that consists of non-invertible transformations and modify it. The easiest basis of $\mathbb{L}(V)$ is the set of all customary **matrix units**: they are the matrices $E(i, j)$ whose $\langle p, q \rangle$ entry is $\delta(i, p)\delta(j, q)$, where δ is the Kronecker delta. (The indices i, j, p, q here run through the values from 1 to n.) In plain language: each $E(i, j)$ has all entries except one equal to 0; the non-zero entry is a 1 in position $\langle i, j \rangle$. Example: if $n = 4$, then

$$E(2, 3) = \begin{pmatrix} 0 & 0 & 0 & 0 \\ 0 & 0 & 1 & 0 \\ 0 & 0 & 0 & 0 \\ 0 & 0 & 0 & 0 \end{pmatrix}.$$

The n^2 matrices $E(i, j)$ constitute a basis for the vector space $\mathbb{L}(V)$, but, obviously, they are not invertible. If

$$F(i, j) = E(i, j) + 1$$

(where the symbol "1" denotes the identity matrix), then the matrices $F(i, j)$ are invertible—that's easy—and they span $\mathbb{L}(V)$—that's not obvious. Since there are n^2 of them, the spanning statement can be proved by showing that the $F(i, j)$'s are linearly independent.

Suppose, therefore, that a linear combination of the F's vanishes:

$$\sum_{ij} \alpha(i, j) F(i, j) = 0,$$

or, in other words,

$$X = \sum_{ij} \alpha(i, j) \cdot 1 + \sum_{ij} \alpha(i, j) E(i, j) = 0.$$

If $p \neq q$, then the $\langle p, q \rangle$ entry of X is $0 + \alpha(p, q)$, and therefore $\alpha(p, q) = 0$. What about the entries $\alpha(p, p)$? The $\langle p, p \rangle$ entry of X is

$$\sum_{ij} \alpha(i, j) + \alpha(p, p),$$

which is therefore 0. But it is already known that $\alpha(i, j) = 0$ when $i = j$, and it follows that

$$\alpha(p, p) + \sum_i \alpha(i, i) = 0$$

for each p. Consequence: the $\alpha(p, p)$'s are all equal (!), and, what's more, their common value is the negative of their sum. The only way that can happen is to have $\alpha(p, p) = 0$ for all p—and that finishes the proof that the F's are linearly independent.

Solution 66.   66

The answer is that on a finite-dimensional vector space every injective linear transformation is surjective, and vice versa.

Suppose, indeed, that $\{u_1, u_2, \ldots, u_n\}$ is a basis of a vector space $\mathbb{V}$ and that T is a linear transformation on $\mathbb{V}$ with kernel $\{0\}$. Look at the transformed vectors $Tu_1, Tu_2, \ldots, Tu_n$: can they be dependent? That is: can there exist scalars $\alpha_1, \alpha_2, \ldots, \alpha_n$ such that

$$\alpha_1 Tu_1 + \alpha_2 Tu_2 + \cdots + \alpha_n Tu_n = 0?$$

If that happened, then (use the linearity of T) it would follow that

$$T(\alpha_1 u_1 + \alpha_2 u_2 + \cdots + \alpha_n u_n) = 0,$$

and hence that

$$\alpha_1 u_1 + \alpha_2 u_2 + \cdots + \alpha_n u_n = 0$$

(here is where the assumption about the kernel of T is used). Since, however, the set

$$\{u_1, u_2, \ldots, u_n\}$$

is independent, it would follow that all the α's are 0—in other words that the transformed vectors

$$Tu_1, Tu_2, \ldots, Tu_n$$

are independent. An independent set of n vectors in an n-dimensional vector space must be a basis (if not, it could be enlarged to become one, but then the number of elements in the enlarged basis would be different from n—see Problem 42). Since a basis of $\mathbb{V}$ spans $\mathbb{V}$, it follows that every vector is a linear combination of the vectors $Tu_1, Tu_2, \ldots, Tu_n$ and hence that the range of T is equal to $\mathbb{V}$. Conclusion: $\ker T = \{0\}$ implies $\operatorname{ran} T = \mathbb{V}$.

The reasoning in the other direction resembles the one just used. Suppose this time that $\{u_1, u_2, \ldots, u_n\}$ is a basis of a vector space $\mathbb{V}$ and that T is a linear transformation on $\mathbb{V}$ such that ran $T = \mathbb{V}$. Assertion: the transformed vectors $Tu_1, Tu_2, \ldots, Tu_n$ span $\mathbb{V}$. Reason: since, by assumption every vector v in $\mathbb{V}$ is the image under T of some vector u, and since every vector u is a linear combination of the form

$$\alpha_1 u_1 + \alpha_2 u_2 + \cdots + \alpha_n u_n,$$

it follows indeed that

$$v = Tu = T(\alpha_1 u_1 + \alpha_2 u_2 + \cdots + \alpha_n u_n)$$

$$= \alpha_1 Tu_1 + \alpha_2 Tu_2 + \cdots + \alpha_n Tu_n.$$

Since a total set of n vectors in an n-dimensional vector space must be a basis (if not, it could be decreased to become one, but then the number of elements in the enlarged basis would be different from n—see Problem 42), it follows that the transformed vectors $Tu_1, Tu_2, \ldots, Tu_n$ are independent. If now u is a vector in $\ker T$, then expand u in terms of the basis $\{u_1, u_2, \ldots, u_n\}$, so that

$$u = \alpha_1 u_1 + \alpha_2 u_2 + \cdots + \alpha_n u_n,$$

infer that

$$0 = Tu = \alpha_1 Tu_1 + \alpha_2 Tu_2 + \cdots + \alpha_n Tu_n,$$

and hence that the α's are all 0. Conclusion: ran $T = \mathbb{V}$ implies $\ker T = \{0\}$.

Comment.   The differentiation operator D on the vector space $\mathbb{P}_5$ is neither injective nor surjective; that's an instance of the result of this section. The differentiation operator D on the vector space $\mathbb{P}$ is surjective (is that right?), but not injective. The integration operator T (see Problem 56) is injective but not surjective. What's wrong?

The answer is that nothing is wrong; the theorem is about finite-dimensional vector spaces, and $\mathbb{P}$ is not one of them.

67   Solution 67.

If the dimension is 2, then there are only two ways a basis (consisting of two elements) can be permuted: leave its elements alone or interchange them. The identity permutation obviously doesn't affect the matrix at all, and the interchange permutation interchanges the two columns.

It is an easy (and familiar?) observation that every permutation can be achieved by a sequence of interchanges of just two objects, and, in the light of the comment in the preceding paragraph, the effect of each such inter-change is the corresponding interchange of the columns of the matrix. It is, however, not necessary to make use of the achievability of permutations by interchanges (technical word: **transpositions**); the conclusion is almost as easy to arrive at directly. If, for instance, the dimension is 3, if a basis is $\{e_1, e_2, e_3\}$, and if the permutation under consideration replaces that basis by $\{e_3, e_1, e_2\}$, then the effect of that replacement on a matrix such as

$$\begin{pmatrix} \alpha_{11} & \alpha_{12} & \alpha_{13} \\ \alpha_{21} & \alpha_{22} & \alpha_{23} \\ \alpha_{31} & \alpha_{32} & \alpha_{33} \end{pmatrix}$$

produces the matrix

$$\begin{pmatrix} \alpha_{13} & \alpha_{11} & \alpha_{12} \\ \alpha_{23} & \alpha_{21} & \alpha_{22} \\ \alpha_{33} & \alpha_{31} & \alpha_{32} \end{pmatrix}.$$

Solution 68.   68

To say that $\{\alpha_{ij}\}$ is a diagonal matrix is the same as saying that $\alpha_{ij} = \alpha_{ij}\delta_{ij}$ for all i and j (where δ_{ij} is the Kronecker delta, equal to 1 or 0 according as $i = j$ or $i \neq j$). If $B = \{\beta_{ij}\}$, then the $\langle i, j \rangle$ entry of AB is

$$\sum_{k=1}^{n} \alpha_{ik}\delta_{ik}\beta_{kj} = \alpha_{ii}\beta_{ij}$$

(because the presence of δ_{ik} makes every term except the one in which $k = i$ equal to 0), and the $\langle i, j \rangle$ entry of BA is

$$\sum_{k=1}^{n} \beta_{ik}\alpha_{kj}\delta_{kj} = \beta_{ij}\alpha_{jj}.$$

If $i \neq j$, then the assumption about the diagonal entries says that $\alpha_{ii} \neq \alpha_{jj}$, and it follows therefore, from the commutativity assumption, that β_{ij} must be 0. Conclusion: B is a diagonal matrix.

Solution 69.   69

If B commutes with every A, then in particular it commutes with every diagonal A with distinct diagonal entries, and it follows therefore, from Problem 68, that B must be diagonal—in the sequel it may be assumed, with no loss of generality, that B is of the form

$$\begin{pmatrix} \beta_1 & 0 & 0 & 0 \\ 0 & \beta_2 & 0 & 0 \\ 0 & 0 & \beta_3 & 0 \\ 0 & 0 & 0 & \beta_4 \end{pmatrix}.$$

At the same time B commutes with the matrices of all those linear transformations that leave fixed all but two entries of the basis. In matrix language those transformations can be described as follows: let p and q be any two distinct indices, and let C be obtained from the identity matrix by replacing the 1's in positions p and q by 0's and replacing the 0's in positions $\langle p, q \rangle$ and $\langle q, p \rangle$ by 1's. Typical example (with $n = 4$, $p = 2$, and $q = 3$):

$$C = \begin{pmatrix} 1 & 0 & 0 & 0 \\ 0 & 0 & 1 & 0 \\ 0 & 1 & 0 & 0 \\ 0 & 0 & 0 & 1 \end{pmatrix}.$$

Since

$$BC = \begin{pmatrix} \beta_1 & 0 & 0 & 0 \\ 0 & 0 & \beta_2 & 0 \\ 0 & \beta_3 & 0 & 0 \\ 0 & 0 & 0 & \beta_4 \end{pmatrix}$$

and

$$CB = \begin{pmatrix} \beta_1 & 0 & 0 & 0 \\ 0 & 0 & \beta_3 & 0 \\ 0 & \beta_2 & 0 & 0 \\ 0 & 0 & 0 & \beta_4 \end{pmatrix}$$

it follows that $\beta_2 = \beta_3$. It's clear (isn't it?) that the method works in general and proves that all the β's are equal.

70   Solution 70.

Consider the linear transformation

$$\begin{pmatrix} 0 & 1 \\ 0 & 0 \end{pmatrix},$$

or, more properly speaking, consider the linear transformation A on $\mathbb{R}^2$ defined by the matrix shown. Note that if $u = (\alpha, \beta)$ is any vector in $\mathbb{R}^2$, then $Au = (\beta, 0)$. Consequence: if M is an invariant subspace that contains a vector (α, β) with $\beta \neq 0$, then M contains $(\beta, 0)$ (and therefore $(1, 0)$), and it follows (via the formation of linear combinations) that M contains $(0, \beta)$ (and therefore $(0, 1)$). In this case $M = \mathbb{R}^2$.

If M is neither $\mathbb{O}$ nor $\mathbb{R}^2$, then every vector in M must be of the form $(\alpha, 0)$, and the set $\mathbb{M}_1$ of all those vectors do in fact constitute an invariant subspace. Conclusion: the only invariant subspaces are $\mathbb{O}$, $\mathbb{M}_1$, and $\mathbb{R}^2$.

Solution 71.   71

If D is the differentiation operator on the space $\mathbb{P}_n$ of polynomials of degree less than or equal to n, and if $m \leq n$, then $\mathbb{P}_m$ is a subspace of $\mathbb{P}_n$, and the subspace $\mathbb{P}_m$ is invariant under D. Does $\mathbb{P}_m$ have an invariant complement in $\mathbb{P}_n$?

The answer is no. Indeed, if p is a polynomial in $\mathbb{P}_n$ that is not in $\mathbb{P}_m$, in other words if the degree k of p is strictly greater than m, then replace p by a scalar multiple so as to justify the assumption that p is **monic** ($p(t) = t^k + a_{k-1}t^{k-1} + \cdots + a_0$). If p belongs to a subspace invariant under D, then $Dp, D^2p, \ldots$ all belong to that subspace, and, therefore, so does the polynomial $D^{k-m}p$, which is of degree m. Consequence: *every* polynomial has the property that if D is applied to it the right number of times, the result is in $\mathbb{P}_m$. Conclusion: $\mathbb{P}_m$ can have no invariant complement.

Comment. If $n = 1$, then $\mathbb{P}_n$ $(= \mathbb{P}_1)$ consists of all polynomials $\alpha + \beta t$ of degree 1 or less, and D sends such a polynomial onto the constant polynomial β $(= \beta + 0 \cdot t)$. That is only trivially (notationally) different from the set of ordered pairs (α, β) with the transformation that sends such a pair onto $(\beta, 0)$—in other words in that case the present solution reduces to Solution 69.

Solution 72.   72

A useful algebraic characterization of projections is **idempotence**. Explanation: to say that a linear transformation A is idempotent means that $A^2 = A$. (The Latin forms "idem" and "potent" mean "same" and "power".) In other words, the assertion is that if E is a projection, then $E^2 = E$, and, conversely, if $E^2 = E$, then E is a projection.

The idempotence of a projection is easy to prove. Suppose, indeed, that E is the projection on M along N. If $z = x + y$ is a vector, with x in M and y in N, then $Ez = x$, and, since $x = x + 0$, so that $Ex = x$, it follows that $E^2 z = Ez$.

Suppose now that E is an idempotent linear transformation, and let M and N be the range and the kernel of E respectively. Both M and N are subspaces; that's known. If z is in M, then, by the definition of range, $z = Eu$ for some vector u, and if z is also in N, then, by the definition of kernel, $Ez = 0$. Since $E = E^2$, the application of E to both sides of the equation $z = Eu$ implies that $Ez = z$; since, at the same time, $Ez = 0$, it follows that $z = 0$. Conclusion: $M \cap N = \mathbb{O}$.

If z is an arbitrary vector in $\mathbb{V}$, consider the vectors

$$Ez \qquad \text{and} \qquad z - Ez \quad (= (1 - E)z);$$

call them x and y. The vector x is in ran E, and, since

$$Ey = Ez - E^2 z = 0,$$

the vector y is in ker E. Since $z = x + y$, it follows that

$$M + N = \mathbb{V}.$$

The preceding two paragraphs between them say exactly that M and N are complementary subspaces and that the projection of any vector z to M along N is equal to Ez—that settles everything. Note, in particular, that the argument answers both questions: projections are just the idempotent linear transformations, and if E is the projection on M along N, then ran $E = M$ and ker $E = N$.

It is sometimes pleasant to know that if E is a projection, then ran E consists exactly of the fixed points of E. That is: if z is in ran E, then $Ez = z$, and, trivially, if $Ez = z$, then z is in ran E.

73   Solution 73.

If E and F are projections such that $E + F$ also is a projection, then

$$(E + F)^2 = E + F,$$

which says, on multiplying out, that

$$EF + FE = 0.$$

Multiply this equation on both left and right by E and get

$$EF + EFE = 0 \qquad \text{and} \qquad EFE + FE = 0.$$

Subtract one of these equations from the other and conclude that

$$EF - FE = 0,$$

and hence (since both the sum and the difference vanish)

$$EF = FE = 0.$$

That's a necessary condition that $E + F$ be a projection.

It is much easier to prove that the condition is sufficient also: if it is known that $EF = FE = 0$, then the cross product terms in $(E + F)^2$ disappear, and, in view of the idempotence of E and F separately, it follows that $E + F$ is idempotent.

Conclusion: the sum of two projections is a projection if and only if their products are 0. (Careful: two products, one in each order.)

Question. Can the product of two projections be 0 in one order but not the other? Yes, and that takes only a little thought and a little experimental search. If

$$E = \begin{pmatrix} 1 & 0 \\ 0 & 0 \end{pmatrix}, \quad F = \begin{pmatrix} \alpha & \beta \\ \gamma & \delta \end{pmatrix},$$

and $EF = 0$, then $\alpha = \beta = 0$. The resulting $F = \begin{pmatrix} 0 & 0 \\ \gamma & \delta \end{pmatrix}$ is idempotent if and only if either $\gamma = \delta = 0$ or else $\delta = 1$. A pertinent example is

$$F = \begin{pmatrix} 0 & 0 \\ 2 & 1 \end{pmatrix};$$

in that case $EF = 0$ and $FE \neq 0$.

Solution 74.

The condition is that $E^2 = E^3$; a strong way for a linear transformation to satisfy that is to have $E^2 = 0$. Is it possible to have $E^2 = 0$ without $E = 0$? Sure; a standard easy example is

$$E = \begin{pmatrix} 0 & 1 \\ 0 & 0 \end{pmatrix}.$$

In that case, indeed, $E^2(1 - E) = 0$, but $E(1 - E) = 0$ is false. That settles the first question.

It is easy to see that the answer to the second question is no—for the E just given it is *not* true that $E(1 - E)^2 = 0$ (because, in fact, $E(1 - E)^2 = E - E^2$).

That answers both questions, but it does not answer all the natural questions that should be asked.

One natural question is this: if $E(1 - E)^2 = 0$, does it follow that E is idempotent? No—how could it? Just replace the E used above by $1 - E$—that is, use

$$\begin{pmatrix} 1 & -1 \\ 0 & 1 \end{pmatrix}$$

as the new E. Then

$$1 - E = \begin{pmatrix} 0 & 1 \\ 0 & 0 \end{pmatrix},$$

so that $(1 - E)^2 = 0$, and therefore $E(1 - E)^2 = 0$, but it is not true that $E^2 = E$.

Another natural question: if both

$$E^2(1 - E) = 0 \qquad \text{and} \qquad E(1 - E)^2 = 0,$$

does it follow that E is idempotent? Sure: add the two equations and simplify to get $E - E^2 = 0$.

Chapter 5. Duality

75   **Solution 75.**

If $\eta = 0$, then $\xi = 0$, everything is trivial and the conclusion is true. In the remaining case, consider a vector x_0 such that $\eta(x_0) \neq 0$, and reason backward. That is, assume for a moment that there does exist a scalar α such that $\xi(x) = \alpha\eta(x)$ for all x, and that therefore, in particular, $\xi(x_0) = \alpha\eta(x_0)$, and infer that

$$\alpha = \frac{\xi(x_0)}{\eta(x_0)}.$$

[Note, not a surprise, but pertinent: it doesn't matter which x_0 was picked —so long as $\eta(x_0) \neq 0$, the fraction gives the value of α. Better said: if there is an α, it is uniquely determined by the linear functionals ξ and η.]

Now start all over again, and go forward (under the permissible assumption that there exists a vector x_0 such that $\eta(x_0) \neq 0$). The linear functional η sends

$$x_0 \quad \text{to} \quad \eta(x_0),$$

and hence it sends

$$\frac{x_0}{\eta(x_0)} \quad \text{to} \quad 1,$$

and hence

$$\frac{\gamma x_0}{\eta(x_o)} \quad \text{to} \quad \gamma$$

for every scalar γ. Special case: η sends

$$\frac{\eta(x)x_0}{\eta(x_0)} \quad \text{to} \quad \eta(x)$$

for every vector x. Consequence:

$$\eta\left(x - \frac{\eta(x)x_0}{\eta(x_0)}\right) = 0$$

for all x. The relation between ξ and η now implies that

$$\xi\left(x - \frac{\eta(x)x_0}{\eta(x_0)}\right) = 0$$

for all x, which says exactly that

$$\xi(x) = \frac{\xi(x_0)}{\eta(x_0)}\eta(x),$$

and that is what was wanted (with $\alpha = \frac{\xi(x_0)}{\eta(x_0)}$).

Solution 76.

Yes, the dual of a finite-dimensional vector space V is finite-dimensional, and the way to prove it is to use a basis of V to construct a basis of V'. Suppose, indeed, that $\{x_1, x_2, \ldots, x_n\}$ is a basis of V. Plan: find linear functionals $\xi_1, \xi_2, \ldots, \xi_n$ that "separate" the x's in the sense that

$$\xi_i(x_j) = \delta_{ij}$$

for each $i, j = 1, 2, \ldots, n$. Can that be done? If it could, then the value of ξ_i at a typical vector

$$x = \alpha_1 x_1 + \alpha_2 x_2 + \cdots + \alpha_n x_n$$

would be α_i—and that shows how ξ_i should be defined when it is not yet known. That is: writing $\xi_i(x) = \alpha_i$ for each i does indeed define a linear functional (verification?).

The linear functionals $\xi_1, \xi_2, \ldots, \xi_n$ are linearly independent. Proof: if

$$\beta_1\xi_1(x) + \beta_2\xi_2(x) + \cdots + \beta_n\xi_n(x) = 0$$

for all x, then, in particular, the linear combination vanishes when $x = x_j$ ($j = 1, \ldots, n$), which says exactly that $\beta_j = 0$ for each j.

Every linear functional is a linear combination of the ξ_i's. Indeed, if ξ is an arbitrary linear functional and $x = \alpha_1 x_1 + \alpha_2 x_2 + \cdots + \alpha_n x_n$ is an arbitrary vector, then

$$\xi(x) = \xi(\alpha_1 x_1 + \alpha_2 x_2 + \cdots + \alpha_n x_n)$$
$$= \alpha_1 \xi(x_1) + \alpha_2 \xi(x_2) + \cdots + \alpha_n \xi(x_n)$$
$$= \xi_1(x)\xi(x_1) + \xi_2(x)\xi(x_2) + \cdots + \xi_n(x)\xi(x_n)$$
$$= \left(\xi_1(x_1)\xi_1 + \xi_2(x_2)\xi_2 + \cdots + \xi_n(x_n)\xi_n\right)(x).$$

The preceding two paragraphs yield the conclusion that the x's constitute a basis of V' (and hence that V' is finite-dimensional).

Corollary. $\dim V' = \dim V = n$.

77   Solution 77.

The answer is yes: the linear transformation T defined by

$$T(x) = T(x_1, \ldots, x_n) = (y_1(x), \ldots, y_n(x))$$

is invertible. One reasonably quick way to prove that is to examine the
kernel of T. Suppose that $x = (x_1, \ldots, x_n)$ is a vector in $\mathbb{F}^n$ that belongs
to the kernel of T, so that

$$\left(y_1(x), \ldots, y_n(x)\right) = (0, \ldots, 0).$$

Since the coordinate projections p_j belong to the span of $y_1, \ldots, y_n$, it follows that for each j there exist scalars $\alpha_1, \ldots, \alpha_n$ such that

$$p_j = \sum_k \alpha_k y_k.$$

Consequence:

$$p_j(x) = \sum_k \alpha_k y_k = 0,$$

for each j, which implies, of course, that $x = 0$; in other words the kernel
of T is $\{0\}$. Conclusion: T is invertible.

Solution 78.

78

The verification that T is linear is the easiest step. Indeed: if x and y are in V and α and β are scalars, then

$$T(\alpha x + \beta y)(u) = u(\alpha x + \beta y) = \alpha u(x) + \beta u(y)$$
$$= \alpha (Tx)(u) + \beta (Ty)(u)$$
$$= (\alpha Tx + \beta Ty)(u).$$

How can it happen (for a vector x in V) that $Tx = 0$ (in V'')? Answer: it happens when

$$u(x) = 0$$

for every linear functional u on V—and that must imply that $x = 0$ (see the discussion preceding Problem 74). Consequence: T is always a one-to-one mapping from V to V''.

  The only question that remains to be asked and answered is whether or not T maps V onto V'', and in the finite-dimensional case the answer is easily accessible. The range of T is a subspace of V'' (Problem 55); since T is an isomorphism from V to $\operatorname{ran} T$, the dimension of $\operatorname{ran} T$ is equal to the dimension of V. The dimension of V'' is equal to the dimension of V also (because $\dim V = \dim V'$ and $\dim V' = \dim V''$). A subspace of dimension n in a vector space of dimension n cannot be a proper subspace. Consequence: $\operatorname{ran} T = V''$. Conclusion: the natural mapping of a finite-dimensional vector space to its double dual is an isomorphism, or, in other words, every finite-dimensional vector space is reflexive.

Solution 79.

79

Some proofs in mathematics require ingenuity, and others require nothing more than remembering and using the definitions—this one begins with a tiny inspiration and then finishes with the using-the-definitions kind of routine.

  Choose a basis $\{x_1, x_2, \ldots, x_n\}$ for V so that its first m elements are in M (and therefore form a basis for M); let $\{u_1, u_2, \ldots, u_n\}$ be the dual basis in V' (see Solution 75). Since $u_i(x_j) = \delta_{ij}$, it follows that the u_i's with $i > m$ annihilate M and with $i \leq m$ do not. In other words, if the span of the u_i's with $i > m$ is called N, then $N \subset M^0$.

  If, on the other hand, u is in M^0, then, just because it is in the space V', the linear functional u is a linear combination of u_i's. Since any such linear combination

$$u = \beta_1 u_1 + \beta_2 u_2 + \cdots + \beta_n u_n$$

applied to one of the x_j's with $j \leq m$ yields 0 (because x_j is in M) and, at the same time, yields β_j (because $u_i(x_j) = 0$ when $i \neq j$), it follows that the coefficients of the early u_i's are all 0. Consequence: u is a linear combination of the latter u_i's, or, in other words, u is in N, or, better still $M^0 \subset N$.

The conclusions of the preceding two paragraphs imply that $M^0 = N$, and hence, since N has a basis of $n - m$ elements, that

$$\dim M^0 = n - m.$$

80   Solution 80.

If the spaces V and V″ are identified (as suggested by Problem 77), then, by definition, M^{00} consists of the set of all those vectors x in M such that $u(x) = 0$ for all u in V′. Since, by the definition of V′, the equation $u(x) = 0$ holds for all x in M and all u in M^0, so that every x in M satisfies the condition just stated for belonging to M^{00}, it follows that $M \subset M^{00}$. If the dimension of V is n and the dimension of M is m, then (see Problem 78) the dimension of M^0 is $n - m$, and therefore, by the same result, the dimension of M^{00} is $n - (n - m)$. In other words M is an m-dimensional subspace of an m-dimensional space M^{00}, and that implies that M and M^{00} must be the same.

81   Solution 81.

Suppose that A is a linear transformation on a finite-dimensional vector space V and A' is its adjoint on V′. If u is an arbitrary vector in ker A', so that $A'u = 0$, then, of course $(A'u)(x) = 0$ for every x in V, and consequently $u(Ax) = 0$ for every x in V. The latter equation says exactly that u takes the value 0 at every vector in the range of A, or, simpler said, that u belongs to $(\operatorname{ran} A)^0$. The argument is reversible: if u belongs to $(\operatorname{ran} A)^0$, so that $u(Ax) = 0$ for every x, then $(A'u)(x) = 0$ for every x, and therefore $A'u = 0$, or, simpler said, u belongs to ker A'. Conclusion:

$$\ker A' = (\operatorname{ran} A)^0.$$

It should not come as too much of a surprise that annihilators enter. The range and the kernel of A are subspaces of V, and the range and the kernel of A' are subspaces of V′—what possible relations can there be between subspaces of $\mathbb{V}$ and subspaces of $\mathbb{V}'$? The only known kind of relation (at least so far) has to do with annihilators.

If A is replaced by A' in the equation just derived, the result is

$$\ker A'' = (\operatorname{ran} A')^0,$$

an equation that seems to give some information about $\operatorname{ran} A'$—that's good. The information is, however, indirect (via the annihilator), and it is expressed indirectly (in terms of A'' instead of A). Both of these blemishes can be removed. If $\mathbb{V}''$ is identified with $\mathbb{V}$ (remember reflexivity), then A'' becomes A, and if the annihilator of both sides of the resulting equation is formed (remember double annihilators), the result is

$$\operatorname{ran} A' = (\ker A)^0.$$

Question. Was finite-dimensionality needed in the argument? Sure: the second paragraph made use of reflexivity. What about the first paragraph —is finite-dimensionality needed there?

Solution 82.   82

What is obvious is that the adjoint of a projection is a projection. The reason is that projections are characterized by idempotence (Problem 71), and idempotence is inherited by adjoints.

Problem 71 describes also what a projection is "on" and "along": it says that if E is the projection on $\mathbb{M}$ along $\mathbb{N}$, then

$$\mathbb{N} = \ker E$$

and

$$\mathbb{M} = \operatorname{ran} E.$$

It is a special case of the result of Solution 80 that

$$\ker E' = (\operatorname{ran} E)^0$$

and

$$\operatorname{ran} E' = (\ker E)^0.$$

Consequence: E' is the projection on $\mathbb{N}^0$ along $\mathbb{M}^0$.

Solution 83.   83

Suppose that A is a linear transformation on a finite-dimensional vector space $\mathbb{V}$, with basis $\{e_1, \ldots, e_1\}$, and consider its adjoint A' on the dual space $\mathbb{V}'$, with the dual basis $\{u_1, \ldots, u_n\}$. What is wanted is to compare the expansion of each $A'u$ in terms of the u's with the expansion of each Ae in terms of the e's. The choice of notation should exercise some alphabetic care; this is a typical case where subscript juggling cannot be avoided, and carelessness with letters can make them step on their own tails in a confusing manner.

The beginning of the program is easy enough to describe: expand $A'u_j$ in terms of the u's, and compare the result with what happens when Ae_i is expanded in terms of the e's. The alphabetic care is needed to make sure that the "dummy variable" used in the summation is a harmless one— meaning, in particular, that it doesn't collide with either j or i. Once that's said, things begin to roll: write

$$A'u_j = \sum_k \alpha'_{kj} u_k,$$

evaluate the result at each e_i, and do what the notation almost seems to force:

$$A'u_j(e_i) = \sum_k \alpha'_{kj} u_k(e_i) = \sum_k \alpha'_{kj} \delta_{ki} = \alpha'_{ij}.$$

All right—that gives an expression for the matrix entries α'_{ij} of A'; what is to be done next? Answer: recall the way the matrix entries are defined for A, and hope that the two expressions together give the desired information. That is: look at

$$Ae_i = \sum_k \alpha_{ki} e_k,$$

apply each u_j, and get

$$u_j(Ae_i) = u_j\left(\sum_k \alpha_{ki} e_k\right) = \sum_k \alpha_{ki} e_k u_j(e_k) = \sum_k \alpha_{ki} e_k \delta_{jk} = \alpha_{ji}.$$

Since

$$u_j(Ae_i) = A'u_j(e_i),$$

it follows that

$$\alpha'_{ij} = \alpha_{ji}$$

for all i and j. Victory: that's a good answer. It says that the matrix entries of A' are the same as the matrix entries of A with the subscripts interchanged. Equivalently: the matrix of A' is the same as the matrix of A with the rows and columns interchanged. Still better (and this is the most popular point of view): the matrix of A' is obtained from the matrix of A by flipping it over the main diagonal. In the customary technical language: the matrix of A' is the **transpose** of the matrix of A.

Chapter 6. Similarity

Solution 84.

84

The interesting and useful feature of the relation between x and y is the answer to this question: how does one go from x to y? To "go" means (and that shouldn't be a surprise) to apply a linear transformation. The natural way to go that offers itself is the unique linear transformation T determined by the equations

$$Tx_1 = y_1,$$
$$\vdots$$
$$Tx_n = y_n.$$

The linear transformation T has the property that $Tx = y$; indeed

$$Tx = T(\alpha_1 x_1 + \cdots + \alpha_1 x_n)$$
$$= \alpha_1 Tx_1 + \cdots + \alpha_1 Tx_n$$
$$= \alpha_1 y_1 + \cdots + \alpha_1 y_n = y.$$

The answer to the original question, expressed in terms of T, is therefore simply this: the relation between x and y is that $Tx = y$. That is: a "change of basis" is effected by the linear transformation that changes one basis to another.

Question. Is T invertible?

Solution 85.

85

The present question compares with the one in Problem 83 the way matrices compare with linear transformations. The useful step in the solution of Problem 83 was to introduce the linear transformation T that sends the x's to the y's. Question: what is the matrix of that transformation (with respect to the basis $(x_1, \ldots, x_n)$)? The answer is obtained (see Solution 59) by applying T to each x_j and expanding the result in terms of the x's. If

$$y_j = Tx_j = \sum_i \alpha_{ij} x_i,$$

then

$$\sum_j \eta_j y_j = \sum_j \eta_j \sum_i \alpha_{ij} x_i = \sum_i \left(\sum_j \alpha_{ij} \eta_j \right) x_i.$$

Since, however, by assumption

$$\sum_i \xi_i x_i = \sum_j \eta_j y_j,$$

it follows that

$$\xi_i = \sum_j \alpha_{ij} \eta_j.$$

That's the answer: the relation that the question asks for is that the ξ's can be calculated from the η's by an application of the matrix (α_{ij}). Equivalently: a change of basis is effected by the matrix that changes one coordinate system to another.

86   Solution 86.

The effective tool that solves the problem is the same linear transformation T that played an important role in Solutions 84 and 85, the one that sends the x's to the y's. If, that is,

$$Tx_j = y_j \qquad (j = 1, \ldots, n)$$

then

$$Cy_j = CTx_j$$

and

$$Cy_j = \sum_i \alpha_{ij} y_i = \sum_i \alpha_{ij} Tx_i = T \left(\sum_i \alpha_{ij} x_i \right) = TBx_j.$$

Consequence:

$$CTx_j = TBx_j$$

for all j, so that

$$CT = TB.$$

That's an acceptable answer to the question, but the usual formulation of the answer is slightly different. Solution 84 ended by a teasing puzzle that asked whether T is invertible. The answer is obviously yes—T sends a basis (namely the x's) onto a basis (namely the y's), and that guarantees invertibility. In view of that fact, the relation between C and B can be written in the form

$$C = TBT^{-1},$$

and that is the usual equation that describes the similarity of B and C.

The last phrase requires a bit more explanation. What it is intended to convey is that B and C are similar if and only if there exists an invertible transformation T such that $C = TBT^{-1}$. The argument so far has proved only the "only if". The other direction, the statement that if an invertible T of the sort described exists, then B and C are indeed similar, is not immediately obvious but it is pretty easy. It is to be proved that if T exists, then B and C do indeed correspond to the same matrix via two different bases. All right; assume the existence of a T, write B as a matrix in terms of an arbitrary basis $\{x_1, \ldots, x_n\}$, so that

$$Bx_j = \sum_i \alpha_{ij} x_i,$$

define a bunch of vectors y by writing $Tx_j = y_j$ ($j = 1, \ldots, n$), and then compute as follows:

$$Cy_j = CTx_j = TBx_j = \sum_i \alpha_{ij} Tx_i = \sum_i \alpha_{ij} y_i.$$

Conclusion: the matrix of B with respect to the x's is the same as the matrix of C with respect to the y's.

Solution 87.   87

Some notation needs to be set up. The assumption is that *one* linear transformation is given, call it B, and *two* bases

$$\{x_1, \ldots, x_n\} \quad \text{and} \quad \{y_1, \ldots, y_n\}.$$

Each basis can be used to express B as a matrix,

$$Bx_j = \sum_i \beta_{ij} x_i \quad \text{and} \quad By_j = \sum_i \gamma_{ij} y_i,$$

and the question is about the relation between the β's and the γ's.

The transformation T that has been helpful in the preceding three problems $(Tx_j = y_j)$ can still be helpful, but this time (because temporarily matrices are at the center of the stage, not linear transformations), it is advisable to express its action in matrix language. The matrix of T with respect to the x's is defined by

$$Tx_j = \sum_k \tau_{kj} x_k,$$

and now the time has come to compute with it. Here goes:

$$By_j = BTx_j = B \sum_k \tau_{kj} x_k$$

$$= \sum_k \tau_{kj} B x_k$$

$$= \sum_k \tau_{kj} \sum_i \beta_{ik} x_i$$

$$= \sum_i \left(\sum_k \beta_{ik} \tau_{kj} \right) x_i$$

and

$$By_j = \sum_k \gamma_{kj} y_k$$

$$= \sum_k \gamma_{kj} T x_k \sum_k \gamma_{kj} \sum_i \tau_{ik} x_i$$

$$= \sum_i \left(\sum_k \tau_{ik} \gamma_{kj} \right) x_i.$$

Consequence:

$$\sum_k \tau_{ik} \gamma_{kj} = \sum_k \beta_{ik} \tau_{kj}.$$

In an abbreviated but self-explanatory form the last equation asserts a relation between the matrices β and χ, namely that

$$\tau\gamma = \beta\tau.$$

The invertibility of τ permits this to be expressed in the form

$$\gamma = \tau^{-1}\beta\tau,$$

and, once again, the word similarity can be used: the matrices β and γ are similar.

Solution 88.   88

Yes, it helps to know that at least one of B and C is invertible; in that case the answer is yes. If, for instance, B is invertible, then

$$BC = BC(BB^{-1}) = B(CB)B^{-1};$$

the argument in case C is invertible is a trivial modification of this one.

If neither B nor C is invertible, the conclusion is false. Example: if

$$B = \begin{pmatrix} 1 & 0 \\ 0 & 0 \end{pmatrix} \quad \text{and} \quad C = \begin{pmatrix} 0 & 1 \\ 0 & 0 \end{pmatrix},$$

then

$$BC = C \quad \text{and} \quad CB = 0.$$

The important part of this conclusion is that $BC \neq 0$ but $CB = 0$.

Comment. There is an analytic kind of argument, usually frowned upon as being foreign in spirit to pure algebra, that can sometimes be used to pass from information about invertible transformations to information about arbitrary ones. An example would be this: if B is invertible, then BC and CB are similar; if B is not invertible then it is the limit (here is the analysis) of a sequence $\{B_n\}$ of invertible transformations. Since $B_n C$ is similar to CB_n, it follows (?) by passage to the limit that BC is similar to CB.

The argument is phony of course—where does it break down? What is true is that there exist invertible transformations T_n such that

$$(B_n C)T_n = T_n(CB_n),$$

and what the argument tacitly assumes is that the sequence of T's (or possibly a subsequence) converges to an invertible limit T. If that were true, then it would indeed follow that $(BC)T = T(BC)$—hence, as the proof above implies, that cannot always be true. Here is a concrete example for the B and C mentioned above: if

$$B_n = T_n = \begin{pmatrix} 1 & 0 \\ 0 & \frac{1}{n} \end{pmatrix},$$

then indeed

$$(B_n C)T_n = T_n(CB_n)$$

for all n, and the sequence $\{T_n^{-1}\}$ converges all right, but its limit refuses to be invertible.

89   Solution 89.

Yes, two real matrices that are complex similar are also real similar. Suppose, indeed, that A and B are real and that

$$SA = BS,$$

where S is an invertible complex matrix. Write S in terms of its real and imaginary parts,

$$S = P + iQ.$$

Since $PA + iQA = BP + iBQ$, and since A, B, P, and Q are all real, it follows that

$$PA = BP \qquad \text{and} \qquad QA = BQ.$$

The problem might already be solved at this stage; it is solved if either P or Q is invertible, because in that case the preceding equations imply that A and B are "real similar". Even if P and Q are not invertible, however, the solution is not far away.

Consider the polynomial

$$p(\lambda) = \det(P + \lambda Q).$$

Since $p(i) = \det(P + iQ) \neq 0$ (because S is invertible), the polynomial p is not identically 0. It follows that the equation $p(\lambda) = 0$ can have only a finite number of roots and hence that there exists a real number λ such that the real matrix $P + \lambda Q$ is invertible. That does it: since

$$(P + \lambda Q)A = PA + \lambda QA = BP + \lambda BQ = B(P + \lambda Q),$$

the matrices A and B are similar over the field of real numbers.

The computation in this elementary proof is surely mild, but it's there just the same. An alternative proof involves no computation at all, but it is much less elementary; it depends on the non-elementary concept of "elementary divisors". They are polynomials associated with a matrix; their exact definition is not important at the moment. What is important is that their coefficients are in whatever field the entries of the matrix belong to, and that two matrices are similar if and only if they have the same elementary divisors. Once these two statements are granted, the proof is finished: if A and B are real matrices that are similar (over whatever field happens to be under consideration, provided only that it contains the entries of A and B), then they have the same elementary divisors, and therefore they must be similar over every possible field that contains their entries—in particular over the field of real numbers.

Solution 90.   **90**

Since

$$\ker A' = (\operatorname{ran} A)^0$$

(by Problem 80) and since

$$\dim(\operatorname{ran} A)^0 = n - \dim \operatorname{ran} A$$

(by Problem 65), it follows immediately that

$$\operatorname{null} A' = n - \operatorname{rank} A,$$

Suppose now that $\{x_1, \ldots, x_m\}$ is a basis for $\ker A$ and extend it to a basis $\{x_1, \ldots, x_m, x_{m+1}, \ldots, x_n\}$ of the entire space $\mathbb{V}$. If

$$x = \alpha_1 x_1 + \cdots + \alpha_m x_m + \alpha_{m+1} x_{m+1} + \cdots + \alpha_n x_n$$

is an arbitrary vector in $\mathbb{V}$, then

$$Ax = \alpha_{m+1} A x_{m+1} + \cdots + \alpha_n A x_n,$$

which implies that $\operatorname{ran} A$ is spanned by the set $\{A x_{m+1}, \ldots, A x_n\}$. Consequence:

$$\dim \operatorname{ran} A \leqq n - m,$$

or, in other words,

$$\operatorname{rank} A \leqq n - \operatorname{null} A.$$

Apply the latter result to A', and make use of equation above connecting null A' and rank A to get

$$\operatorname{rank} A' \leqq \operatorname{rank} A.$$

That almost settles everything. Indeed: apply it to A' in place of A to get

$$\operatorname{rank} A'' \leqq \operatorname{rank} A';$$

in view of the customary identification of A'' and A, the last two inequalities together imply that

$$\operatorname{rank} A = \operatorname{rank} A'.$$

Consequence:

$$\operatorname{null} A' = n - \operatorname{rank} A',$$

and that equation, with A' in place of A (and the same identification argument as just above) yields

$$\text{rank } A + \text{null } A = n.$$

The answer to the first question of the problem as stated is that if rank $A = r$, then there is only one possible value of rank A', namely the same r. The answer to the second question is that there is only one possible value of null A, namely $n - r$.

Comment. A special case of the principal result above (rank + nullity = dimension) is obvious: if rank $A = 0$ (which means exactly that ran $A = \mathbb{O}$), then A must be the transformation 0, and therefore null A must be n. A different special case, not quite that trivial, has already appeared in this book, in Problem 65. The theorem there says, in effect, that the nullity of a linear transformation on a space of dimension n is 0 if and only if its rank is n—and that says exactly that the sum formula is true in case null $A = 0$.

91   Solution 91.

Yes, similar transformations have the same rank. Suppose, indeed, that B and C are linear transformations and T is an invertible linear transformation such that

$$CT = TB.$$

If y is a vector in ran B, so that $y = Bx$ for some vector x, then the equation

$$CTx = Ty$$

implies that Ty is in ran C, and hence that y belongs to $T^{-1}(\text{ran } C)$. What this argument proves is that

$$\text{ran } B \subset T^{-1}(\text{ran } C).$$

Since the invertibility of T implies that $T^{-1}(\text{ran } C)$ has the same dimension as ran C, it follows that

$$\text{rank } B \leqq \text{rank } C.$$

The proof can be completed by a lighthearted call on symmetry. The assumption that B and C are similar is symmetric in B and C; if that assumption implies that rank $B \leqq \text{rank } C$, then it must also imply that

$$\text{rank } C \leqq \text{rank } B,$$

and from the two inequalities together it follows that

$$\operatorname{rank} B = \operatorname{rank} C.$$

Solution 92.

The answer is yes, every 2×2 matrix is similar to its transpose, and a surprisingly simple computation provides most of the proof:

$$\begin{pmatrix} \gamma & 0 \\ 0 & \beta \end{pmatrix}\begin{pmatrix} \alpha & \beta \\ \gamma & \delta \end{pmatrix} = \begin{pmatrix} \alpha\gamma & \beta\gamma \\ \beta\gamma & \beta\delta \end{pmatrix} = \begin{pmatrix} \alpha & \gamma \\ \beta & \delta \end{pmatrix}\begin{pmatrix} \gamma & 0 \\ 0 & \beta \end{pmatrix}.$$

If neither β nor γ is 0, then $\begin{pmatrix} \gamma & 0 \\ 0 & \beta \end{pmatrix}$ is invertible and that's that: $\begin{pmatrix} \alpha & \beta \\ \gamma & \delta \end{pmatrix}$ is indeed similar to $\begin{pmatrix} \alpha & \gamma \\ \beta & \delta \end{pmatrix}$.

If $\beta = 0$ and $\gamma \neq 0$, the proof is still easy, but it is not quite so near the surface. If worse comes to worst, computation is bound to reveal it: just set

$$\begin{pmatrix} \xi & \eta \\ \zeta & \theta \end{pmatrix}\begin{pmatrix} \alpha & 0 \\ \gamma & \delta \end{pmatrix} = \begin{pmatrix} \alpha & \gamma \\ 0 & \delta \end{pmatrix}\begin{pmatrix} \xi & \eta \\ \zeta & \theta \end{pmatrix},$$

and solve the implied system of four equations in four unknowns. It is of course not enough just to find numbers ξ, η, ζ, and θ that satisfy the equations—for instance $\xi = \eta = \zeta = \theta = 0$ always works—it is necessary also to find them so that the matrix is invertible. One possible solution is indicated by the equation

$$\begin{pmatrix} 0 & \gamma \\ \gamma & \delta - \alpha \end{pmatrix}\begin{pmatrix} \alpha & 0 \\ \gamma & \delta \end{pmatrix} = \begin{pmatrix} \gamma^2 & \gamma\delta \\ \gamma\delta & \delta(\delta - \alpha) \end{pmatrix} = \begin{pmatrix} \alpha & \gamma \\ 0 & \delta \end{pmatrix}\begin{pmatrix} 0 & \gamma \\ \gamma & \delta - \alpha \end{pmatrix}.$$

That works (meaning that $\begin{pmatrix} 0 & \gamma \\ \gamma & \delta - \alpha \end{pmatrix}$ is invertible) because $\gamma \neq 0$.

The case in which $\gamma = 0$ and $\beta \neq 0$, that is, the problem of the similarity of

$$\begin{pmatrix} \alpha & \beta \\ 0 & \delta \end{pmatrix} \qquad \text{and} \qquad \begin{pmatrix} \alpha & 0 \\ \beta & \delta \end{pmatrix},$$

is the same as the one just discussed: just replace β by γ and interchange the order in which the matrices were written.

(Does the assertion that similarity is a symmetric relation deserve explicit mention? If B and C are similar, via T, that is if

$$CT = TB,$$

then

$$T^{-1}(CT)T^{-1} = T^{-1}(TB)T^{-1}.$$

Replace TT^{-1} and $T^{-1}T$ by I and interchange the two sides of the equation to get

$$BT^{-1} = T^{-1}C,$$

and that is exactly the similarity of C and B via T^{-1}.)

If both β and γ are 0, the matrix $\begin{pmatrix} \alpha & \beta \\ \gamma & \delta \end{pmatrix}$ degenerates to a diagonal one, and the question of similarity to its transpose degenerates to a triviality—it is equal to its transpose.

Comment. That settles 2×2 matrices; what happens with matrices of size 3 and greater? The answer is that the same result is true for every size—every matrix is similar to its transpose—but even for 3×3 matrices the problem of generalizing the computations of the 2×2 case becomes formidable. New ideas are needed, more sophisticated methods are needed. They exist, but they will come only later.

93   Solution 93.

The answer is

$$\operatorname{rank}(A + B) \leqq \operatorname{rank} A + \operatorname{rank} B.$$

For the proof, observe first that

$$\operatorname{ran} A + \operatorname{ran} B$$

(in the sense of sums of subspaces, defined in Problem 28) consists of all vectors of the form

$$Ax + By,$$

and that

$$\operatorname{ran}(A + B)$$

consists of all vectors of the form

$$Az + Bz.$$

Consequence:

$$\operatorname{ran}(A + B) \subset \operatorname{ran} A + \operatorname{ran} B,$$

and, as a consequence of that,

$$\text{rank}(A + B) \leq \dim(\text{ran } A + \text{ran } B).$$

How is the right side of this inequality related to

$$\dim \text{ran } A + \dim \text{ran } B?$$

In general: if $\mathbb{M}$ and $\mathbb{N}$ are subspaces, what is the relation between

$$\dim(\mathbb{M} + \mathbb{N}) \qquad \text{and} \qquad \dim \mathbb{M} + \dim \mathbb{N}?$$

The answer is a natural one to guess and an easy one to prove, as follows. If $\{x_1, \ldots, x_m\}$ is a basis for $\mathbb{M}$ and $\{y_1, \ldots, y_n\}$ is a basis for $\mathbb{N}$, then the set

$$\{x_1, \ldots, x_m, y_1, \ldots, y_n\}$$

is surely big enough to span $\mathbb{M} + \mathbb{N}$. Consequence: the dimension of $\mathbb{M} + \mathbb{N}$ is not more than $m + n$, or, in other words,

$$\dim(\mathbb{M} + \mathbb{N}) \leq \dim \mathbb{M} + \dim \mathbb{N},$$

The proof of the rank sum inequality is complete.

Solution 94.   94

Since $(AB)x = A(Bx)$, it follows that

$$\text{ran } AB \subset \text{ran } A,$$

and hence that

$$\text{rank } AB \leq \text{rank } A.$$

Words are more useful here than formulas: what was just proved is that the rank of a product is less than or equal to the rank of the left-hand factor. That formulation implies that

$$\text{rank}(B'A') \leq \text{rank } B'.$$

Since, however,

$$\text{rank}(B'A') = \text{rank}\big((AB)'\big) = \text{rank } AB,$$

and

$$\text{rank } B' = \text{rank } B$$

(Problem 88), it follows that rank AB is less than or equal to both rank A and rank B, and hence that

$$\text{rank } AB \leq \min\{\text{rank } A, \text{rank } B\}.$$

That's it; that's the good relation between the rank of a product and the ranks of its factors.

Comment. If B happens to be invertible, so that rank B is equal to the dimension of the space, then the result just proved implies that

$$\text{rank } AB \leq \text{rank } A$$

and at the same time that

$$\text{rank } A = \text{rank}(AB)B^{-1} \leq \text{rank } AB,$$

so that, in fact,

$$\text{rank } AB = \text{rank } A.$$

It follows that

$$\text{rank}(BA) = \text{rank}(BA)' = \text{rank}(A'B') = \text{rank } A' = \text{rank } A.$$

In sum: the product of a given transformation with an invertible one (in either order) always has the same rank as the given one.

95   Solution 95.

The range of a transformation A is the image under A of the entire space V, and its dimension is an old friend by now—that's just the rank. What can be said about the dimension of the image of a proper subspace of V? The question is pertinent because

$$\text{ran}(AB) = (AB)V = A(BV) = A(\text{ran } B),$$

so that

$$\text{rank}(AB) = \dim\big(A(\text{ran } B)\big).$$

If M is a subspace of dimension m, say, and if N is any complement of M, so that

$$V = M + N,$$

then

$$\text{ran } A = AV = AM + AN.$$

It follows that

$$\text{rank } A \leqq \dim(A\mathbb{M}) + \dim(A\mathbb{N}) \leqq \dim(A\mathbb{M}) + \dim(\mathbb{N})$$

(because the application of a linear transformation can never *increase* dimension), and hence that

$$n - \text{null } A \leqq \dim(A\mathbb{M}) + n - m$$

(where $n = \dim \mathbb{V}$, of course). If in particular

$$\mathbb{M} = \text{ran } B,$$

then the last inequality implies that

$$\text{rank } B - \text{null } A \leqq \text{rank}(AB),$$

or, equivalently, that

$$n - \text{null } A - \text{null } B \leqq n - \text{null}(AB).$$

Conclusion:

$$\text{null}(AB) \leqq \text{null } A + \text{null } B.$$

Together the two inequalities about products, namely the one just proved about nullity and the one (Problem 89) about rank,

$$\text{rank } AB \leqq \min\{\text{rank } A, \text{rank } B\},$$

are known as **Sylvester's law of nullity**.

Solution 96.

(a) The "natural" basis vectors

$$e_1 = (1,0,0), \quad e_2 = (0,1,0), \quad e_3 = (0,0,1)$$

have a curious and special relation to the transformation C: it happens that

$$Ce_1 = e_1, \quad Ce_2 = 2e_2, \quad Ce_3 = 3e_3.$$

If B and C were similar,

$$CT = TB,$$

or, equivalently,

$$BT^{-1} = T^{-1}C,$$

then the vectors

$$T^{-1}e_1, \quad T^{-1}e_2, \quad T^{-1}e_3$$

would have the same relation to B:

$$BT^{-1}e_j = T^{-1}Ce_j = T^{-1}(je_j) = jT^{-1}e_j$$

for $j = 1, 2, 3$. Is that possible? Are there any vectors that are so related to B?

There is no difficulty about $j = 1$: the vector f_1 can be chosen to be the same as e_1. What about f_2? Well, that's not hard either. Since

$$Be_1 = e_1, \quad Be_2 = e_1 + 2e_2,$$

it follows that

$$B(e_1 + e_2) = e_1 + (e_1 + 2e_2) = 2(e_1 + e_2);$$

in other words if $f_2 = e_1 + e_2$, then

$$Bf_2 = 2f_2.$$

Is this the beginning of a machine? Yes, it is. Since

$$Be_3 = e_1 + e_2 + e_3,$$

it follows that

$$B(e_1 + e_2 + e_3) = B(e_1 + e_2) + Be_3$$
$$= 2(e_1 + e_2) + (e_1 + e_2 + 3e_3) = 3(e_1 + e_2 + e_3);$$

to get

$$Bf_3 = 3f_3,$$

just set

$$f_3 = e_1 + e_2 + e_3.$$

What good does all that do? Answer: it proves that B and C are similar. Indeed: the vectors f_1, f_2, f_3, expressed in coordinate forms in terms of the e's as

$$f_1 = (1, 0, 0)$$
$$f_2 = (1, 1, 0)$$
$$f_3 = (1, 1, 1),$$

constitute a basis. The matrix of B with respect to that basis is the matrix C. Isn't that clear from the definition of the matrix of B with respect to the basis $\{f_1, f_2, f_3\}$? What that phrase means is this: form Bf_j (for each j), and express it as a linear combination of the f's; the resulting coefficients are the entries of column number j. Conclusion: B and C are similar.

(b) The reasoning here is similar to the one used above. The linear transformation C has the property that

$$Ce_1 = 0, \quad Ce_2 = e_1, \quad Ce_3 = e_2.$$

If B and C are similar,

$$BT^{-1} = T^{-1}C,$$

then the vectors

$$f_j = T^{-1}e_j \qquad (j = 1, 2, 3)$$

are such that $Bf_1 = 0$, and, for $j > 0$,

$$Bf_j = BT^{-1}e_j = T^{-1}Ce_j = T^{-1}e_{j-1} = f_{j-1}.$$

At this moment it may not be known that B and C are similar, but it makes sense to ask whether there exist vectors f_1, f_2, f_3 that B treats in the way just described.

Yes, such vectors exist, and the proof is not difficult. Just set f_1 equal to e_1, set f_2 equal to e_2, and then start looking for f_3. Since $Be_3 = e_1 + e_2$, it follows that

$$B(e_3 - e_2) = (e_1 + e_2) - e_1 = e_2;$$

in other words, if $f_3 = e_3 - e_2$, then

$$Bf_3 = f_2.$$

Once that's done, the problem is solved. The vectors f_1, f_2, f_3, expressed in coordinate forms in terms of the e's as

$$f_1 = (1, 0, 0)$$
$$f_2 = (0, 1, 0)$$
$$f_3 = (0, -1, 1),$$

constitute a basis, and the matrix of B with respect to that basis is equal to C.

(c) The most plausible answer to both (a) and (b) is no—how could a similarity kill all the entries above the diagonal? Once, however, the answers have been shown to be yes, most people approaching (c) would probably be ready to guess yes—but this time the answer is no.

What is obvious is that

$$Ce_1 = 2e_1, \quad Ce_2 = 3e_2, \quad Ce_3 = 3e_3.$$

What is one millimeter less obvious is that every linear combination of e_2 and e_3 is mapped onto 3 times itself by C. What must therefore be asked (in view of the technique established in (a)) is whether or not there exist vectors f_1, f_2, f_3 that B treats the way C treats the e's. The answer turns out to be no.

Suppose indeed that $Bf = 3f$, where f is a vector whose coordinate form in terms of the e's is, say $(\alpha_1, \alpha_2, \alpha_3)$. Since

$$Bf = (2\alpha_1 + \alpha_2 + \alpha_3, 3\alpha_2 + \alpha_3, 3\alpha_3),$$

the only way that can be equal to

$$3f = (3\alpha_1, 3\alpha_2, 3\alpha_3),$$

is to have $\alpha_3 = 0$ (look at the second coordinates). From that in turn it follows that $\alpha_2 = \alpha_1$ (look at the first coordinates). To sum up: f must look like $(\tau, \tau, 0)$, or simpler said, every solution of the vector equation $Bf = 3f$ is of the form $(\tau, \tau, 0)$. Consequence: the set of solutions of that vector equation is a subspace of dimension 1, not 2. For C the corresponding dimension was 2, and that distinction settles the argument—B and C cannot be similar.

(d) In view of all this, what would a reasonable person guess about (d) by now? Is it imaginable that a similarity can *double* a linear transformation?

Yes, it is. The action of B on the natural basis $\{e_1, e_2, e_3\}$ can be described this way: the first basis vector is killed, and the other two are shifted backward to their predecessors. The question is this: is there a basis such that the first of its vectors is killed and each of the others is shifted backward to *twice* its predecessor? In that form the answer is easy to see: put

$$f_1 = (1, 0, 0)$$

$$f_2 = (0, 2, 0)$$

$$f_3 = (0, 0, 4).$$

That solves the problem, and nothing more needs to be said, but it might be illuminating to see a linear transformation that sends the e's to the f's and, therefore, actually transforms B into C. That's not hard: if

$$T = \begin{pmatrix} 1 & 0 & 0 \\ 0 & \frac{1}{2} & 0 \\ 0 & 0 & \frac{1}{4} \end{pmatrix},$$

then

$$T^{-1} = \begin{pmatrix} 1 & 0 & 0 \\ 0 & 2 & 0 \\ 0 & 0 & 4 \end{pmatrix}$$

and painless matrix multiplication proves that $TBT^{-1} = 2B$.

(e) The matrix of B with respect to the natural basis $\{e_1, e_2, e_3\}$ is the one exhibited in the question; what is the matrix of B with respect to the basis given by

$$f_1 = e_3,$$
$$f_2 = e_2,$$
$$f_3 = e_1?$$

The answer is as easy as any matrix determination can ever be. Since

$$Bf_1 = Be_3 = e_1 + e_2 + e_3 = f_3 + f_2 + f_1$$
$$Bf_2 = Be_2 = e_1 + e_2 = f_3 + f_2$$
$$Bf_3 = Be_1 = f_3,$$

it follows that the matrix of B with respect to the f's is exactly C.
Note that C is the transpose of B; compare with Problem 92.

Solution 97.

97

Define a linear transformation P by

$$P\widehat{x}_j = \widehat{y}_j \qquad (j = 1, \ldots, n)$$

and compute:

$$Cy_j = \sum_i \alpha_{ij} \widehat{y}_i = \sum_i \alpha_{ij} P\widehat{x}_i = P \sum_i \alpha_{ij} \widehat{x}_i = PBx_j.$$

The result almost forces the next step: to make the two extreme terms of this chain of equalities comparable, it is desirable to introduce the linear transformation Q for which

$$x_j = Qy_j \qquad (j = 1, \ldots, n).$$

The result is that $Cy_j = PBQy_j$ for all j, and hence that

$$C = PBQ.$$

That's the answer: B and C are equivalent if and only if there exist invertible transformations P and Q such that $C = PBQ$. The argument just given proves "only if", but, just as for Problem 85, the "if" must be proved too. The proof is a routine imitation of the proof in Problem 85, except it is not quite obvious how to set up the notation: what can be chosen arbitrarily to begin with and what should be defined in terms of it? Here is one way to answer those questions.

Assume that $C = PBQ$, and choose, arbitrarily, two bases

$$\{\widehat{x}_1, \ldots, \widehat{x}_n\} \qquad \text{and} \qquad \{y_1, \ldots, y_n\}.$$

Write

$$x_j = Qy_j \qquad \text{and} \qquad \widehat{y}_j = P\widehat{x}_j,$$

and write B as a matrix with respect to the x's and $\widehat{x}$'s:

$$Bx_j = \sum_i \alpha_{ij}\widehat{x}_i.$$

It follows that

$$Cy_j = PBQy_j = PBx_j = P\sum_i \alpha_{ij}\widehat{x}_i = \sum_i \alpha_{ij}P\widehat{x}_i = \sum_i \alpha_{ij}\widehat{y}_i.$$

Comparison of the last two displayed equations shows the matrix $C(\mathbb{Y}, \widehat{\mathbb{Y}})$ of C with respect to the y's is the same as the matrix $B(\mathbb{X}, \widehat{\mathbb{X}})$ of B with respect to the x's.

Question. If B is equivalent to C, does it follow that B^2 is equivalent to C^2? The first attempt at answering the question, without using the following problem, is not certain to be successful.

98   Solution 98.

Suppose that A is a linear transformation of rank r, say, on a finite-dimensional vector space $\mathbb{V}$. Since the kernel of A is a subspace of dimension $n - r$, standard techniques of extending bases show that there exists a basis

$$x_1, \ldots, x_r, x_{r+1}, \ldots, x_n$$

of $\mathbb{V}$ such that $\{x_{r+1}, \ldots, x_n\}$ is a basis for ker A. Assertion: the vectors

$$y_1 = Ax_1, \ldots, y_r = Ax_r$$

are linearly independent. Indeed: the only way it can happen that

$$\gamma_1 Ax_1 + \cdots + \gamma_r Ax_r = 0,$$

is to have $\gamma_1 x_1 + \cdots + \gamma_r x_r$ in the kernel of A. Reason: since the x's form a basis for $\mathbb{V}$, the only way a linear combination of the first r of them can be equal to a linear combination of the last $n - r$, is to have all coefficients equal to 0.

Once that is known, then, of course, the set $\{y_1, \ldots, y_r\}$ can be extended to a basis

$$y_1, \ldots, y_r, y_{r+1}, \ldots, y_n$$

of $\mathbb{V}$. What is the matrix of A with respect to the pair of bases (the x's and the y's) under consideration here? Answer: it is the $n \times n$ diagonal matrix the first r of whose diagonal terms are 1's and the last $n - r$ are 0's.

That remarkable conclusion should come as a surprise. It implies that every matrix of rank r is equivalent to a projection of rank r, and hence that any two matrices of rank r are equivalent to one another.

Chapter 7. Canonical Forms

Solution 99.   99

If E is a projection, and if λ is an eigenvalue of E with eigenvector x, so that $Ex = \lambda x$, then

$$Ex = E^2 x = E(Ex) = E(\lambda x) = \lambda Ex = \lambda(\lambda x) = \lambda^2 x.$$

Since $\lambda x = \lambda^2 x$ and $x \neq 0$ (by the definition of eigenvector), it follows that $\lambda = \lambda^2$. Consequence: the only possible eigenvalues of E are 0 and 1. Since the roots of the characteristic equation are exactly the eigenvalues, it follows that the only possible factors of the characteristic polynomial can be λ and $1 - \lambda$, and hence that the characteristic polynomial must be of the form $\lambda^k(\lambda - 1)^{n-k}$, with $k = 0, 1, \ldots, n$.

Question. If rank $E = 0$ (that is, if $E = 0$), then $k = n$; if rank $E = n$ (that is, if $E = 1$), then $k = 0$; what is k for other values of rank E?

Solution 100.   100

The sum of the roots of a (monic) polynomial equation

$$\lambda^n + \alpha_1 \lambda^{n-1} + \cdots + \alpha_{n-1} \lambda + \alpha_n = 0$$

is equal to $-\alpha_1$, and the product of the roots is equal to plus or minus α_n (depending on whether n is even or odd). To become convinced of these statements, just write the polynomial in factored form

$$(\lambda - \lambda_1) \cdots (\lambda - \lambda_n).$$

It follows that the sum and the product of the eigenvalues of a matrix A belong to the field in which the entries of A lie.

The product of the eigenvalues is equal to the determinant (think about triangularization), and that observation yields an alternative proof of the assertion about the product. The sum of the eigenvalues is also easy to read off the matrix A: it is the coefficient of $(-\lambda)^{n-1}$ in the expansion of $\det(A - \lambda)$, and hence it is the sum of the diagonal entries. That sum has a name: it is called the **trace** of the matrix A.

The answer to the question as it was asked is a strong NO: even though the eigenvalues of a rational matrix can be irrational, their sum and their product must be rational.

101   Solution 101.

The answer is yes: AB and BA always have the same eigenvalues.

It is to be proved that if $\lambda \neq 0$ and $AB - \lambda$ is not invertible, then neither is $BA - \lambda$, or, contrapositively, that if $AB - \lambda$ is invertible, then so is $BA - \lambda$. Change signs (that is surely harmless), divide by λ, and then replace A, say, by $\frac{A}{\lambda}$. These manipulations reduce the problem to proving that if $1 - AB$ is invertible, then so is $1 - BA$.

At this point it turns out to be clever to do something silly. Pretend that the classical infinite series formula for $\frac{1}{1-x}$ is applicable,

$$\frac{1}{1-x} = 1 + x + x^2 + x^3 + \cdots,$$

and apply it, as a matter of purely formal juggling, to BA in place of x. The result is

$$(1 - BA)^{-1} = 1 + BA + BABA + BABABA + \cdots$$
$$= 1 + B(1 + AB + ABAB + \cdots)A$$
$$= 1 + B(1 - AB)^{-1}A.$$

Granted that this is all meaningless, it suggests that, maybe, if $1 - AB$ is invertible, with, say

$$(1 - AB)^{-1} = X,$$

then $1 - BA$ is also invertible, with

$$(1 - BA)^{-1} = 1 + BXA.$$

Now *that* statement may be true or false, but it is in any case meaningful—it can be checked. Assume, that is, that $(1 - AB)X = 1$ (which can be written as $ABX = X - 1$) and calculate:

$$
\begin{aligned}
(1 - BA)(1 + BXA) &= (1 + BXA) - BA(1 + BXA) \\
&= 1 + BXA - BA - BABXA \\
&= 1 + BXA - BA - B(X - 1)A \\
&= 1 + BXA - BA - BXA + BA = 1.
\end{aligned}
$$

Victory !

Solution 102.   102

If λ is an eigenvalue of A with eigenvector x,

$$Ax = \lambda x,$$

then

$$A^2 x = A(Ax) = A(\lambda x) = \lambda(Ax) = \lambda(\lambda x) = \lambda^2 x,$$

and, by an obvious inductive repetition of this argument,

$$A^n x = \lambda^n x$$

for every positive integer n. (For the integer 0 the equation is if possible even truer.) This in effect answers the question about monomials. A linear combination of a finite number of these true equations yields a true equation. That statement is a statement about polynomials in general: it says that if p is a polynomial, then $p(\lambda)$ is an eigenvalue of $p(A)$.

Solution 103.   103

Since the matrix of A is

$$
\begin{pmatrix}
0 & 1 & 0 \\
0 & 0 & 1 \\
1 & 0 & 0
\end{pmatrix},
$$

the characteristic equation is

$$\lambda^3 = 1.$$

Since

$$\lambda^3 - 1 = (\lambda - 1)(\lambda^2 - \lambda + 1),$$

it follows that the roots of the characteristic equation are the three cube roots of unity:

$$1, \qquad \omega = -\frac{1}{2} + \frac{i}{2}\sqrt{3} \qquad \text{and} \qquad \omega^2 = -\frac{1}{2} - \frac{i}{2}\sqrt{3}.$$

The corresponding eigenvectors are easy to calculate (and even easier to guess and to verify); they are

$$u = (1,1,1), \qquad v = (1,\omega,\omega^2), \qquad \text{and} \qquad w = (1,\omega^2,\omega).$$

Comment. It is easy and worth while to generalize the question to dimensions greater than 3 and to permutations more complicated than the simple cyclic permutation that sends $(1,2,3)$ to $(2,3,1)$. The most primitive instance of this kind occurs in dimension 2. The eigenvalues of the transformation A defined on $\mathbb{C}^2$ by

$$A(x_1, x_2) = (x_2, x_1)$$

are of course the eigenvalues

$$1 \qquad \text{and} \qquad -1$$

of the matrix

$$A = \begin{pmatrix} 0 & 1 \\ 1 & 0 \end{pmatrix},$$

with corresponding eigenvectors

$$(1,1) \qquad \text{and} \qquad (1,-1).$$

These two vectors constitute a basis; the matrix of A with respect to that basis is

$$\begin{pmatrix} 1 & 0 \\ 0 & -1 \end{pmatrix}.$$

The discussion of the 3×3 matrix A of the problem can also be regarded as solving a diagonalization problem; its result is that that matrix is similar to

$$\begin{pmatrix} 1 & 0 & 0 \\ 0 & \omega & 0 \\ 0 & 0 & \omega^2 \end{pmatrix}.$$

The next higher dimension, $n = 4$, is of interest. There the matrix becomes

$$\begin{pmatrix} 0 & 1 & 0 & 0 \\ 0 & 0 & 1 & 0 \\ 0 & 0 & 0 & 1 \\ 1 & 0 & 0 & 0 \end{pmatrix}.$$

Its characteristic equation is

$$\lambda^4 = 1,$$

and, therefore, its eigenvalues are the fourth roots of unity:

$$1, \ i, \ -1, \ -i.$$

Consequence: the diagonalized version of the 4×4 matrix is

$$\begin{pmatrix} 1 & 0 & 0 & 0 \\ 0 & i & 0 & 0 \\ 0 & 0 & -1 & 0 \\ 0 & 0 & 0 & -i \end{pmatrix}.$$

Solution 104.   104

Yes, every eigenvalue of $p(A)$ is of the form $p(\lambda)$ for some eigenvalue λ of A. Indeed, if μ is an eigenvalue of $p(A)$, consider the polynomial $p(\lambda) - \mu$. By the fundamental theorem of algebra that polynomial can be factored into linear pieces. If

$$p(\lambda) - \mu = (\lambda - \lambda_1) \cdots (\lambda - \lambda_n),$$

then

$$p(A) - \mu = (A - \lambda_1) \cdots (A - \lambda_n).$$

The assumption about μ implies that there exists a non-zero vector x such that $p(A)x = \mu x$, and from that it follows that

$$(A - \lambda_1) \cdots (A - \lambda_n)x = 0.$$

Consequence: the product of the linear transformations

$$(A - \lambda_1), \ldots, (A - \lambda_n)$$

is not invertible, and from that it follows that $A - \lambda_j$ is not invertible for at least one j. That means that λ_j is an eigenvalue of A for at least one j.

Conclusion: since

$$p(\lambda_j) - \mu = 0,$$

the eigenvalue μ does indeed have the form $p(\lambda)$ for some eigenvalue λ of A (namely $\lambda = \lambda_j$).

Comment. The set of all eigenvalues of a linear transformation A on a finite-dimensional vector space is called the **spectrum** of A and is often referred to by the abbreviation spec A. With that notation Solution 102 can be expressed by saying that if A is a linear transformation, then

$$p(\text{spec } A) \subset \text{spec}\big(p(A)\big),$$

and the present solution strengthens that to

$$p(\text{spec } A) = \text{spec}\big(p(A)\big).$$

Another comment deserves to be made, one about the factorization technique used in the proof above. Is spec(A) always non-empty? That is: does every linear transformation on a finite-dimensional vector space have an eigenvalue? The answer is yes, of course, and the shortest proof of the answer (the one used till now) uses determinants (the characteristic equation of A). The factorization technique provides an alternative proof, one without determinants, as follows.

Given A, on a space of dimension n, take any non-zero vector x and form the vectors

$$x, \; Ax, \; A^2x, \ldots, A^nx.$$

Since there are $n + 1$ of them, they cannot be linearly independent. It follows that there exist scalars $\alpha_0, \alpha_1, \alpha_2, \ldots, \alpha_n$, not all 0, such that

$$\alpha_0 x + \alpha_1 Ax + \alpha_2 A^2x + \cdots + \alpha_n A^nx = 0.$$

It simplifies the notation and it loses no information to assume that if k is the largest index for which $\alpha_k \neq 0$, then, in fact, $\alpha_k = 1$—just divide through by α_k. A different language for saying what the preceding equation says (together with the normalization of the preceding sentence) is to say that there exists a monic polynomial p of degree less than or equal to n such that $p(A)x = 0$. Apply the fundamental theorem of algebra to factor p:

$$p(\lambda) = (\lambda - \lambda_1) \cdots (\lambda - \lambda_k).$$

Since $p(A)x = 0$ it is possible to reason as in the solution above to infer that $A - \lambda_j$ is not invertible for at least one j—and there it is!—λ_j is an eigenvalue of A.

The fundamental theorem of algebra is one of the deepest and most useful facts of mathematics—its repeated use in linear algebra should not come as a surprise. The need to use it is what makes it necessary to work with complex numbers instead of only real ones.

Solution 105.   105

The polynomials $1, x, x^2, x^3$ form a basis of $\mathbb{P}_3$; the matrix of D with respect to that basis is

$$\begin{pmatrix} 0 & 1 & 0 & 0 \\ 0 & 0 & 2 & 0 \\ 0 & 0 & 0 & 3 \\ 0 & 0 & 0 & 0 \end{pmatrix}.$$

Consequences: the only eigenvalue of D is 0, and the characteristic polynomial of D is λ^4. The algebraic multiplicity of the eigenvalue 0 is the exponent 4. What about the geometric multiplicity? The question is about the solutions p of the equation

$$Dp = 0;$$

in other words, the question is about the most trivial possible differential equation. Since the only functions (polynomials) whose derivative is 0 are the constants, the geometric multiplicity of 0 (the dimension of the eigenspace corresponding to 0) is 1.

Solution 106.   106

The answer is good: every transformation on an n-dimensional space with n distinct eigenvalues is diagonalizable.

Suppose, to begin with, that $n = 2$. If A is a linear transformation on a 2-dimensional space, with distinct eigenvalues λ_1 and λ_2 and corresponding eigenvectors x_1 and x_2, then (surprise?) x_1 and x_2 are linearly independent. Reason: if

$$\alpha_1 x_1 + \alpha_2 x_2 = 0,$$

apply $A - \lambda_1$ to that equation. Since $A - \lambda_1$ kills x_1, the result is

$$\alpha_2(\lambda_2 - \lambda_1)x_2 = 0,$$

and since $\lambda_2 - \lambda_1 \neq 0$ (assumption) and $x_2 \neq 0$ (eigenvector), it follows that $\alpha_2 = 0$. That in turn implies that $\alpha_1 = 0$, or, alternatively, an application of $A - \lambda_2$ to the assumed equation yields the same conclusion.

If $n = 3$, and if the three distinct eigenvalues in question are λ_1, λ_2, and λ_3, with eigenvectors x_1, x_2, and x_3, the same conclusion holds: the x's are linearly independent. Reason: if

$$\alpha_1 x_1 + \alpha_2 x_2 + \alpha_3 x_3 = 0,$$

apply $A - \lambda_1$ to infer

$$\alpha_2(\lambda_2 - \lambda_1)x_2 + \alpha_3(\lambda_3 - \lambda_1)x_3 = 0,$$

and then apply $A - \lambda_2$ to infer

$$\alpha_3(\lambda_3 - \lambda_1)(\lambda_3 - \lambda_2)x_3 = 0.$$

That implies $\alpha_3 = 0$ (because $(\lambda_3 - \lambda_1)(\lambda_3 - \lambda_2) \neq 0$ and $x_3 \neq 0$). Continue the same way: apply first $A - \lambda_1$ and then $A - \lambda_3$ to get $\alpha_2 = 0$, and by obvious small modifications of these steps get $\alpha_1 = 0$.

The general case, for an arbitrary n, should now be obvious, and from it diagonalization follows. Indeed, once it is known that a transformation on an n-dimensional space has n linearly independent eigenvectors, then its matrix with respect to the basis those vectors form is diagonal—and in this last step it no longer even matters that the eigenvalues are distinct.

Comment. Here is a minute but enchanting corollary of the result: a 2×2 real matrix with negative determinant is diagonalizable. Reason: since the characteristic polynomial is a quadratic real polynomial with leading coeffcient 1 and negative constant term, the quadratic formula implies that the two eigenvalues are distinct.

107   Solution 107.

Suppose that A is a linear transformation on a finite-dimensional vector space and that λ_0 is one of its eigenvalues with eigenspace $\mathbb{M}_0$. If x belongs to $\mathbb{M}_0$,

$$Ax = \lambda_0 x,$$

then

$$A(Ax) = A^2 x = \lambda_0^2 x = \lambda_0(\lambda_0 x) = \lambda_0 \cdot Ax$$

—which says that Ax belongs to $\mathbb{M}_0$. In other words, the subspace $\mathbb{M}_0$ is invariant under A. If A_0 is the linear transformation A considered on $\mathbb{M}_0$

only (the restriction of A to $\mathbb{M}_0$), then the polynomial $\det(A_0 - \lambda)$ is a factor of the polynomial $\det(A - \lambda)$. (Why?). If the dimension of $\mathbb{M}_0$ ($=$ the geometric multiplicity of λ_0) is m_0, then

$$\det(A_0 - \lambda) = (\lambda_0 - \lambda)^{m_0},$$

and it follows that $(\lambda_0 - \lambda)$ occurs as a factor of $\det(A - \lambda)$ with an exponent m greater than or equal to m_0. That's it: the assertion $m \geq m_0$ says exactly that geometric multiplicity is always less than or equal to algebraic multiplicity.

Comment.   What can be said about a transformation for which the algebraic multiplicity of every eigenvalue is equal to 1? In view of the present result the answer is that the geometric multiplicity of every eigenvalue is equal to 1, and hence that the number of eigenvalues is equal to the dimension. Conclusion (see Problem 106): the matrix is diagonalizable.

Solution 108.   108

The calculation of the characteristic polynomials is easy enough:

$$\det(A - \lambda) = (1 - \lambda)(2 - \lambda)(6 - \lambda) + 3 + 6(1 - \lambda) + (6 - \lambda)$$

and

$$\det(B - \lambda) = (1 - \lambda)(5 - \lambda)(3 - \lambda) + 4(3 - \lambda).$$

These both work out to

$$\lambda^3 - 9\lambda^2 + 27\lambda - 27,$$

which is equal to

$$(\lambda - 3)^3.$$

It follows that both A and B have only one eigenvalue, namely $\lambda = 3$, of algebraic multiplicity 3, and, on the evidence so far available, it is possible to guess that A and B are similar.

What are the eigenvectors of A? To have $Au = 3u$, where $u = (\alpha, \beta, \gamma)$, means having

$$\begin{aligned}
\alpha + \beta \qquad &= 3\alpha \\
-\alpha + 2\beta + \gamma &= 3\beta \\
3\alpha - 6\beta + 6\gamma &= 3\gamma.
\end{aligned}$$

These equations are easy to solve; it turns out that the only solutions are the vector $u = (1, 2, 3)$ and its scalar multiples. Consequence: the eigenvalue 3 of A has geometric multiplicity 1.

For B the corresponding equations are

$$\begin{aligned} \alpha + \beta \quad\quad\quad &= 3\alpha \\ -4\alpha + 5\beta + \quad &= 3\beta \\ -6\alpha - 3\beta + 3\gamma &= 3\gamma. \end{aligned}$$

The eigenspace of the eigenvalue 3 is 2-dimensional this time; it is the set of all vectors of the form $(\alpha, 2\alpha, \alpha)$. Consequence: the eigenvalue 3 of B has geometric multiplicity 2. Partial conclusion: A and B are not similar.

The upper triangular form of both A and B must be something like

$$\begin{pmatrix} 3 & \alpha & \gamma \\ 0 & 3 & \beta \\ 0 & 0 & 3 \end{pmatrix}.$$

Even a little experience with similarity indicates that that form is not uniquely determined—the discussion of Problem 94 shows that similarity can effect radical changes in the stuff above the diagonal (see in particular part (b)). Here is a pertinent special example that fairly illustrates the general case:

$$\begin{pmatrix} 1 & -1 & 0 \\ 0 & 1 & 0 \\ 0 & 0 & 1 \end{pmatrix} \begin{pmatrix} 3 & 1 & 1 \\ 0 & 3 & 1 \\ 0 & 0 & 3 \end{pmatrix} \begin{pmatrix} 1 & 1 & 0 \\ 0 & 1 & 0 \\ 0 & 0 & 1 \end{pmatrix} = \begin{pmatrix} 3 & 1 & 0 \\ 0 & 3 & 1 \\ 0 & 0 & 3 \end{pmatrix}.$$

In view of these comments it is not unreasonable to restrict the search for triangular forms to those whose top right corner entry is 0.

For A, the search is for vectors u, v, and w such that

$$Au = 3u, \quad Av = u + 3v, \quad \text{and} \quad Aw = v + 3w.$$

As for u, that's already at hand—that's the eigenvector $(1, 2, 3)$ found above.

If $v = (\alpha, \beta, \gamma)$ (the notation of the calculation that led to u is now abandoned), then the equation for v says that

$$\begin{aligned} \alpha + \beta \quad\quad\quad &= 3\alpha + 1 \\ -\alpha + 2\beta + \gamma &= 3\beta + 2 \\ 3\alpha - 6\beta + 6\gamma &= 3\gamma + 3. \end{aligned}$$

These equations are just as easy to solve as the ones that led to u. Their solutions are the vectors of the form $(\alpha, 2\alpha + 1, 3\alpha + 3)$—a space of dimension 1. One of them (one is enough) is $(0, 1, 3)$—call that one v.

If $w = (\alpha, \beta, \gamma)$ (another release of old notation), then the equation for w becomes

$$\begin{aligned}
\alpha + \beta \qquad &= 3\alpha \\
-\alpha + 2\beta + \gamma &= 3\beta + 1 \\
3\alpha - 6\beta + 6\gamma &= 3\gamma + 3.
\end{aligned}$$

The solutions of these equations are the vectors of the form $(\alpha, 2\alpha, 3\alpha + 1)$; a typical one of which (with $\alpha = 1$) is $w = (1, 2, 4)$.

The vectors u, v, and w so obtained constitute the basis; the matrix of A with respect to that basis is

$$\begin{pmatrix} 3 & 1 & 0 \\ 0 & 3 & 1 \\ 0 & 0 & 3 \end{pmatrix}$$

as it should be.

The procedure for B is entirely similar. Begin with the eigenvector $u = (1, 2, 3)$ with eigenvalue 1, and then look for a vector v such that

$$Bv = u + 3v.$$

If $v = (\alpha, \beta, \gamma)$, this equation becomes

$$\begin{aligned}
\alpha + \beta \qquad &= 3\alpha + 1 \\
-4\alpha + 5\beta \qquad &= 3\beta + 2 \\
-6\alpha + 3\beta + 3\gamma &= 3\gamma + 3,
\end{aligned}$$

and the solutions of that are the vectors of the form $(\alpha, 2\alpha + 1, 3)$. If w is the one with $\alpha = 0$, so that $w = (0, 1, 3)$, then the vectors u, v, and w constitute a basis, and the matrix of B with respect to that basis is

$$\begin{pmatrix} 3 & 1 & 0 \\ 0 & 3 & 0 \\ 0 & 0 & 3 \end{pmatrix}.$$

Solution 109.   109

If n is 2, the answer is trivially yes. If the question concerned $\mathbb{C}^n$ instead of $\mathbb{R}^n$ (with the understanding that in the complex case the dimension being asked about is the complex dimension), the answer would be easily yes again; just triangularize and look. One way of proving that the answer to the original question is yes for every n is to "complexify" $\mathbb{R}^n$ and the linear transformations that act on it. There are sophisticated ways of doing that for completely general real vector spaces, but in the case of $\mathbb{R}^n$ there is hardly anything to do. Just recall that if A is a linear transformation on

$\mathbb{R}^n$, then A can be defined by a matrix (with real entries, of course), and such a matrix defines at the same time a linear transformation (call it A^+) on $\mathbb{C}^n$.

The linear transformation A^+ on $\mathbb{C}^n$ has an eigenvalue and a corresponding eigenvector; that is

$$A^+ z = \lambda z$$

for some complex number λ and for some vector z in $\mathbb{C}^n$. Consider the real and imaginary parts of the complex number λ and, similarly, separate out the real and imaginary parts of the coordinates of the vector z. Some notation would be helpful; write

$$\lambda = \alpha + i\beta,$$

with α and β real, and

$$z = x + iy,$$

with x and y in $\mathbb{R}^n$. Since

$$A^+(x + iy) = (\alpha + i\beta)(x + iy),$$

it follows that

$$Ax = \alpha x - \beta y$$

and

$$Ay = \beta x + \alpha y.$$

There it is—that implies the desired conclusion: the subspace of $\mathbb{R}^n$ spanned by x and y is invariant under A.

110   Solution 110.

Yes, if a linear transformation A on a finite-dimensional (complex) vector space is such that $A^k = 1$ for some positive integer k, then A is diagonalizable. Here is the reasoning.

The assumption implies that every eigenvalue λ of A is a kth root of unity. Consequence: each block in a triangularization of A is of the form $\lambda + T$, where $\lambda^k = 1$ and where

$$T = \begin{pmatrix} 0 & * & * & \\ 0 & 0 & * & \mathbb{O} \\ 0 & 0 & 0 & \mathbb{O} \\ & & & \mathbb{O} \end{pmatrix}$$

is strictly upper triangular. By the binomial theorem,

$$(\lambda + T)^k = 1 + kT + \cdots,$$

where the possible additional terms do not contribute to the lowest non-zero diagonal of T. Conclusion: $(\lambda + T)^k$ can be 1 only when $T = 0$, that is, only when each block in the triangularization is diagonal.

Solution 111.   111

Since $\mathbb{M}(x)$ is spanned by the q vectors

$$x, \; Ax, \; A^2 x, \ldots, A^{q-1} x,$$

its dimension cannot be more than q; the answer to the question is that for an intelligently chosen x that dimension can actually attain the value q. The intelligent choice is not too difficult. Since the index of A is q, there must exist at least one vector x_0 such that

$$A^{q-1} x_0 \neq 0,$$

and each such vector constitutes an intelligent choice.

The assertion is that if

$$\alpha_0 x_0 + \alpha_1 A x_0 + \alpha_2 A^2 x_0 + \cdots + \alpha_q A^{q-1} x_0 = 0,$$

then each α_j must be 0. If that is not true, then choose the smallest index j such that $\alpha_j \neq 0$. (If $\alpha_0 \neq 0$, then of course $j = 0$.) It makes life a little simpler now to normalize the assumed linear dependence equation: divide through by α_j and transpose all but $A^j x_0$ to the right side. The result is an equation that expresses $A^j x_0$ as a linear combination of vectors obtained from x_0 by applying the higher powers of A (that is, the powers A^k with $k \geq j + 1$). Consequence:

$$A^j x_0 = A^{j+1} y$$

for some y. Since

$$A^{q-1} x_0 = A^{q-1-j} A^j x_0 = A^{q-1-j} A^{j+1} y = A^q y = 0$$

(the last equal sign is justified by the assumption that A is nilpotent of index q), a contradiction has arrived. (Remember the choice of x_0.) Since the only possibly shaky step that led here was the choice of j, the forced conclusion is that that choice is not possible. In other words, there is no smallest index j for which $\alpha_j \neq 0$—which says that $\alpha_j = 0$ for all j.

Corollary. *The index of nilpotence of a transformation on an space of dimension n can never be greater than n.*

112   Solution 112.

Perhaps somewhat surprisingly, the answer depends on size. If the dimension of the underlying space is 2, or, equivalently, if A and B denote 2×2 matrices, then AB and BA always have the same characteristic polynomial, and it follows that if AB is nilpotent, then so is BA. If a matrix of size 2 is nilpotent, then its index of nilpotence is less than or equal to 2.

For 3×3 matrices the conclusion is false. If, for instance,

$$A = \begin{pmatrix} 1 & 0 & 0 \\ 0 & 1 & 0 \\ 0 & 0 & 0 \end{pmatrix} \quad \text{and} \quad B = \begin{pmatrix} 0 & 0 & 0 \\ 1 & 0 & 0 \\ 0 & 1 & 0 \end{pmatrix}$$

then

$$AB = \begin{pmatrix} 0 & 0 & 0 \\ 1 & 0 & 0 \\ 0 & 0 & 0 \end{pmatrix}$$

is nilpotent of index 2, but $BA = A$ is not; it is nilpotent of index 3.

113   Solution 113.

The result of applying M to a vector $(\alpha, \beta, \gamma, \delta, \varepsilon)$ is $(\beta + \delta, \gamma - \varepsilon, 0, \varepsilon, 0)$. When is that 0—or, in other words, which vectors are in the kernel of M? Answer: ε must be 0, hence γ must be 0, and $\beta + \delta$ must be 0. So: the kernel consists of all vectors of the form

$$(\alpha, \beta, 0, -\beta, 0),$$

a subspace of dimension 2. In view of this observation, and in view of the given form of M, a reasonable hope is to begin the desired basis with

$$(1, 0, 0, 0, 0)$$
$$(0, 1, 0, 0, 0)$$
$$(0, 0, 1, 0, 0)$$

and

$$(0, -1, 0, 1, 0).$$

What is wanted for a fifth vector is one whose image under M is

$(0, -1, 0, 1, 0)$. Since the image of $(\alpha, \beta, \gamma, \delta, \varepsilon)$ is $(\beta+\delta, \gamma-\varepsilon, 0, \varepsilon, 0)$, what is wanted is to have $\beta + \delta = 0$, $\gamma - \varepsilon = -1$, and $\varepsilon = 1$. These equations have many solutions; the simplest among them is

$$(0, 0, 0, 0, 1).$$

That's it: the last five displayed vectors do the job.

Solution 114.

114

The answer is no but yes. No, not every matrix has a square root, but the reason is obvious (once you see it), and there is a natural way to get around the obstacle.

An example of a matrix with no square root is

$$A = \begin{pmatrix} 0 & 1 & 0 \\ 0 & 0 & 1 \\ 0 & 0 & 0 \end{pmatrix}.$$

(So is $\begin{pmatrix} 0 & 1 \\ 0 & 0 \end{pmatrix}$, but the larger example gives a little more of an idea of why it works.) If, indeed, it were true that $A = B^2$, then (since $A^3 = 0$) it would follow that $B^6 = 0$, and hence that B is nilpotent. A nilpotent matrix of size 3×3 must have index less than or equal to 3 (since the index is always less than or equal to the dimension)—and that implies $B^3 = 0$, and since $B^4 = A^2 \neq 0$, that is a contradiction.

What's wrong? The answer is 0. People familiar with the theory of multivalued analytic functions know that the point $z = 0$ is one at which the function defined by $\sqrt{z}$ misbehaves; the better part of valor dictates that in the study of square roots anything like 0 should be avoided. What in matrix theory is "anything like 0"? Reasonable looking answer: matrices that have 0 in their spectrum. How are they to be avoided? Answer: by sticking to invertible matrices. Very well then: does every invertible matrix have a square root?

Here is where the Jordan form can be used to good advantage. Every invertible matrix is similar to a direct sum of matrices such as

$$\begin{pmatrix} \lambda & 1 & 0 & 0 \\ 0 & \lambda & 1 & 0 \\ 0 & 0 & \lambda & 1 \\ 0 & 0 & 0 & \lambda \end{pmatrix},$$

with $\lambda \neq 0$, and, consequently, it is sufficient to decide whether or not every matrix of that form has a square root.

The computations are somewhat easier in case $\lambda = 1$, and it is possible to reduce to that case simply by dividing by λ. When that is done, the 1's above the diagonal turn into $\frac{1}{\lambda}$'s, to be sure, but in that position they cause no trouble. So the problem is to find a square root for something like

$$\begin{pmatrix} 1 & \alpha & 0 & 0 \\ 0 & 1 & \alpha & 0 \\ 0 & 0 & 1 & \alpha \\ 0 & 0 & 0 & 1 \end{pmatrix}.$$

One way to do that is to look for a square root of the form

$$\begin{pmatrix} 1 & \xi & \eta & \zeta \\ 0 & 1 & \xi & \eta \\ 0 & 0 & 1 & \xi \\ 0 & 0 & 0 & 1 \end{pmatrix}.$$

Set the square of that matrix equal to the given one and look for solutions x, y, z of the resulting equations. That works!

There is a more sophisticated approach. Think of the given matrix as $I + M$, where

$$M = \begin{pmatrix} 0 & \alpha & 0 & 0 \\ 0 & 0 & \alpha & 0 \\ 0 & 0 & 0 & \alpha \\ 0 & 0 & 0 & 0 \end{pmatrix}.$$

The reason that's convenient is that it makes possible the application of facts about the function $\sqrt{1 + \zeta}$.

As is well known, Professor Moriarty "wrote a treatise upon the binomial theorem, which has had a European vogue"; the theorem asserts that the power series expansion of $(1 + \zeta)^\xi$, is

$$(1 + \zeta)^\xi = 1 + \binom{\xi}{1} \zeta + \binom{\xi}{2} \zeta^2 + \binom{\xi}{3} \zeta^3 + \cdots.$$

(Here a binomial coefficient such as, for instance, $\binom{\xi}{3}$ denotes

$$\frac{\xi(\xi - 1)(\xi - 2)}{3!},$$

and the parameter ξ can be any real number.) The series converges for some values of ζ and does not converge for others, but, for the moment, none of that matters. What does matter is that the equation is "formally" right. That means, for instance, that if the series for $\xi = \frac{1}{2}$ is multiplied by itself, then the constant term and the coefficient of ζ turn out to be 1 and all other coefficients turn out to be 0—the product is exactly $1 + \zeta$. In the application that is about to be made the variable ζ will be replaced by a nilpotent matrix, so that only a finite number of non-zero terms will appear—and in that case convergence is not a worry.

All right: consider the series with $k = \frac{1}{2}$, and replace the variable ζ by the matrix M. The result is

$$
\begin{pmatrix}
1 & \dfrac{\alpha}{2} & -\dfrac{\alpha^2}{4} & \dfrac{\alpha^3}{16} \\
0 & 1 & \dfrac{\alpha}{2} & -\dfrac{\alpha^2}{4} \\
0 & 0 & 1 & \dfrac{\alpha}{2} \\
0 & 0 & 0 & 1
\end{pmatrix}
$$

(check?), and that works—meaning that its square is $1 + M$ (check?). So, one way or another, it is indeed true that every invertible matrix has a square root.

Solution 115.

115

The differentiation transformation D is nilpotent of index 4 (the dimension of the space). Consequence: both the minimal polynomial and the characteristic polynomial are equal to λ^4.

As for T, its only eigenvalue is 1. Indeed: if

$$
\alpha + \beta(t+1) + \gamma(t+1)^2 + \delta(t+1)^3 = \lambda(\alpha + \beta t + \gamma t^2 + \delta t^3)
$$

then

$$
\begin{aligned}
\alpha + \beta + \gamma + \delta &= \lambda\alpha, \\
\beta + 2\gamma + 3\delta &= \lambda\beta, \\
\gamma + 3\delta &= \lambda\gamma, \\
\delta &= \lambda\delta.
\end{aligned}
$$

It follows that if $\lambda \neq 1$, then

$$
\delta = 0, \quad \gamma = 0, \quad \beta = 0,
$$

and therefore

$$
\alpha = 0.
$$

On the other hand if $\lambda = 1$, then

$$\beta + \gamma + \delta = 0,$$
$$2\gamma + 3\delta = 0,$$
$$3\delta = 0,$$

and therefore

$$\delta = \gamma = \beta = 0.$$

(Another way to get here is to look at the matrix in Solution 108.) Conclusion: both the minimal polynomial and the characteristic polynomial are $(\lambda - 1)^4$.

116   Solution 116.

Yes, it's always true that one polynomial can do on each of n prescribed transformations what n prescribed polynomials do. The case $n = 2$ is typical and notationally much less cumbersome; here is how it goes.

Given: two linear transformations A and B with disjoint spectra, and two polynomials p and q. Wanted: a polynomial r such that

$$r(A) = p(A)$$

and

$$r(B) = q(B).$$

If there is such a polynomial r, then the difference $r - p$ annihilates A. The full annihilator of A, that is the set of all polynomials f such that $f(A) = 0$, is an ideal in the ring of all complex polynomials; every such polynomial is a multiple of the minimal polynomial p_0 of A. Consequence: if there is an r of the kind sought, then

$$r = sp_0 + p$$

for some polynomial p, and, similarly,

$$r = tq_0 + q,$$

where q is the minimal polynomial of B. Conversely, clearly, any $sp_0 + p$ maps A onto $p(A)$, and any $tq_0 + q$ maps B onto $q(B)$; the problem is to find an r that is simultaneously an $sp_0 + p$ and a $tq_0 + q$. In other words, the problem is to find polynomials s and t such that

$$sp_0 - tq_0 = q - p.$$

Since p_0 and q_0 are relatively prime (this is the step that uses the assumed disjointness of the spectra of A and B), it is a standard consequence of the Euclidean algorithm that such polynomials s and t do exist.

The general case ($n > 2$) can be treated either by imitating the special case or else by induction. Here is how the induction argument goes.

Assume the conclusion for n, and pass to $n + 1$ as follows. By the induction hypothesis, there is a polynomial p such that

$$p(A_j) = p_j(A_j)$$

for $j = 1, \ldots, n$. Write

$$A = A_1 \oplus \cdots \oplus A_n$$

(direct sum),

$$B = A_{n+1},$$

and

$$q = p_{j+1}.$$

Note that the spectra of A and B are disjoint (because the spectrum of A is the union of the spectra of the A_j's, $j = 1, \ldots, n$), and therefore, by the case $n = 2$ of the theorem, there exists a polynomial r such that

$$r(A) = p(A)$$

and

$$r(B) = q(B).$$

Once the notation is unwound, these equations become

$$r(A_1) \oplus \cdots \oplus r(A_n) = p_1(A_1) \oplus \cdots \oplus p_n(A_n)$$

and

$$r(A_{n+1}) = p_{n+1}(A_{n+1}).$$

The first of these equations implies that

$$r(A_j) = p_j(A_j)$$

for $j = 1, \ldots, n$, and that concludes the proof.

The result holds for all fields, not only $\mathbb{C}$, provided that the hypothesis of the disjointness of spectra is replaced by its algebraically more usable version, namely the pairwise relative primeness of the minimal polynomials of the given transformations.

Chapter 8. Inner Product Spaces

117   Solution 117.

An orthogonal set of non-zero vectors is always linearly independent. (The case in which one of them is zero is degenerate—then, of course, they are dependent.) Indeed, if

$$\alpha_1 x_1 + \cdots + \alpha_n x_n = 0,$$

form the inner product of both sides of the equation with any x_j and get

$$\alpha_j(x_j, x_j) = 0.$$

The reason is that if $i \neq j$, then the inner product (x_i, x_j) is 0; that's what the assumed orthogonality says. Since $(x_j, x_j) \neq 0$ (by the assumed non-zeroness), it follows that $\alpha_j = 0$—every linear dependence relation must be trivial.

118   Solution 118.

The answer to the question as posed is no: different inner products must yield different norms. The proof is a hard one to discover but a boring one to verify—the answer is implied by the equation

$$(x, y) = \left\| \frac{1}{2}(x + y) \right\|^2 - \left\| \frac{1}{2}(x - y) \right\|^2 + i \left\| \frac{1}{2}(x + iy) \right\|^2 - i \left\| \frac{1}{2}(x - iy) \right\|^2,$$

which is called the **polarization** formula. It might be somewhat frightening when first encountered, but it doesn't take long to understand, and once it's absorbed it is useful—it is worth remembering, or, at the very least, its existence is worth remembering.

119   Solution 119.

What is always true is that

$$\|x + y\|^2 = \|x\|^2 + (x, y) + (y, x) + \|y\|^2.$$

For real vector spaces the two cross product terms are equal; the equation

$$\|x + y\|^2 = \|x\|^2 + \|y\|^2$$

is equivalent to $(x, y) = 0$, and all is well.

In complex vector spaces, however, (y, x) is the complex conjugate of (x, y); the sum of the two cross product terms is $2Re(x, y)$. The equation between norms is equivalent to $Re(x, y) = 0$, and that is not the same as orthogonality. An obvious way to try to construct a concrete counterexample is to start with an arbitrary vector x and set $y = ix$. In that case

$$||x + y||^2 = ||(1 + i)x||^2 = 2||x||^2 = ||x||^2 + ||y||^2,$$

but (except in the degenerate case $x = 0$) the vectors x and y are not orthogonal.

Solution 120.   120

Multiply out $||x + y||^2 + ||x - y||^2$, get

$$||x||^2 + (x, y) + (y, x) + ||y||^2 + ||x||^2 - (x, y) - (y, x) + ||y||^2,$$

and conclude that the equation in the statement of the problem is in fact an identity, true for all vectors x and y in all inner product spaces.

Solution 121.   121

Yes, every inner product space of dimension n has an orthonormal set of n elements. Indeed consider, to begin with, an arbitrary orthonormal set. If no larger one jumps to the eye, a set with one element will do: take an arbitrary non-zero vector x, and normalize it (that is replace it by $\frac{x}{||x||}$). If the orthonormal set on hand is not maximal, enlarge it, and if the resulting orthonormal set is still not maximal, enlarge it again, and proceed in this way by induction. Since an orthonormal set can contain at most n elements (Problem 111), this process leads to a complete orthonormal set in at most n steps.

Assertion: such a set spans the whole space. Reason: if the set is $\{x_1, x_2, \ldots\}$, and if some vector x is *not* a linear combination of the x_j's, then form the vector

$$y = x - \sum_j (x, x_j)x_j.$$

The assumption about x implies that $y \neq 0$. Since, moreover,

$$(y, x_i) = (x, x_i) - \sum_j (x, x_i)\delta_{ij} = (x, x_i) - (x, x_i) = 0,$$

so that y is orthogonal to each of the x_i's, the normalized vector $\frac{y}{||y||}$, when adjoined to the x_i's, leads to a larger orthonormal set. That's a contradiction, and, therefore, the x_j's do indeed span the space. Since they are also linearly independent (Problem 111 keeps coming up), it follows that they constitute a basis, and hence that there must be n of them.

Comment. There is a different way to express the proof, a more constructive way. The idea is to start with a basis $\{x_1, \ldots, x_n\}$ and by continued modifications convert it to an orthonormal set. Here is an outline of how that goes. Since $x_1 \neq 0$, it is possible to form $y_1 = \frac{x_1}{||x_1||}$. Once $y_1, \ldots, y_r$ have been found so that each y_j is a linear combination of $x_1, \ldots, x_j$, form

$$x_{r+1} - \sum_{j=1}^{r} (x_{r+1}, y_j) y_j,$$

verify that it is linearly independent of $y_1, \ldots, y_r$, and normalize it. These steps are known as the **Gram-Schmidt orthogonalization process**.

122   Solution 122.

If x and y are the same vector, then both sides of the Schwarz inequality are equal to $||x||^2$. More generally if one of x and y is a scalar multiple of the other (in that case there is no loss of generality in assuming that $y = \alpha x$), then both sides of the inequality are equal to $|\alpha| \cdot ||x||^2$. If x and y are linearly dependent, then one of them is a scalar multiple of the other. In all these cases the Schwarz inequality becomes an equation—can the increasing generality of this sequence of statements be increased still further? The answer is no: the Schwarz inequality can become an equation for linearly dependent pairs of vectors only.

One proof of that assertion is by black magic, as follows. If

$$|(x, y)| = ||x|| \cdot ||y||,$$

replace x by γx, where γ is a complex number of absolute value 1 chosen so that $\gamma(x, y)$ is real. The assumed "Schwarz equation" is still true, but with the new x (and the same old y) it takes the form

$$(x, y) = ||x|| \cdot ||y||.$$

This is not an important step—it just makes the black magic that follows a tiny bit more mysterious still. Once that step has been taken, evaluate the expression

$$\left|\left| \, ||y||x - ||x||y \, \right|\right|^2.$$

Since it is equal to

$$(||y||x - ||x||y, ||y||x - ||x||y) = ||x||^2||y||^2 - 2||x||^2||y||^2 + ||x||^2||y||^2 = 0,$$

it follows that $||y||x - ||x||y = 0$, which is indeed a linear dependence between x and y.

One reason why the Schwarz inequality is true, and why equality happens only in the presence of linear dependence, can be seen by looking at simple special cases. Look, for instance, at two vectors in $\mathbb{R}^2$, say $x = (\alpha, \beta)$ and $y = (1, 0)$. Then

$$||x|| = \sqrt{|\alpha|^2 + |\beta|^2}, \qquad ||y|| = 1, \qquad \text{and} \qquad (x, y) = \alpha;$$

the Schwarz inequality reduces to the statement

$$|\alpha| \leqq \sqrt{|\alpha|^2 + |\beta|^2},$$

which becomes an equation just when $\beta = 0$.

An approach to the theorem that is neither black magic nor overly simplistic could go like this. Assume that $|(x, y)| = ||x|| \cdot ||y||$ and, temporarily fixing a real parameter α, consider

$$||x - \alpha y||^2 = (x - \alpha y, x - \alpha y) = ||x||^2 - 2Re(x, \alpha y) + |\alpha|^2||y||^2.$$

This indicates why changing x so as to make (x, y) real is a helpful thing to do; if that's done, then the right term becomes

$$\left(||x|| - |\alpha| \cdot ||y||\right)^2.$$

Inspiration: choose the parameter α so as to make that term equal to 0 (which explains the reason for writing down the black magic expression)— the possibility of such a choice proves that $x - \alpha y$ can be made equal to 0, which is a statement of linear dependence.

Solution 123.   123

If M is a subspace of a finite-dimensional inner product space V, then M and $M^\perp$ are complements (Problem 28), and $M^{\perp\perp} = M$. For the proof, consider an orthonormal basis $\{x_1, \ldots, x_m\}$ for the subspace M. If z is an arbitrary vector in V, form the vectors

$$x = \sum_{i=1}^{m}(z, x_i)x_i \qquad \text{and} \qquad y = z - x.$$

Since x is a linear combination of the x_j's, it belongs to M, and since y is orthogonal to each x_j it belongs to $M^\perp$. Consequence:

$$V = M + M^\perp.$$

If a vector u belongs to both M and $M^\perp$, then $(u, u) = 0$ (by the definition of $M^\perp$): that implies, of course, that $u = 0$, that is that the subspaces M and $M^\perp$ are disjoint. Conclusion (in the language of Problem 50): V is the direct sum of M and $M^\perp$, and that's as good a relation between M and $M^\perp$ as can be hoped for.

The definitions of x and y imply that

$$(z, x) = (x + y, x) = ||x||^2 + (y, x) = ||x||^2,$$

and, similarly,

$$(z, y) = (x + y, y) = (x, y) + ||y||^2 = ||y||^2.$$

It follows that if z is in $M^{\perp\perp}$, so that $(z, y) = 0$, then $||y||^2 = 0$, so that $z = x$ and therefore z is in M; in other words $M^{\perp\perp} \subset M$. Since the reverse inclusion $M \subset M^{\perp\perp}$ is already known, it now follows that $M = M^{\perp\perp}$, and that's as good a relation between M and $M^{\perp\perp}$ as can be hoped for.

124   Solution 124.

The answer is yes: every linear functional ξ on an inner product space V is induced as an inner product. For the proof it is good to look at the vectors x for which $\xi(x) = 0$. If every x is like that, then $\xi = 0$, and there is nothing more to say. In any case, the kernel of ξ is a subspace of V, and it is pertinent to consider its orthogonal complement, which it is convenient to denote by $\ker^\perp \xi$. If $\ker \xi \neq V$ (and that may now be assumed), then $\ker^\perp \xi$ contains at least one non-zero vector y_0. It is true in fact (even though for present purposes it is not strictly needed) that $\ker^\perp \xi$ consists of all scalar multiples of any such vector y_0; in other words, the subspace $\ker^\perp \xi$ has dimension 1. Indeed: if y is in $\ker^\perp \xi$, then so is every vector of the form $y - \alpha y_0$. The value of ξ at such a vector, that is

$$\xi(y - \alpha y_0) = \xi(y) - \alpha \xi(y_0),$$

can be made equal to 0 by a suitable choice of the scalar α (namely,

$$\alpha = \frac{\xi(y)}{\xi(y_0)})$$

which means, for that value of ξ, that $y - \xi y_0$ belongs to both $\ker \xi$ and $\ker^\perp \xi$. Conclusion: $y - \alpha y_0 = 0$, that is $y = \alpha y_0$.

The vector y_0 "works" just fine for the vectors in $\ker \xi$, meaning that if x is in $\ker \xi$, then

$$\xi(x) = (x, y_0)$$

(because both sides are equal to 0), and the same thing is true for every scalar multiple of y_0. Does the vector y_0 work for the vectors in $\ker^\perp \xi$ also? That is: is it true for an arbitrary element αy_0 in $\ker^\perp \xi$ that

$$u(\alpha y_0) = (\alpha y_0, y_0)?$$

The equation is equivalent to

$$\xi(y_0) = \|y_0\|^2,$$

and there is no reason why that *must* be true, but, obviously, it *can* be true if y_0 is replaced by a suitable scalar multiple of itself. Indeed: if y_0 is replaced by γy_0, the desired equation reduces to

$$\gamma \xi(y_0) = |\gamma|^2 \cdot \|y_0\|^2,$$

which can be satisfied by choosing γ so that

$$\xi(y_0) = \overline{\gamma} \|y_0\|^2.$$

Solution 125.   125

(a) Since by the very definition of adjoints $(U^*\eta, \zeta)$ is always equal to $(\eta, U\zeta)$, the way to determine U^* is to calculate with $(U^*\eta, \zeta)$. That's not inspiring, but it is doable. The way to do it is to begin with

$$\left(U^*\langle x_1, y_1 \rangle, \langle x_2, y_2 \rangle\right)$$

and juggle till it becomes an inner product with the same second term $\langle x_2, y_2 \rangle$ and a pleasant, simple first term that does not explicitly involve U. The beginning is natural enough:

$$\left(U^*\langle x_1, y_1 \rangle, \langle x_2, y_2 \rangle\right) = \left(\langle \xi_1, \eta_1 \rangle, U\langle x_2, y_2 \rangle\right)$$

$$= \left(\langle x_1, y_1 \rangle, \langle y_2, -x_2 \rangle\right) = (x_1, y_2) - (y_1, x_2).$$

That's an inner product all right, but it is one whose second term is $\langle y_2, x_2 \rangle$ instead of $\langle x_2, y_2 \rangle$. Easy to fix:

$$\left(U^*\langle x_1, y_1 \rangle, \langle x_2, y_2 \rangle\right) = (-y_1, x_2) + (x_1, y_2) = \left(\langle -y_1, x_1 \rangle, \langle x_2, y_2 \rangle\right).$$

That does it: the identity so derived implies that

$$U^*\langle x, y \rangle = \langle -y, x \rangle,$$

and that's the sort of thing that is wanted. What it shows is a surprise: it shows that

$$U^* = -U.$$

The calculation of U^*U is now trivial:

$$U^*U\langle x, y\rangle = U^*\langle y, -x\rangle = \langle x, y\rangle,$$

so that U^*U is equal to the identity transformation. The verification that UU^* is the same thing is equally trivial.

(b) Yes, a graph is always a subspace. The verification is direct: if $\langle x_1, y_1\rangle$ and $\langle x_2, y_2\rangle$ are in the graph of A, so that

$$y_1 = Ax_1 \qquad \text{and} \qquad y_2 = Ax_2,$$

then $\alpha_1\langle x_1, y_1\rangle + \alpha_2\langle x_2, y_2\rangle$ is in the graph of A, because

$$\alpha_1 y_1 + \alpha_2 y_2 = Ax_1 + Ax_2.$$

(c) The graph of A^* is the orthogonal complement of the image under U of the graph of A. To prove that, note that the graph of A is the set of all pairs of the form $\langle x, Ax\rangle$, and hence the U image of that graph is the set of all pairs of the form $\langle -Ax, x\rangle$. The orthogonal complement (in $V \oplus V$) of that image is the set of all those pairs $\langle u, v\rangle$ for which

$$(-Ax, u) + (x, v) = 0$$

identically in x. That means that

$$(x, -A^*u + v) = 0$$

for all x, and hence that $A^*u = v$. The set of all pairs $\langle u, v\rangle$ for which $A^*u = v$ is the set of all pairs of the form $\langle u, A^*u\rangle$, and that's just the graph of A^*.

That wasn't bad to verify—was it?—but how could it have been discovered? That sort of question is always worth thinking about.

126   Solution 126.

(a) Yes, congruence is an equivalence relation. Indeed, clearly,

$$A = P^*AP \quad \text{(with } P = 1\text{)};$$

$$\text{if} \quad B = P^*AP, \text{ then } A = Q^*BQ \quad \text{(with } Q = P^{-1}\text{)};$$

and

$$\text{if} \quad B = P^*AP \quad \text{and} \quad C = Q^*BQ, \text{ then } C = R^*AR \quad \text{(with } R = PQ\text{)}.$$

(b) Yes: if $B = P^*AP$, then $B^* = P^*A^*P$.

(c) No: a transformation congruent to a scalar doesn't have to be a scalar. Indeed, if P is an arbitrary invertible transformation such that P^*P is not a scalar (such things abound), then $P^*P (= P^* \cdot 1 \cdot P)$ is congruent to the scalar 1 without being equal to it.

(d) The answer to this one is not obvious—some head scratching is needed. The correct answer is yes: it is possible for A and B to be congruent without A^2 and B^2 being congruent. Here is one example:

$$A = \begin{pmatrix} 0 & 1 \\ 0 & 0 \end{pmatrix} \quad \text{and} \quad B = \begin{pmatrix} 0 & 1 \\ 1 & \xi \end{pmatrix}.$$

The computation is easy. If

$$P = \begin{pmatrix} 1 & \xi \\ 0 & 1 \end{pmatrix},$$

then

$$P^*AP = \begin{pmatrix} 1 & 0 \\ \xi & 1 \end{pmatrix}\begin{pmatrix} 0 & 1 \\ 0 & 0 \end{pmatrix}\begin{pmatrix} 1 & \xi \\ 0 & 1 \end{pmatrix} = \begin{pmatrix} 0 & 1 \\ 0 & \xi \end{pmatrix}\begin{pmatrix} 1 & \xi \\ 0 & 1 \end{pmatrix} = \begin{pmatrix} 0 & 1 \\ 0 & \xi \end{pmatrix}$$

so that, indeed, A is congruent to B. Since, however,

$$A^2 = \begin{pmatrix} 0 & 0 \\ 0 & 0 \end{pmatrix} \quad \text{and} \quad B^2 = \begin{pmatrix} 0 & \xi \\ 0 & \xi^2 \end{pmatrix},$$

it follows that A^2 cannot be congruent to B^2. (Is a microsecond's thought necessary? Can the transformation 0 be congruent to a transformation that is not 0? No: since $P^* \cdot 0 \cdot P = 0$, it follows that being congruent to 0 implies being equal to 0.)

Solution 127.

The desired statement is the converse of a trivial one: if $A = 0$, then $(Ax, x) = 0$ for all x. In the non-trivial direction the corresponding statement about sesquilinear forms (in place of quadratic ones) is accessible: if $(Ax, y) = 0$ for all x and y, then $A = 0$. Proof: set $y = Ax$. A possible approach to the quadratic result, therefore, is to reduce it to the sesquilinear one—try to prove that if $(Ax, x) = 0$ for all x, then $(Ax, y) = 0$ for all x and y.

What is wanted is (or should be?) reminiscent of polarization (Solution 118). What that formula does is express the natural sesquilinear form (x, y) in terms of the natural quadratic form $\|x\|^2$. Can that expression be generalized? Yes, it can, and the generalization is no more troublesome than the original version. It looks like this:

$$(Ax, y) = A\left(\frac{1}{2}(x+y), \frac{1}{2}(x+y)\right) - A\left(\frac{1}{2}(x-y), \frac{1}{2}(x-y)\right)$$
$$+ iA\left(\frac{1}{2}(x+iy), \frac{1}{2}(x+iy)\right) - iA\left(\frac{1}{2}(x-iy), \frac{1}{2}(x-iy)\right).$$

Once that's done, everything is done: if (Az, z) is identically 0, then so is (Ax, y).

128   Solution 128.

The product of two Hermitian transformations is not always Hermitian— or, equivalently, the product of two conjugate symmetric matrices is not always conjugate symmetric. It is hard *not* to write down an example. Here is one:

$$\begin{pmatrix} 0 & 1 \\ 1 & 0 \end{pmatrix}\begin{pmatrix} 1 & 0 \\ 0 & 2 \end{pmatrix} = \begin{pmatrix} 0 & 2 \\ 1 & 0 \end{pmatrix}.$$

Does the order matter? Yes, it matters in the sense that if the same two matrices are multiplied in the other order, then they give a different answer,

$$\begin{pmatrix} 1 & 0 \\ 0 & 2 \end{pmatrix}\begin{pmatrix} 0 & 1 \\ 1 & 0 \end{pmatrix} = \begin{pmatrix} 0 & 1 \\ 2 & 0 \end{pmatrix},$$

but the answer "no" does not change to the answer "yes".

How likely is the product of two Hermitian transformations to be Hermitian? If A and B are Hermitian, and if AB also is Hermitian, then

$$(AB)^* = AB,$$

which implies that $BA = AB$. What this proves is that for the product of two Hermitian transformations to be Hermitian it is necessary that they commute. Is the condition sufficient also? Sure—just read the argument backward.

129   Solution 129.

(a) If $B = P^*AP$ and $A^* = -A$, then

$$B^* = P^*(-A)P = -P^*AP = -B.$$

Conclusion: a transformation congruent to a skew one is skew itself.
    (b) If $A^* = -A$, then

$$(A^2)^* = (A^*)^2 = (-A)^2 = A^2,$$

which is not necessarily the same as $-A^2$. Conclusion: the square of a skew transformation doesn't have to be skew. Sermon: this is an incomplete proof. For perfect honesty it should be accompanied by a concrete example of a skew transformation A such that $A^2 \neq -A^2$. One of the simplest such transformations is given by the matrix

$$\begin{pmatrix} 0 & 1 \\ -1 & 0 \end{pmatrix}.$$

As for A^3, since $(-1)^3 = -1$, it follows that $A^* = -A$ implies $(A^3)^* = -A^3$, so that A^* is skew along with A.

(c) Write

$$S \text{ (for sum)} = AB + BA$$

and

$$D \text{ (for difference)} = AB - BA.$$

The question is: what happens to

$$S^* = B^*A^* + A^*B^*$$

and

$$D^* = B^*A^* - A^*B^*$$

when A^* and B^* are replaced by A and B, possibly with changes of sign? The answer is that if the number of sign changes is even (0 or 2), then S remains Hermitian and D remains skew, but if the number of sign changes is odd (which has to mean 1), then S becomes skew and D becomes Hermitian.

Solution 130.   130

If $A = A^*$, then

$$(Ax, x) = (x, A^*x) = (x, Ax) = \overline{(Ax, x)},$$

so that (Ax, x) is equal to its own conjugate and is therefore real. If, conversely, (Ax, x) is always real, then

$$(Ax, x) = \overline{(Ax, x)} = (x, A^*x) = (A^*x, x),$$

so that $((A - A^*)x, x) = 0$ for all x, and, by Problem 127, $A = A^*$.

131   **Solution 131.**

(a) The entries not on the main diagonal influence positiveness less than the ones on it. So, for example, from the known positiveness of

$$\begin{pmatrix} 2 & 1 \\ 1 & 1 \end{pmatrix}$$

it is easy to infer the positiveness of

$$\begin{pmatrix} 2 & -1 \\ -1 & 1 \end{pmatrix}.$$

(b) Yes, and an example has already been seen, namely $\begin{pmatrix} 2 & 2 \\ 2 & 1 \end{pmatrix}$.

(c) A careful look at

$$\begin{pmatrix} 1 & 1 & 1 \\ 1 & 1 & 1 \\ 1 & 1 & 1 \end{pmatrix}$$

shows that the quadratic form associated with it is

$$|\xi_1 + \xi_2 + \xi_3|^2,$$

and that answers the question: yes, the matrix is positive.

(d) The quadratic form associated with

$$\begin{pmatrix} 1 & 0 & 1 \\ 0 & 1 & 0 \\ 1 & 0 & 1 \end{pmatrix}$$

is

$$|\xi_1 + \xi_3|^2 + |\xi_2|^2,$$

and that settles the matter; yes, the matrix is positive.

(e) The quadratic form associated with

$$\begin{pmatrix} \alpha & 1 & 1 \\ 1 & 0 & 0 \\ 1 & 0 & 0 \end{pmatrix}$$

is

$$\alpha|\xi_1|^2 + 2Re\bar{\xi}_1\xi_2 + 2Re\bar{\xi}_1\xi_3$$

and the more one looks at that, the less positive it looks. It doesn't really matter what ξ_3 is—it will do no harm to set it equal to 0. The enemy is the coefficient α, and it can be conquered. No matter what α is, choose ξ_1 to be 1, and then choose ξ_2 to be a gigantic negative number—the resulting value of the quadratic form will be negative. The answer to the question as posed is: none.

Solution 132.   132

Yes, if a positive transformation is invertible, then its inverse also is positive. The proof takes one line, but a trick has to be thought of.

How does it follow from $(Ax, x) \geq 0$ for all x that $(A^{-1}y, y) \geq 0$ for all y? Answer: put $y = Ax$. Indeed, then

$$(A^{-1}y, y) = (A^{-1}Ax, Ax) = (x, Ax) = (Ax, x),$$

and the proof is complete.

(Is the reason for the last equality sign clear? Since A^{-1} is positive, A^{-1} is Hermitian, and therefore A is Hermitian.)

Solution 133.   133

If E is the perpendicular projection onto $\mathbb{M}$, so that E is the projection onto $\mathbb{M}$ along $\mathbb{M}^{\perp}$, then Problem 82 implies that E^* is the perpendicular projection onto $(\mathbb{M}^{\perp})^{\perp}$ along $\mathbb{M}^{\perp}$. (Problem 82 talks about annihilators instead of orthogonal complements, but the two languages can be translated back and forth mechanically.) That means that E^* is the perpendicular projection onto $\mathbb{M}$ (along $\mathbb{M}^{\perp}$)—and that is exactly E.

If, conversely, $E = E^2 = E^*$, then the idempotence of E guarantees that E is the projection onto ran E along ker E (Problem 72). If x is in ran E and y is in ker E, then

$(x, y) = (x, y)$ (because the vectors in the range of a

projection are fixed points of it—see Problem 72)

$= (x, E^*y)$ (just by the definition of adjoints)

$= (x, Ey)$ (because E was assumed to be Hermitian)

$= 0$ (because y is in ker E).

Consequence: ran E and ker E are not only complements—they are orthogonal complements, and, therefore, E is a perpendicular projection.

Summary. Perpendicular projections are exactly those linear transformations that are both Hermitian and idempotent.

134   **Solution 134.**

Since a perpendicular projection is Hermitian, the matrix of a projection on $\mathbb{C}^2$ must always look like

$$\begin{pmatrix} \alpha & \beta \\ \bar\beta & \gamma \end{pmatrix},$$

where α and γ must be real, and must, in fact, be in the unit interval. (Why?) The question then is just this: which of the matrices that look like that are idempotent?

To get the answer, compute. If

$$\begin{pmatrix} \alpha & \beta \\ \bar\beta & \gamma \end{pmatrix}\begin{pmatrix} \alpha & \beta \\ \bar\beta & \gamma \end{pmatrix} = \begin{pmatrix} \alpha^2 + |\beta|^2 & \alpha\beta + \beta\gamma \\ \alpha\bar\beta + \bar\beta\gamma & |\beta|^2 + \gamma^2 \end{pmatrix} = \begin{pmatrix} \alpha & \beta \\ \bar\beta & \gamma \end{pmatrix},$$

then (top right corner) $\alpha + \gamma = 1$, so that

$$\gamma = 1 - \alpha.$$

Consequence (lower right corner): $|\beta|^2 + (1-\alpha)^2 = 1-\alpha$, which simplifies to

$$|\beta|^2 = \alpha(1 - \alpha).$$

Conclusion: the matrices of projections on $\mathbb{C}^2$ are exactly the ones of the form

$$\begin{pmatrix} \alpha & \theta\sqrt{\alpha(1-\alpha)} \\ \bar\theta\sqrt{\alpha(1-\alpha)} & 1-\alpha \end{pmatrix},$$

where

$$0 \leq \alpha \leq 1 \qquad \text{and} \qquad |\theta| = 1.$$

Comment.   The case $\beta = 0$ seems to be more important than any other; in any event it is the one we are most likely to bump into.

135   **Solution 135.**

If E and F are projections, with ran $E = \mathsf{M}$ and ran $F = \mathsf{N}$, then the statements

$$E \leqq F \qquad \text{and} \qquad \mathsf{M} \subset \mathsf{N}$$

are equivalent.

Suppose, indeed, that $E \leq F$. If x is in M, then

$$(Fx, x) \geq (Ex, x) = (x, x),$$

Since the reverse inequality

$$(x, x) \geqq (Fx, x)$$

is always true, it follows that

$$((1 - F)x, x) = 0,$$

and hence that

$$\|(1 - F)x\|^2 = 0.$$

(Why is the last "hence" true?) Conclusion: $Fx = x$, so that x is in $\mathbb{M}$.

If, conversely, $\mathbb{M} \subset \mathbb{N}$, the $FEx = Ex$ (because Ex is in $\mathbb{M}$ for all x), so that $FE = E$. It follows (from adjoints) that $EF = E$, and that justifies a small computational trick:

$$(Ex, x) = \|Ex\|^2 = \|EFx\|^2 \leqq \|Fx\|^2 = (Fx, x).$$

Conclusion: $E \leqq F$.

Solution 136.   136

If E and F are projections with $\operatorname{ran} E = \mathbb{M}$ and $\operatorname{ran} F = \mathbb{N}$, then the statements

$$\mathbb{M} \perp \mathbb{N} \qquad \text{and} \qquad EF = 0$$

are equivalent.

Suppose indeed that $EF = 0$. If x is in $\mathbb{M}$ and y is in $\mathbb{N}$, then

$$(x, y) = (Ex, Fy) = (x, E^*Fy) = (x, EFy) = 0,$$

If, conversely, $\mathbb{M} \perp \mathbb{N}$, so that

$$\mathbb{N} \subset \mathbb{M}^\perp,$$

then, since Fx is in $\mathbb{N}$ for all x, it follows that Fx is in $\mathbb{M}^\perp$ for all x. Conclusion: $EFx = 0$ for all x.

Solution 137.   137

If A is Hermitian, and if x is a non-zero vector such that $Ax = \lambda x$, then, of course,

$$(Ax, x) = \lambda(x, x);$$

since (Ax, x) is real (Problem 130), it follows that λ is real. If, in addition, A is positive, so that (Ax, x) is positive, then it follows that λ is positive.

Note that these conditions on the eigenvalues of Hermitian and positive transformations are necessary, but nowhere near sufficient.

138   **Solution 138.**

The answer is that for Hermitian transformations eigenvectors belonging to distinct eigenvalues must be orthogonal.

Suppose, indeed, that

$$Ax_1 = \lambda_1 x_1 \quad \text{and} \quad Ax_2 = \lambda_2 x_2,$$

with $\lambda_1 \neq \lambda_2$. If A is Hermitian, then

$$\lambda_1(x_1, x_2) = (Ax_1, x_2) = (x_1, Ax_2) \quad \text{(why?)}$$
$$= \lambda_2(x_1, x_2) \quad \text{(why?)}.$$

Since $\lambda_1 \neq \lambda_2$, it must follow that $(x_1, x_2) = 0$.

Comment. Since the product of the eigenvalues of a transformation on a finite-dimensional complex vector space is equal to its determinant (remember triangularization), these results imply that the determinant of a Hermitian transformation is real. Is there an obvious other way to get the same result?

Chapter 9. Normality

139   **Solution 139.**

Caution: the answer depends on whether the underlying vector space is of finite or infinite dimension.

For finite-dimensional spaces the answer is yes. Indeed, if $U^*U = 1$, then U must be injective, and therefore surjective (Problem 66), and therefore invertible (definition), and once that's known the equation $U^*U = 1$ can be multiplied by U^{-1} on the right to get $U^* = U^{-1}$.

For infinite-dimensional spaces the answer may be no. Consider, indeed, the set V of all finitely non-zero infinite sequences

$$\{\xi_1, \xi_2, \xi_3, \dots\}$$

of complex numbers. The phrase "finitely non-zero" means that each sequence has only a finite number of non-zero terms (though that finite number might vary from sequence to sequence). With the obvious way of adding sequences and multiplying them by complex scalars, V is a complex vector space. With the definition

$$(\{\xi_1, \xi_2, \xi_3, \ldots\}, \{\eta_1, \eta_2, \eta_3, \ldots\}) = \sum_{n=1}^{\infty} \xi_n \overline{\eta_n},$$

the space V becomes an inner product space. If U and W are defined by

$$U\{\xi_1, \xi_2, \xi_3, \ldots\} = \{0, \xi_1, \xi_2, \xi_3, \ldots\}$$

and

$$W\{\xi_1, \xi_2, \xi_3, \ldots\} = \{\xi_2, \xi_3, \xi_4, \ldots\},$$

then U and W are linear transformations on V, and a simple computation establishes that the equation

$$(Ux, y) = (x, Wy)$$

is true for every pair of vectors x and y in V. In other words W is exactly the adjoint U^* of U, and, as another, even simpler, computation shows

$$U^*U = 1.$$

(Caution: it is essential to keep in mind that when U^*U is applied to a vector x, the transformation U is applied first.) It is, however, not true that $UU^* = 1$. Not only does U^* fail to be the inverse of U, but in fact U has no inverse at all. The range of U contains only those vectors whose first coordinate is 0, so that the range of U is not the entire space V—that's what rules out invertibility.

Solution 140.   140

When is $\begin{pmatrix} \alpha & \beta \\ \gamma & \delta \end{pmatrix}$ the matrix of a unitary transformation on $\mathbb{C}^2$? Answer: if and only if the product of

$$\begin{pmatrix} \alpha & \beta \\ \gamma & \delta \end{pmatrix}^* \qquad \text{and} \qquad \begin{pmatrix} \alpha & \beta \\ \gamma & \delta \end{pmatrix}$$

is the identity matrix. Since

$$\begin{pmatrix} \alpha & \beta \\ \gamma & \delta \end{pmatrix}^* \begin{pmatrix} \alpha & \beta \\ \gamma & \delta \end{pmatrix} = \begin{pmatrix} |\alpha|^2 + |\gamma|^2 & \overline{\alpha}\beta + \overline{\gamma}\delta \\ \alpha\overline{\beta} + \gamma\overline{\delta} & |\beta|^2 + |\delta|^2 \end{pmatrix},$$

that condition says that

$$|\alpha|^2 + |\gamma|^2 = |\beta|^2 + |\delta|^2 = 1 \qquad \text{and} \qquad \overline{\alpha}\beta + \overline{\gamma}\delta = 0,$$

or, in other words that the vectors (α, β) and (γ, δ) in $\mathbb{C}^2$ constitute an orthonormal set.

This 2×2 calculation extends to the general case. If U is a linear transformation on a finite-dimensional inner product space, and if the matrix of U with respect to an orthonormal basis is (u_{ij}), then a necessary and sufficient condition that U be unitary is that

$$\sum_k \overline{u}_{ki} u_{kj} = \delta_{ij}.$$

That matrix equation is, in fact, just the equation $U^*U = 1$ in matrix notation.

These comments make it easy to answer the questions about the special matrices in (a), (b), and (c).

For (a): since the second row is not a unit vector, it doesn't matter what α is, the matrix can never be unitary.

For (b): the rows must be orthonormal unit vectors. Since the norm of each row is $|\alpha|^2 + \frac{1}{4}$, the condition of normality is equivalent to $|\alpha| = \frac{\sqrt{3}}{2}$. Since the inner product of the two rows is $\frac{1}{2}(-\alpha + \overline{\alpha})$, their orthogonality is equivalent to α being real. Conclusion:

$$\begin{pmatrix} \alpha & \frac{1}{2} \\ -\frac{1}{2} & \alpha \end{pmatrix}$$

is unitary if and only if $\alpha = \pm \frac{\sqrt{3}}{2}$.

For (c): the question is an awkward way of asking whether or not a multiple of $(1, 1, 1)$ can be the first term of an orthonormal set. The answer is: why not? In detail: if ω is a complex cube root of 1, then the vectors

$$(1, 1, 1), \qquad (1, \omega, \omega^2), \qquad \text{and} \qquad (1, \omega^2, \omega)$$

all have the same norm $(\frac{\sqrt{3}}{2})$; normalization yields an explicit answer to the question.

141   Solution 141.

None of the three conditions $U^* = U$, $U^*U = 1$, and $U^2 = 1$ implies any of the others. Indeed,

$$\begin{pmatrix} 1 & 0 \\ 0 & 2 \end{pmatrix}$$

is Hermitian but neither unitary nor involutory;

$$\begin{pmatrix} 0 & 1 \\ -1 & 0 \end{pmatrix}$$

is unitary but neither Hermitian nor involutory; and

$$\begin{pmatrix} 1 & -2 \\ 0 & 1 \end{pmatrix}$$

is involutory but neither Hermitian nor unitary.

The implicative power of pairs of these conditions is much greater than that of each single one; indeed it turns out that any two together imply the third. That's very easy; here is how it goes.

If $U^* = U$, then the factor U^* in U^*U can be replaced by U, and, consequently, $U^*U = 1$ implies $U^2 = 1$.

If $U^*U = 1$ and $U^2 = 1$, then of course $U^*U = U^2$; multiply by U^{-1} $(= U^*)$ on the right and get $U^* = U$.

If, finally, $U^* = U$, then one of the factors U in U^2 can be replaced by U^*, and consequently, $U^2 = 1$ implies $U^*U = 1$.

Solution 142.   142

Each row and each column of a unitary matrix is a unit vector. If, in particular, a unitary matrix is triangular (upper triangular, say), then its first *column* is of the form

$$(*, 0, 0, 0, \ldots),$$

and, consequently, those entries in the first *row* that come after the first can contribute nothing—they must all be 0. Proceed inductively: now it's known that the second column is of the form

$$(0, *, 0, 0, \ldots),$$

and, consequently, those entries in the second row that come after the first two can contribute nothing—etc., etc. Conclusion: a triangular unitary matrix must be diagonal.

Comment. This solution tacitly assumed that the matrices in question correspond to unitary transformations via *orthonormal bases*. A similar comment applies in the next problem, about Hermitian diagonalizability.

Solution 143.   143

The answer is yes; every Hermitian matrix is unitarily similar to a diagonal one. This result is one of the cornerstones of linear algebra (or, perhaps more modestly, of the part of linear algebra known as unitary geometry). Its proof is sometimes considered recondite, but with the tools already available here it is easy.

Suppose, indeed, that A is a Hermitian transformation with the distinct eigenvalues

$$\lambda_1, \ldots, \lambda_r$$

and corresponding eigenspaces M_i:

$$M_i = \{x : Ax = \lambda_i x\},$$

$i = 1, \ldots, r$. If $i \neq j$ (so that $\lambda_i \neq \lambda_j$), then

$$M_i \perp M_j$$

(by Problem 138). The M_i's must span the entire space. Reason: the restriction of A to the orthogonal complement of their span is still a Hermitian transformation and, as such, has eigenvalues and corresponding eigenspaces.

That settles everything. Just choose an orthonormal basis within each M_i, and note that the union of all those little bases is an orthonormal basis for the whole space. Otherwise said: there exists an orthonormal basis

$$x_1, \ldots, x_n$$

of eigenvectors; the matrix of A with respect to that basis is diagonal.

144   Solution 144.

The answer is 1: every positive transformation has a unique positive square root. A quick proof of existence goes via diagonalization. If $A \geq 0$, then, in particular, A is Hermitian, and, consequently, A can be represented by a diagonal matrix such as

$$\begin{pmatrix} \alpha & 0 & 0 & \\ 0 & \beta & 0 & \\ 0 & 0 & \gamma & \\ & & & \ddots \end{pmatrix}.$$

The diagonal entries $\alpha, \beta, \gamma, \ldots$ are the eigenvalues of A, and, therefore, they are real; since, moreover, A is positive, it follows that they are positive.

Write

$$\begin{pmatrix} \sqrt{\alpha} & 0 & 0 & \\ 0 & \sqrt{\beta} & 0 & \\ 0 & 0 & \sqrt{\gamma} & \\ & & & \ddots \end{pmatrix}$$

(where the indicated numerical square roots are the positive ones), and jump happily to the conclusions that (i) $B \geq 0$ and (ii) $B^2 = A$.

What about uniqueness? If $C \geq 0$ and $C^2 = A$, then C can be diagonalized,

$$C = \begin{pmatrix} \xi & 0 & 0 & \\ 0 & \eta & 0 & \\ 0 & 0 & \zeta & \\ & & & \ddots \end{pmatrix}.$$

The numbers $\xi, \eta, \zeta, \ldots$ are positive and their squares are the numbers $\alpha, \beta, \gamma, \ldots$ — Q.E.D.

Solution 145.   145

Every linear transformation A on a finite-dimensional inner product space is representable as

$$A = UP$$

with U unitary and P positive; if A is invertible, the representation is unique.

To get a clue to a way of constructing U and P when only A is known, think backward: assume the result and try to let it suggest the method. If $A = UP$, then $A^* = PU^*$, and therefore

$$A^*A = P^2.$$

That's a big hint: since A^*A and P^2 are positive linear transformations, they have positive square roots; the equation $P^2 = A^*A$ implies the square root equation

$$P = \sqrt{A^*A}.$$

That's enough of a hint: given A, define P by the preceding equation, and then ask where U can come from. If A is to be equal to UP, then it's tempting to "divide through" by P—which would make sense if P were invertible. All right: assume for a moment that A is invertible; in that case A^* is invertible, and so are A^*A and P. If U is defined by

$$U = AP^{-1},$$

then

$$U^*U = P^{-1}A^*AP = P^{-1}P^2P = 1,$$

and victory has been achieved: U is indeed unitary.

Uniqueness is not hard. If

$$U_1P_1 = U_2P_2,$$

with U_1 and U_2 unitary and P_1 and P_2 invertible and positive, then

$$P_1^2 = (U_1P_1)^*(U_1P_1) = (U_2P_2)^*(U_2P_2) = P_2^2,$$

and therefore (by the uniqueness of positive square roots)

$$P_1 = P_2.$$

"Divide" the equation $U_1P_1 = U_2P_2$ through by P_1 ($= P_2$) and conclude that $U_1 = U_2$.

If A is not invertible, the argument becomes a little more fussy. What is wanted is $Ax = UPx$ for all x, or, writing $y = Px$, what is wanted is

$$Uy = Px$$

whenever y is in the range of P. Can that equation be used as a definition of U—is it an unambiguous definition? That is: if one and the same y is in the range of P for two reasons,

$$y = Px_1 \quad \text{and} \quad y = Px_2,$$

must it then be true that $Ax_1 = Ax_2$? The answer is yes: write

$$x = x_1 - x_2$$

and note the identity

$$\|Px\|^2 = (Px, Px) = (P^2x, x) = (A^*Ax, x) = \|Ax\|^2.$$

It implies that if $Px = 0$, then $Ax = 0$, or, in other words, that if

$$Px_1 = Px_2,$$

then

$$Ax_1 = Ax_2;$$

the proposed definition of U is indeed unambiguous.

Trouble: the proposed definition works on the range of P only; it defines a linear transformation U with domain equal to ran P and range equal

to ran A. Since that linear transformation preserves lengths (and therefore distances), it follows that ran A and ran P have the same dimension.
Consequence: $\text{ran}^\perp A$ and $\text{ran}^\perp P$ have the same dimension, and, consequently, there exists a linear transformation V that maps $\text{ran}^\perp P$ onto
$\text{ran}^\perp A$ and that preserves lengths. Extend the transformation U (already
defined on ran P) to the entire space by defining it to be equal to V on
$\text{ran}^\perp P$. The enlarged U has the property that $\|Ux\| = \|x\|$ for all x, which
implies that it is unitary; since $A = UP$, everything falls into place.

In the non-invertible case there is no hope of uniqueness and the arbi-
trariness of the definition of U used in the proof shows why. For a concrete
counterexample consider

$$A = \begin{pmatrix} 0 & 1 \\ 0 & 0 \end{pmatrix};$$

both the equations

$$A = \begin{pmatrix} 0 & 1 \\ 1 & 0 \end{pmatrix} \begin{pmatrix} 0 & 1 \\ 0 & 0 \end{pmatrix}$$

and

$$A = \begin{pmatrix} 0 & 1 \\ -1 & 0 \end{pmatrix} \begin{pmatrix} 0 & 0 \\ 0 & 1 \end{pmatrix}$$

are polar decompositions of A.

Solution 146.

Yes, eigenvectors belonging to distinct eigenvalues of a normal transfor-
mation (on a finite-dimensional inner product space) must be orthogonal.
The natural way to try to prove that is to imitate the proof that worked for
Hermitian (and unitary) transformations. That is: assume that

$$Ax_1 = \lambda_1 x_1 \qquad \text{and} \qquad Ax_2 = \lambda_2 x_2,$$

with $\lambda_1 \neq \lambda_2$, and look at

$$(Ax_1, x_2) = (x_1, A^* x_2).$$

The left term is equal to $\lambda_1(x_1, x_2)$—so far, so good—but there isn't any
grip on the right term. Or is there? Is there a connection between the eigenvalues of a normal transformation and its adjoint? That is: granted that
$Ax = \lambda x$, can something intelligent be said about $A^* x$? Yes, but it's a bit
tricky.

The normality of A implies that

$$\|Ax\|^2 = (Ax, Ax) = (A^*Ax, x)$$
$$= (AA^*x, x) = (A^*x, A^*x) = \|A^*x\|^2.$$

Since $A - \lambda$ is just as normal as A, and since

$$(A - \lambda)^* = A^* - \bar{\lambda},$$

it follows that

$$\|(A - \lambda)x\| = \|(A^* - \bar{\lambda})x\|.$$

Consequence: if λ is an eigenvalue of A with eigenvector x, then $\bar{\lambda}$ is an eigenvalue of A^* with the same eigenvector x.

The imitation of the proof that worked in the Hermitian case can now be comfortably resumed: since

$$(Ax_1, x_2) = \lambda_1(x_1, x_2)$$

and

$$(x_1, A^*x_2) = \lambda_2(x_1, x_2),$$

the distinctness of λ_1 and λ_2 implies the vanishing of (x_1, x_2), and the proof is complete.

147   Solution 147.

The answer is yes—normal transformations are diagonalizable. The key preliminary question is whether or not every restriction of a normal transformation is normal. That is: if A is normal on $\mathbb{V}$, if $\mathbb{M}$ is a subspace of $\mathbb{V}$, and if $A_\mathbb{M}$ is the restriction $A|\mathbb{M}$ of A to $\mathbb{M}$ (which means that $A_\mathbb{M}x = Ax$ whenever x is in $\mathbb{M}$), does it follow that $A_\mathbb{M}$ is normal? The trouble with the question is that it doesn't quite make sense—and it doesn't quite make sense even for Hermitian transformations. The reason is that the restriction is rigorously defined, but it may not be a linear transformation *on* $\mathbb{M}$—that is, it may fail to send vectors in $\mathbb{M}$ to vectors in $\mathbb{M}$. For the question to make sense it must be assumed that the subspace is invariant under the transformation. All right, what if that is assumed?

One good way to learn the answer is to write the transformation A under consideration as a 2×2 matrix according to the decomposition of the space into $\mathbb{M}$ and $\mathbb{M}^\perp$. The result looks like

$$A = \begin{pmatrix} P & * \\ 0 & * \end{pmatrix},$$

where P is the linear transformation A_M on M and the asterisks are linear transformations from M to $M^\perp$ (top right corner) and from $M^\perp$ to $M^\perp$ (bottom right corner). It doesn't matter what linear transformations they are, and there is no point in spending time inventing a notation for them—what is important is the 0 in the lower left corner. The reason for that 0 is the assumed invariance of M under A.

Once such a matrix representation is known for A, one for A^* can be deduced:

$$A^* = \begin{pmatrix} P^* & 0 \\ * & * \end{pmatrix}.$$

Now use the normality of A in an easy computation: since

$$A^* A = \begin{pmatrix} P^* & 0 \\ * & * \end{pmatrix} \begin{pmatrix} P & * \\ 0 & * \end{pmatrix} = \begin{pmatrix} P^* P & * \\ * & * \end{pmatrix}$$

and

$$AA^* = \begin{pmatrix} P & * \\ 0 & * \end{pmatrix} \begin{pmatrix} P^* & 0 \\ * & * \end{pmatrix} = \begin{pmatrix} PP^* & * \\ * & * \end{pmatrix},$$

normality implies that

$$P^* P = PP^*,$$

that is, it implies that P is normal—in other words A_M is normal.

That's all the hard work that has to be done—at this point the diagonalizability theorem for normal transformations can be abandoned in good conscience. The point is that intellectually the proof resembles the one for Hermitian transformations in every detail. There might be some virtue in checking the technical details, and the ambitious reader is encouraged to do so—examine the proof of diagonalizability for Hermitian transformations, replace the word "Hermitian" by "normal", delete all references to reality, and insist that the action take place on a complex inner product space, and note, happily, that the remaining parts of the proof remain unchanged.

Language. The diagonalizability of normal (and, in particular, Hermitian) transformations is sometimes called the **spectral theorem**.

Solution 148.   **148**

If A and B are defined on $\mathbb{C}^2$ by

$$A = \begin{pmatrix} 0 & 1 \\ 0 & 0 \end{pmatrix} \quad \text{and} \quad B = \begin{pmatrix} 1 & 0 \\ 0 & 0 \end{pmatrix},$$

then B is normal and every eigenspace of A is invariant under B, but A and B do not commute.

If, however, A is normal, and every eigenspace of A is invariant under B, then A and B do commute. The most obvious approach to the proof is to use the spectral theorem (Problem 147); the main purpose of that theorem is, after all, to describe the relation between a normal transformation and its eigenspaces. The assertion of the theorem can be formulated this way: if A is normal with distinct eigenvalues $\lambda_1, \ldots, \lambda_r$, and if E_j is, for each j, the (perpendicular) projection on the eigenspace corresponding to λ_j, then $A = \sum_j \lambda_j E_j$. The assumption that the eigenspace corresponding to λ_j is invariant under B can be expressed in terms of E_j as the equation

$$BE_j = E_j BE_j.$$

From the assumption that every eigenspace of A is invariant under B it follows that the orthogonal complement of the eigenspace corresponding to λ_j is invariant under B (because it is spanned by the other eigenspaces), and hence that

$$B(1 - E_j) = (1 - E_j)B(1 - E_j).$$

The two equations together simplify to

$$BE_j = E_j B,$$

and that, in turn implies the desired commutativity $BA = AB$.

149   Solution 149.

There are three ways for two of three prescribed linear transformations A, B, and C to be adjoints of one another; the adjoint pairs can be $\langle A, B \rangle$, or $\langle B, C \rangle$, or $\langle A, C \rangle$. There are, therefore, except for notational differences, just three possible commutativity hypotheses:

$$
\begin{array}{lllll}
A & \text{with} & A^* & \text{and} & A^* & \text{with} & C, \\
A & \text{with} & B & \text{and} & B & \text{with} & B^*, \\
A & \text{with} & B & \text{and} & B & \text{with} & A^*.
\end{array}
$$

The questioned conclusion from the last of these is obviously false; for a counterexample choose A so that it is not normal and choose $B = 0$. The implications associated with the first two differ from one another in notation only; both say that if something commutes with a normal transformation, then it commutes with the adjoint of that normal transformation. That implication is true.

The simplest proof uses the fact that if A is normal, then a necessary and sufficient condition for $AB = BA$ is that each of the eigenspaces of A is invariant under B (see Solution 148). Consequence: if A is normal and $AB = BA$, then the eigenspaces of A are invariant under B. The normality of A implies that the eigenspaces of A are exactly the same as the eigenspaces of A^*. Consequence: the eigenspaces of A^* are invariant under B. Conclusion: $A^*B = BA^*$, and the proof is complete.

Solution 150.   150

Almost every known proof of the adjoint commutativity theorem (Solution 143) can be modified to yield the intertwining generalization: it is indeed true that if A and B are normal and $AS = SB$, then $A^*S = SB^*$. Alternatively, there is a neat derivation, via matrices whose entries are linear transformations, of the intertwining version from the commutative one. Write

$$A^\wedge = \begin{pmatrix} A & 0 \\ 0 & B \end{pmatrix} \qquad \text{and} \qquad S^\wedge = \begin{pmatrix} 0 & S \\ 0 & 0 \end{pmatrix}$$

The transformation $A^\wedge$ is normal, and a straightforward verification proves that $B^\wedge$ commutes with it. The adjoint commutativity theorem implies that $B^\wedge$ commutes with $A^{\wedge*}$ also. To get the desired conclusion from this fact, just multiply the matrices $A^{\wedge*}$ and $B^\wedge$ in both orders and compare corresponding entries.

Solution 151.   151

Yes; if A, B, and AB are all normal, then BA is normal too. One good way to prove that statement is a splendid illustration of what is called a **trace argument**. In general terms, a trace argument can sometimes be used to prove an equation between linear transformations, or, what comes to the same thing, to prove that some linear transformation C is equal to 0, by proving that the trace of C^*C is 0. Since C^*C is positive, the only way it can have trace 0 is to be 0, and once C^*C is known to be 0 it is immediate that C itself must be 0. The main techniques available to prove that the trace of something is 0 are the additivity of trace,

$$\text{tr}(X + Y) = \text{tr}\,X + \text{tr}\,Y,$$

and the invariance of the trace of a product under cyclic permutations of its factors,

$$\text{tr}(XYZ) = \text{tr}(ZXY).$$

If it could be proved that A and B must commute, then all would be well (see the discussion preceding the statement of the problem), but that is not necessarily true (see the discussion preceding the statement of the problem). A step in the direction of commutativity can be taken anyway: the assumptions do imply that B commutes with A^*A. That is: if $C = BA^*A - A^*AB$, then $C = 0$. That's where the trace argument comes in.

A good way to study C^*C is to multiply out

$$(A^*AB^* - B^*A^*A)(BA^*A - A^*AB),$$

getting

$$A^*AB^*BA^*A - B^*A^*ABA^*A - A^*AB^*A^*AB + B^*A^*AA^*AB,$$

and then examine each of the four terms. As a device in that examination, introduce an ad hoc equivalence relation, indicated by $X \sim Y$ for any two products X and Y, if they can be obtained from one another by a cyclic permutation of factors. A curious thing happens: the assumptions (A, B, and AB are normal) and the cyclic permutation property of trace imply that all four terms are equivalent to one another. Indeed:

$$A^*AB^*BA^*A = A^*ABB^*A^*A \quad \text{(because } B \text{ is normal)}$$
$$\sim A^*B^*A^*ABA \quad \text{(because } AB \text{ is normal)},$$
$$B^*A^*ABA^*A = B^*A^*ABAA^* \quad \text{(because } A \text{ is normal)}$$
$$\sim A^*B^*A^*ABA,$$
$$A^*AB^*A^*AB = AA^*B^*A^*AB \quad \text{(because } A \text{ is normal)}$$
$$\sim A^*B^*A^*ABA,$$
$$B^*A^*AA^*AB \sim AA^*ABB^*A^*$$
$$= AA^*B^*A^*AB \quad \text{(because } AB \text{ is normal)}$$
$$\sim A^*B^*A^*ABA.$$

Consequence: all four terms have the same trace, and, therefore, the trace of C^*C is 0.

The result of the preceding paragraph implies that B commutes with A^*A. If $A = UP$ is the polar decomposition of A, then U commutes with P (because A is normal), and, since B commutes with P^2 ($= A^*A$), it follows that B also commutes with P. These commutativities imply that

$$U^*(AB)U = U^*(UP)BU = (U^*U)(BP)U = B(UP) = BA.$$

Conclusion: BA is unitarily equivalent to the normal transformation AB, and, consequently, BA itself must be normal.

Solution 152.   152

(a) The adjoint of a matrix is its conjugate transpose. Polynomials are not clever enough to transpose matrices. If, for instance,

$$A = \begin{pmatrix} 0 & 0 \\ 1 & 0 \end{pmatrix},$$

then every polynomial in A is of the form

$$\begin{pmatrix} \alpha & 0 \\ \beta & \alpha \end{pmatrix},$$

which has no chance of being equal to

$$A^* = \begin{pmatrix} 0 & 1 \\ 0 & 0 \end{pmatrix}.$$

Question. What made this A work? Would any non-normal A work just as well?

(b) This time the answer is yes; the inverse of an invertible matrix A can always be obtained via a polynomial. For the proof, consider the characteristic polynomial

$$\lambda^n + a_{n-1}\lambda^{n-1} + \cdots + a_1\lambda + a_0$$

of A and observe that a_0 cannot be 0. Reason: the assumed invertibility of A implies that 0 is not an eigenvalue. Multiply the Hamilton-Cayley equation

$$A^n + a_{n-1}A^{n-1} + \cdots + a_1 A + a_0 = 0$$

by A^{-1} to get

$$A^{n-1} + a_{n-1}A^{n-2} + \cdots + a_1 + a_0 A^{-1} = 0.$$

Conclusion: if

$$p(\lambda) = -\frac{1}{a_0} \left(\lambda^{n-1} + a_{n-1}\lambda^{n-2} + \cdots + a_1 \right),$$

then $p(A) = A^{-1}$.

Solution 153.   153

The answer is that all positive matrices are Gramians. Suppose, indeed, that $A \geq 0$ and infer (Problem 144) that there exists a positive matrix B

such that $B^2 = A$. If $A = (\alpha_{ij})$, then the equations

$$\alpha_{ij} = (Ae_j, e_i) \qquad \text{(why?)}$$
$$= (B^2 e_j, e_i) = (Be_j, Be_i)$$

imply that A is a Gramian (the Gramian of the vectors $Be_1, Be_2, \ldots$), and that's all there is to it.

154   Solution 154.

Squaring is not monotone; a simple counterexample is given by the matrices

$$A = \begin{pmatrix} 1 & 0 \\ 0 & 0 \end{pmatrix} \qquad \text{and} \qquad \begin{pmatrix} 2 & 1 \\ 1 & 1 \end{pmatrix}$$

The relation $A \leq B$ can be verified by inspection. Since

$$A^2 = \begin{pmatrix} 1 & 0 \\ 0 & 0 \end{pmatrix} (= A) \qquad \text{and} \qquad B^2 = \begin{pmatrix} 5 & 3 \\ 3 & 2 \end{pmatrix},$$

so that

$$B^2 - A^2 = \begin{pmatrix} 4 & 3 \\ 3 & 2 \end{pmatrix},$$

it is also easy to see that the relation $A^2 \leq B^2$ is false; indeed, the determinant of $B^2 - A^2$ is negative.

Is it a small blemish that not both the matrices in this example are invertible? That's easy to cure (at the cost of an additional small amount of computation): the matrices $A + 1$ and $B + 1$ are also a counterexample.

That's the bad news; for square roots the news is good. That is: if

$$0 \leq A \leq B,$$

then it is true that

$$\sqrt{A} \leq \sqrt{B}.$$

Various proofs of that conclusion can be constructed, but none of them jumps to the eye—the only way to go is by honest toil. The idea of one proof is to show that every eigenvalue of $\sqrt{B} - \sqrt{A}$ is non-negative; for invertible Hermitian transformations that property is equivalent to positiveness. All right: suppose then that λ is an eigenvalue of $\sqrt{B} - \sqrt{A}$, with corresponding (non-zero) eigenvector x, so that

$$\sqrt{A}x = \sqrt{B}x - \lambda x;$$

it is to be shown that $\lambda \geq 0$.

If it happens that $\sqrt{B}x = 0$, then, of course, $Bx = 0$, and therefore it follows from the assumed relation between A and B that $(Ax, x) = 0$. Consequence: $\sqrt{A}x = 0$. Reason:

$$0 = (Ax, x) = (\sqrt{A}\sqrt{A}x, x) = (\sqrt{A}x, \sqrt{A}x) = ||\sqrt{A}x||^2.$$

Once that's known, then the assumed eigenvalue equation implies that $\lambda x = 0$, and hence that $\lambda = 0$.

If $\sqrt{B}x \neq 0$, then $(\sqrt{B}x, x) \neq 0$—to see that apply the chain of equations displayed just above to B instead of A. Consequence:

$$\begin{aligned}
(\sqrt{B}x, \sqrt{B}x) &= ||\sqrt{B}x||^2 \\
&\geq ||\sqrt{B}x|| \cdot ||\sqrt{A}x|| \quad \text{(why?)} \\
&\geq (\sqrt{B}x, \sqrt{A}x) \quad \text{(why?)} \\
&= (\sqrt{B}x, \sqrt{B}x - \lambda x) \\
&= (\sqrt{B}x, \sqrt{B}x) - \lambda(\sqrt{B}x, x).
\end{aligned}$$

Conclusion: $\lambda > 0$, because the contrary possibility yields the contradiction,

$$(\sqrt{B}x, \sqrt{B}x) > (\sqrt{B}x, \sqrt{B}x).$$

The proof is complete.

Solution 155.   155

In some shallow combinatorial sense there are 32 cases to examine: 16 obtained via combining the four constituents ran A, ker A, ran A^*, and ker A^* with one another by spans, and 16 others via combining them by intersections. Consideration of the duality given by orthogonal complements (and other, even simpler eliminations) quickly reduce the possibilities to two, namely

$$\text{ker } A \cap \text{ker } A^* \qquad \text{and} \qquad \text{ran } A \cap \text{ran } A^*.$$

The first of these is always a reducing subspace; indeed both A and A^* map it into $\{0\}$. An explicit look at the duality can do no harm: since the orthogonal complement of a reducing subspace is a reducing subspace, it follows that ran $A +$ ran A^* is always a reducing subspace. This corollary is just as easy to get directly: A maps everything, and therefore in particular ran $A +$ ran A^*, into ran A (which is included in ran $A +$ ran A^*), and a similar statement is true about A^*.

The second possibility, ran $A \cap$ ran A^*, is not always a reducing subspace. One easy counterexample is given by

$$A = \begin{pmatrix} 0 & 1 & 0 \\ 0 & 0 & 1 \\ 0 & 0 & 0 \end{pmatrix}.$$

Its range consists of all vectors of the form $\langle \alpha, \beta, 0 \rangle$, and the range of its adjoint consists of all vectors of the form $\langle 0, \beta, \gamma \rangle$. The intersection of the two ranges is the set of all vectors of the form $\langle 0, \beta, 0 \rangle$, which is not only not invariant under both A and A^*, but, in fact, is invariant under neither. The dual is ker A + ker A^*, which in the present case consists of the set of all vectors of the form $\langle \alpha, 0, \gamma \rangle$, not invariant under either A or A^*.

156   Solution 156.

The only eigenvalue of A is 0 (look at the diagonal of the matrix). If $x = \langle \alpha_1, \alpha_2, \ldots, \alpha_n \rangle$, then

$$Ax = \langle 0, \alpha_1, \alpha_2, \ldots, \alpha_{n-1} \rangle;$$

it follows that $Ax = 0$ if and only if x is a multiple of

$$x_n = \langle 0, 0, \ldots, 0, 1 \rangle.$$

That is: although the algebraic multiplicity of 0 as an eigenvalue of A is n (the characteristic polynomial of A is $(-\lambda)^n$), the geometric multiplicity is only 1. One way to emphasize the important one of these facts is to say that the subspace M_1 consisting of all multiples of x_n is the only 1-dimensional subspace invariant under A.

Are there any 2-dimensional subspaces invariant under A? Yes; one of them is the subspace M_2 spanned by the last two basis vectors x_{n-1} and x_n (or, equivalently, the subspace consisting of all vectors whose first $n-2$ coordinates vanish). That, moreover, is the only possibility. Reason: every such subspace has to contain x_n (because it has to contain an eigenvector), and, since A is nilpotent, the restriction of A to each such subspace must be nilpotent (of index 2). It follows that each such subspace must contain at least one vector y in M_2 that is not in M_1, and hence (consider the span of y and x_n) must coincide with M_2.

The rest of the proof climbs up an inductive ladder. If M_k is the subspace spanned by the last k vectors of the basis $\{x_1, x_2, \ldots, x_n\}$ (or, equivalently, the subspace consisting of all vectors whose first $n-k$ coordinates vanish), then it is obvious that each M_k is invariant under the truncated shift A, and by a modification of the argument of the preceding paragraph (just keep raising the dimensions by 1) it follows that $\mathbb{M}_k$ is in fact the only invariant subspace of dimension k. (Is it permissible to interpret $\mathbb{M}_0$ as $\{0\}$?)

Conclusion: the number of invariant subspaces is $n+1$, and the number of reducing subspaces is 2; the truncated shift is irreducible.

Solution 157.   157

The matrix A is the direct sum of the 2×2 matrix

$$B = \begin{pmatrix} 0 & 1 \\ 0 & 0 \end{pmatrix}$$

and the 1×1 matrix 0. A few seconds' reflection should yield the conclusion that the same direct sum statement can be made about

$$A^{\wedge} = \begin{pmatrix} 0 & 0 & 1 \\ 0 & 0 & 0 \\ 0 & 0 & 0 \end{pmatrix};$$

the only difference between A and $A^{\wedge}$ is that for $A^{\wedge}$ the third column plays the role that the second column played for A. A more formal way of saying that is to say that the permutation matrix that interchanges the second and the third columns effects a similarity between A and $A^{\wedge}$:

$$\begin{pmatrix} 1 & 0 & 0 \\ 0 & 0 & 1 \\ 0 & 1 & 0 \end{pmatrix} \cdot \begin{pmatrix} 0 & 1 & 0 \\ 0 & 0 & 0 \\ 0 & 0 & 0 \end{pmatrix} \cdot \begin{pmatrix} 1 & 0 & 0 \\ 0 & 0 & 1 \\ 0 & 1 & 0 \end{pmatrix} = \begin{pmatrix} 0 & 0 & 1 \\ 0 & 0 & 0 \\ 0 & 0 & 0 \end{pmatrix}.$$

Since $A^{\wedge}$ does have a square root, namely the 3×3 truncated shift, so does A. Since in fact, more generally,

$$\begin{pmatrix} 0 & \xi & \eta \\ 0 & 0 & \frac{1}{\xi} \\ 0 & 0 & 0 \end{pmatrix}$$

is a square root of $A^{\wedge}$, it follows that A too has many square roots, namely the matrices of the form

$$\begin{pmatrix} 0 & \eta & \xi \\ 0 & 0 & 0 \\ 0 & \frac{1}{\xi} & 0 \end{pmatrix}$$

obtained from the square roots of $A^{\wedge}$ by the permutation similarity.

158   **Solution 158.**

Similar normal transformations are unitarily equivalent. Suppose, indeed, that A_1 and A_2 are normal and that

$$A_1 B = B A_2,$$

where B is invertible. Let $B = UP$ be the polar decomposition of B (Problem 145), so that U is unitary and $P = \sqrt{B^* B}$, and compute as follows:

$$A_2(B^* B) = (A_2 B^*)B$$
$$= (B^* A_1)B$$

(by the facts about adjoint intertwining, Solution 150)

$$= B^*(A_1 B) = B^*(B A_2) \quad \text{(by assumption)}$$
$$= (B^* B)A_2.$$

The result is that

$$A_2 P^2 = P^2 A_2,$$

from which it follows (since P is a polynomial in P^2) that

$$A_2 P = P A_2.$$

Consequence:

$$A_1 UP = UP A_2 \quad \text{(by assumption)}$$
$$= U A_2 P \quad \text{(by what was just proved)},$$

and therefore, since P is invertible,

$$A_1 U = U A_2.$$

That completes the proof of the unitary equivalence of A_1 and A_2.

159   **Solution 159.**

Are the matrices

$$A = \begin{pmatrix} 0 & 1 & 0 \\ 0 & 0 & 2 \\ 0 & 0 & 0 \end{pmatrix} \qquad \text{and} \qquad B = \begin{pmatrix} 0 & 0 & 0 \\ 2 & 0 & 0 \\ 0 & 1 & 0 \end{pmatrix}$$

unitarily equivalent? The answer is yes, and it's not especially surprising; if

$$U = \begin{pmatrix} 0 & 0 & 1 \\ 0 & 1 & 0 \\ 1 & 0 & 0 \end{pmatrix},$$

then $U^*AU = B$.

Are the matrices

$$A = \begin{pmatrix} 0 & 1 & 0 \\ 0 & 0 & 2 \\ 0 & 0 & 0 \end{pmatrix} \quad \text{and} \quad B = \begin{pmatrix} 0 & 2 & 0 \\ 0 & 0 & 1 \\ 0 & 0 & 0 \end{pmatrix}$$

unitarily equivalent? The surprising answer is no. More or less sophisticated proofs for that negative answer are available, but the quickest proof is a simple computation that is not sophisticated at all. What can be said about a 3×3 matrix S with the property that

$$SA = BS?$$

Written down in terms of matrix entries, the question becomes a system of nine equations in nine unknowns. The general solution of the system is easy to find; the answer is that the matrix S must have the form

$$S = \begin{pmatrix} 2\xi & 0 & \eta \\ 0 & \xi & 0 \\ 0 & 0 & 2\xi \end{pmatrix}.$$

A matrix like that cannot possibly be unitary, and that settles that.

An alternative proof is based on the observation that

$$A^2 = B^2 = \begin{pmatrix} 0 & 0 & 2 \\ 0 & 0 & 0 \\ 0 & 0 & 0 \end{pmatrix}.$$

Since $SA = BS$ implies that $SA^2 = B^2S$, it becomes pertinent to find out which matrices commute with A^2. That's another simple computation, which leads to the same conclusion.

These comments seem not to address the main issue (unitary equivalence of transposes), but in fact they come quite close to it. The A's in the two pairs of examples are the same, but the B's are not: the first B is the transpose of the second. Since the first B is unitarily equivalent to A but the second one is not, since, in fact, the second B is unitarily equivalent to the transpose of A, it follows that A is not unitarily equivalent to its own transpose, and that settles the issue.

Yes, it settles the issue, but not very satisfactorily. How could one possibly discover such examples, and, having discovered them, how could one give a conceptual proof that they work instead of an unenlightening computational one?

Here is a possible road to discovery. What is sought is a matrix A that is not unitarily equivalent to the transpose A. Write A in polar form $A = UP$, with U unitary and P positive (Problem 145), and assume for the time being that P is invertible. There is no real loss of generality in that assumption; if there is any example at all, then there are both invertible and non-invertible examples. Proof: the addition of a scalar doesn't change the unitary equivalence property in question. Since, moreover, transforming every matrix in sight by a fixed unitary one doesn't change the unitary equivalence property in question either, there is no loss of generality in assuming that the matrix P is in fact diagonal.

If $A = UP$, then $A' = PU'$, so that to say that A and A' are unitarily equivalent is the same as saying that there exists a unitary matrix W such that

$$W^*(UP)W = PU', \qquad (*)$$

or, equivalently, such that

$$(W^*U)P(W\overline{U}) = P.$$

(The symbol $\overline{U}$ here denotes the complex conjugate of the matrix U.) Assume then that $(*)$ is true, and write $Q = W^*U$, and $R = W\overline{U}$; note that Q and R are unitary and that

$$QPR = P.$$

It follows that

$$P^2 = PP^* == QPRR^*PQ^* = QP^2Q^*,$$

so that Q commutes with P^2; since P is a polynomial in P^2, it follows that Q commutes with P (and similarly that R commutes with P).

To get a powerful grip on the argument, it is now a good idea to make a restrictive assumption: assume that the diagonal entries (the eigenvalues) of P are all distinct. In view of the commutativity of Q and P, that assumption implies that Q too is diagonal and hence, incidentally, that $W = UP^2$. The equation $(*)$ yields

$$PQU\overline{Q} = QPU\overline{Q} = PU',$$

and hence, since P is invertible, that

$$QU\overline{Q} = U'.$$

Since the entries of the unitary diagonal matrix Q are complex numbers of absolute value 1, it follows that the absolute values of the matrix U constitute a symmetric matrix.

That last result is unexpected but does not seem to be very powerful; in fact, it solves the problem. The assumption of the existence of W has implied that U must satisfy a condition. The matrix U, however, has not yet been specified; it could have been chosen to be a quite arbitrary unitary matrix. If it is chosen so as not to satisfy the necessary condition that the existence of W imposes, then it follows that no W can exist, and victory is achieved.

The simplest example of a unitary matrix whose absolute values do not form a symmetric matrix is

$$U = \begin{pmatrix} 0 & 1 & 0 \\ 0 & 0 & 1 \\ 1 & 0 & 0 \end{pmatrix}.$$

A simple P that can be used (positive, diagonal, invertible, with distinct eigenvalues) is given by

$$P = \begin{pmatrix} 1 & 0 & 0 \\ 0 & 2 & 0 \\ 0 & 0 & 3 \end{pmatrix}.$$

Since, however, invertibility is an unnecessary luxury, an even simpler one is

$$P = \begin{pmatrix} 0 & 0 & 0 \\ 0 & 1 & 0 \\ 0 & 0 & 2 \end{pmatrix};$$

if that one is used, then the resulting counterexample is

$$A = UP = \begin{pmatrix} 0 & 1 & 0 \\ 0 & 0 & 1 \\ 1 & 0 & 0 \end{pmatrix} \cdot \begin{pmatrix} 0 & 0 & 0 \\ 0 & 1 & 0 \\ 0 & 0 & 2 \end{pmatrix} = \begin{pmatrix} 0 & 1 & 0 \\ 0 & 0 & 2 \\ 0 & 0 & 0 \end{pmatrix},$$

and the process of "discovery" is complete.

Solution 160.

If A and B are real, U is unitary, and $U^* A U = B$, then there exists a real orthogonal V such that $V^* A V = B$.

A surprisingly important tool in the proof is the observation that the unitary equivalence of A and B via U implies the same result for A^* and B^*. Indeed, the adjoint of the assumed equation is $U^* A^* U = B^*$.

Write U in terms of its real and imaginary parts (compare Solution 89): $U = E + iF$. It follows from $AU = UB$ that $AE = EB$ and $AF = FB$, and hence that $A(E + \lambda F) = (E + \lambda F)B$ for every scalar λ. If λ is real and different from a finite number of troublesome scalars (the ones for which $\det(E + \lambda F) = 0$), the real matrix $S = E + \lambda F$ is invertible, and, of course, has the property that $AS = SB$.

Proceed in the same way from $U^*A^*U = B^*$: deduce that $A^*(E + \lambda F) = (E + \lambda F)B^*$ for all λ, and, in particular, for the ones for which $E + \lambda F$ is invertible, and infer that $A^*S = SB^*$ (and hence that $S^*A^* = BS^*$).

From here on in the technique of Solution 158 works. Let $S = VP$ be the polar decomposition of S (that theorem works just as well in the real case as in the complex one, so that V and P are real). Since

$$BP^2 = BS^*S = S^*A^*S = S^*SB = P^2B,$$

so that P^2 commutes with B, it follows that P commutes with B. Since

$$AVP = AS = SB = VPB = VBP$$

and P is invertible, it follows that $AV = VB$, and the proof is complete.

161   Solution 161.

It is a worrisome fact that eigenvalues of absolute value 1 can not only stop the powers of a matrix from tending to 0, they can even make those powers explode to infinity. Example: if

$$A = \begin{pmatrix} 1 & 1 \\ 0 & 1 \end{pmatrix},$$

then

$$A^n = \begin{pmatrix} 1 & n \\ 0 & 1 \end{pmatrix}.$$

Despite this bad omen, *strict* inequalities do produce the desired convergence.

An efficient way to prove convergence is to use the Jordan canonical form. (Note that $A^n \to 0$ if and only if $(S^{-1}AS)^n \to 0$.) The relevant part of Jordan theory is the assertion that (the Jordan form of) A is the direct sum of matrices of the form $\lambda + B$, where B is nilpotent (of some index k). Since

$$(\lambda + B)^n = \lambda^n + \binom{n}{1}\lambda^{n-1}B + \cdots + \binom{n}{k-1}\lambda^{n-k+1}B^{k-1}$$

as soon as $n \geq k - 1$, and since the assumption $|\lambda| < 1$ (*strict* inequality) implies that the coefficients tend to 0 as $n \to \infty$, the proof is complete.

162   Solution 162.

Yes, every power bounded transformation is similar to a contraction.

Note first that if A is power bounded, then every eigenvalue of A is less than or equal to 1 in absolute value. (Compare the reasoning preceding the statement of Problem 161.) To get more powerful information, use the Jordan form to write (the matrix of) A as the direct sum of matrices of one of the forms

$$E = \begin{pmatrix} \lambda & 0 & 0 & 0 \\ 1 & \lambda & 0 & 0 \\ 0 & 1 & \lambda & 0 \\ 0 & 0 & 1 & \lambda \end{pmatrix} \quad \text{or} \quad F = \begin{pmatrix} \lambda & 0 & 0 & 0 \\ 0 & \lambda & 0 & 0 \\ 0 & 0 & \lambda & 0 \\ 0 & 0 & 0 & \lambda \end{pmatrix},$$

where, for typographical convenience, 4×4 matrices are used to indicate the general $n \times n$ case. It is then enough to prove that each such direct summand that can actually occur in a power bounded matrix is similar to a contraction.

Since $|\lambda| \leq 1$, the matrix F is a contraction, and nothing else needs to be said about it. As far as E is concerned, two things must be said: first, $|\lambda|$ cannot be equal to 1, and, second, when $|\lambda| < 1$, then E is similar to a contraction.

As for the first, a direct computation shows that the entry in row 2, column 1 of E^n is $n\lambda^{n-1}$; if $|\lambda| = 1$, that is inconsistent with power boundedness. As for the second, E is similar to

$$E_\varepsilon = \begin{pmatrix} \lambda & 0 & 0 & 0 \\ \varepsilon & \lambda & 0 & 0 \\ 0 & \varepsilon & \lambda & 0 \\ 0 & 0 & \varepsilon & \lambda \end{pmatrix},$$

where ε can be any number different from 0. There are two ways to prove that similarity: brute force and pure thought. For brute force, form

$$S = \begin{pmatrix} 1 & 0 & 0 & 0 \\ 0 & \varepsilon & 0 & 0 \\ 0 & 0 & \varepsilon^2 & 0 \\ 0 & 0 & 0 & \varepsilon^3 \end{pmatrix},$$

and verify that $SES^{-1} = E_\varepsilon$. For pure thought, check, by inspection, that E and E_ε have the same elementary divisors, and therefore, by abstract similarity theory, they must be similar.

The proof can now be completed by observing that if $|\lambda| < 1$ and ε is sufficiently small, then E_ε is a contraction. The quickest way of establishing that observation is to recall that $\|X\|$ is a continuous function of X, and that, therefore, $\|E_\varepsilon\|$ is a continuous function of ε. Since $\|E_0\| = |\lambda| < 1$, it follows that $\|E_\varepsilon\| < 1$ when ε is sufficiently small, and that settles everything.

163   Solution 163.

What is obvious is that some nilpotent transformations of index 2 *can* be reducible: just form direct sums. That can be done even in spaces of dimension 3; the direct sum of a 0 (of size 1) and a nilpotent of index 2 (of size 2) is nilpotent of index 2 (and size 3). What is not obvious is that, in fact, on a space V of dimension greater than 2 every nilpotent transformation A of index 2 must be reducible. In the proof it is permissible to assume that $A \neq 0$ (for otherwise the conclusion is trivial).

(1) $V = \ker A + \ker A^*$. Reason: $V = \operatorname{ran} A + \operatorname{ran}^\perp A$; nilpotence of index 2 implies that $\operatorname{ran} A \subset \ker A$, and always $\operatorname{ran}^\perp A = \ker A^*$. In the rest of the proof it is permissible to assume that

$$\ker A \cap \ker A^* = \{0\}$$

(for if $x \neq 0$ but $Ax = A^*x = 0$, then the span of x is a 1-dimensional reducing subspace).

(2) The dimension of $\ker A^*$ (the **nullity** of A^*, abbreviated null A^*) is strictly greater than 1. Since A and A^* play completely symmetric roles in all these considerations, it is sufficient to prove that null $A > 1$ (and that way there is less notational fuss). Suppose, indeed, that rank $A \leq 1$. Since $A \neq 0$ by assumption, rank A must be 1 (not 0). Since $\operatorname{ran} A = A(\ker^\perp A)$ and the restriction of A to $\ker^\perp A$ is a one-to-one transformation, it follows that

$$\dim \ker^\perp A = \dim \operatorname{ran} A = \operatorname{rank} A = 1.$$

Thus both $\ker A$ and $\ker A^*$ have dimension 1, and hence V has dimension 2 (see (1) above), contradicting the assumption that $\dim V > 2$. This contradiction destroys the hypothesis null $A \leq 1$.

(3) If $x \in \ker A^*$, then $A^*Ax \in \operatorname{ran} A^* \subset \ker A^*$; in other words, the subspace $\ker A^*$ is invariant under the Hermitian transformation A^*A. It follows that $\ker A^*$ contains an eigenvector of A^*A, or, equivalently, that $\ker A^*$ has a subspace N of dimension 1 that is invariant under A^*A.

(4) Consider the subspace $M = N + AN$. Since A maps N to AN and AN to $\{0\}$ (recall that $A^2 = 0$), the subspace M is invariant under A. Since A^* maps N to $\{0\}$ (recall that $N \subset \ker A^*$) and AN to N, the subspace M is invariant under A^*. Consequence: M reduces A. Since $M \supset N$, the dimension of M is not less than 1, and since $M = N + AN$, the dimension of M is not more than $1 + 1$. Conclusion: M is a non-trivial proper reducing subspace for A.

Solution 164.

Yes, a nilpotent transformation of index 3 can be irreducible on $\mathbb{C}^4$. One example, in a sense "between" the truncated shift and its square, is given by the matrix

$$A = \begin{pmatrix} 0 & 0 & 0 & 0 \\ 1 & 0 & 0 & 0 \\ 1 & 0 & 0 & 0 \\ 0 & 1 & 0 & 0 \end{pmatrix} \quad \text{with adjoint} \quad A^* = \begin{pmatrix} 0 & 1 & 1 & 0 \\ 0 & 0 & 0 & 1 \\ 0 & 0 & 0 & 0 \\ 0 & 0 & 0 & 0 \end{pmatrix}.$$

The kernel of A is the set of all vectors of the form $x = \langle 0, 0, \gamma, \delta \rangle$. These being the only eigenvectors (the only possible eigenvalue being 0), every non-trivial invariant subspace for A must contain one of them (other than 0). One way to establish that A is irreducible is to show that, for any x of the indicated form, the set consisting of x together with all its images under repeated applications of A and A^* necessarily spans $\mathbb{C}^4$. Consider, indeed, the following vectors:

$$\begin{aligned}
y_1 &= & x &= \langle 0, 0, \gamma, \delta \rangle, \\
y_2 &= & A^* x &= \langle \gamma, \delta, 0, 0 \rangle, \\
y_3 &= & A^{*2} x &= \langle \delta, 0, 0, 0 \rangle, \\
y_4 &= & AA^* x &= \langle 0, \gamma, \gamma, \delta \rangle, \\
y_5 &= & AA^{*2} x &= \langle 0, \delta, \gamma, 0 \rangle, \\
y_6 &= & A^2 A^* x &= \langle 0, 0, 0, \gamma \rangle.
\end{aligned}$$

It is true that no matter what γ and δ are, so long as not both are 0, these vectors span the space. If $\gamma = 0$, then y_1, y_2, y_3, and y_4 form a basis; if $\delta = 0$, then y_1, y_2, y_6, and a simple linear combination of y_4 and y_2 form a basis; if neither γ nor δ is 0, then y_3, y_6, and simple linear combinations of y_1 and y_6 for one and of y_2 and y_3 for another form a basis.

The question as asked is now answered, but the answer gives only a small clue to the more general facts (about possible irreducibility) for nilpotent transformations of index k on spaces of dimension n when $k < n$. The case $k = 3$ and $n = 5$ hints at the sort of thing that has to be looked at; the matrix

$$\begin{pmatrix} 0 & 0 & 0 & 0 & 0 \\ 1 & 0 & 0 & 0 & 0 \\ 1 & 0 & 0 & 0 & 0 \\ 0 & 1 & 0 & 0 & 0 \\ 0 & 0 & 2 & 0 & 0 \end{pmatrix}$$

does the job in that case.

It should be emphasized that these considerations have to do with inner product spaces, where reduction is defined in terms of adjoints (or, equivalently, in terms of orthogonal complements). There is a purely algebraic theory of reduction (the existence for an invariant subspace of an invariant complement), and in that theory the present question is much easier to answer in complete generality. The structure theory of nilpotent transformations (in effect, the Jordan normal form), implies that the only chance a nilpotent transformation of index k on a space of dimension n has to be irreducible (that is: one of two complementary invariant subspaces must always be $\{0\}$) is to have $k = n$.